Universal Joints and Driveshafts

H. Chr. Seherr-Thoss · F. Schmelz · E. Aucktor

H. Chr. Seherr-Thoss · F. Schmelz · E. Aucktor

Universal Joints and Driveshafts

Analysis, Design, Applications

Second, enlarged edition with 267 Figures and 72 Tables

Translated by J. A. Tipper and S. J. Hill

Authors:

HANS CHRISTOPH SEHERR-THOSS, Dipl.-Ing.
Holder of the Graf Seherr Archives, Unterhaching

FRIEDRICH SCHMELZ, Dipl.-Ing.
Test- and Computing Engineer, Ingolstadt

ERICH AUCKTOR, Dipl.-Ing.
Development- and Design-Engineer, Offenbach a.M.

Translators:

MRS. JENNIFER A. TIPPER
B.A. HONS., LICHFIELD

DR. STUART J. HILL, B. A. (Eng.)
HONS. PH. D., British Railways Board, London

ISBN-10 3-540-30169-0 Springer-Verlag Berlin Heidelberg New York
ISBN-13 978-3-540-30169-1 Springer-Verlag Berlin Heidelberg New York

Library of Congress Cataloging-in-Publication Data
[Gelenke and Gelenkwellen. English]
Universal joints and driveshafts: analysis, design, applications/F. Schmelz
H. Chr. Seherr-Thoss, E. Aucktor: translated by S. J. Hill and
J. A. Tipper. p. cm.
Translation of: Gelenke und Gelenkwellen. Includes indexes.
ISBN 3-540-41759-1
1. Universal joints. 2. Automobiles – Powertrains.
I. Seherr-Thoss, H.-Chr. (Hans-Christoph), Count, II. Schmelz, F.
(Friedrich), III. Aucktor, E. (Erich), 1913–98.
TJ1059.S3613 1992   621.8'25—dc20   90-28614

This work is subject to copyright. All rights are reserved, whether the whole or part of the material is concerned, specifically the rights of translation, reprinting, reuse of illustrations, recitation, broadcasting, reproduction on microfilm or in any other way, and storage in data banks. Duplication of this publication or parts thereof is permitted only under the provisions of the German Copyright Law of September 9, 1965, in its current version, and permission for use must always be obtained from Springer-Verlag. Violations are liable for prosecution under the German Copyright Law.

Springer is a part of Springer Science+Business Media
springer.com
© Springer-Verlag Berlin Heidelberg 2006
Printed in Germany

The use of registered names, trademarks, etc. in this publication does not imply, even in the absence of a specific statement, that such names are exempt from the relevant protective laws and regulations and therefore free for general use.

Typesetting: Fotosatz-Service Köhler GmbH, Würzburg
Projectmanagement: Reinhold Schöberl, Würzburg
Cover design: medionet AG, Berlin
Printed on acid-free paper – 62/3020 – 543210

Preface to the second English edition

An important date in the history of automotive engineering was celebrated at the start of the 1980s: 50 years of front-wheel drives in production vehicles. This bicentennial event aroused interest in the development, theory and future of driveshafts and joints. The authors originally presented all the available knowledge on constant-velocity and universal-joint driveshafts in German as long ago as 1988, followed by English in 1992 and Chinese in 1997.

More than ten years have passed since then, in which time technology has also made major progress in the field of driveshafts. Driveshaft design and manufacturing process has kept pace with the constantly growing demands of the various users. More powerful engines with higher torques, new fields of application with increased stresses, e.g. off-road and heavy goods vehicles or rolling mills, improved materials, new production processes and advanced experimental and test methods have imposed completely new requirements on the driveshaft as a mechanical component.

GKN Driveline has made a major contribution to the further development of the driveshaft and will maintain this effort in the future. At our Research & Product Development Centres, GKN engineers have defined basic knowledge and conceived product improvements for the benefit of the automotive, agricultural, and machinery industry of mechanically engineered products for the world. In close cooperation with its customers GKN has created low-noise, vibration and maintenance-free driveshafts.

The cumulative knowledge acquired has been compiled in this second edition book that has been updated to reflect the latest state of the art. It is intended to serve both as a textbook and a work of reference for all driveline engineers, designers and students who are in some way involved with constant-velocity and universal-joint driveshafts.

Redditch, England
July 2005

ARTHUR CONNELLY
Chief Executive Officer
GKN Automotive Driveline Driveshafts

Preface to the second German edition

1989 saw the start of a new era of the driveline components and driveshafts, driven by changing customer demands. More vehicles had front wheel drive and transverse engines, which led to considerable changes in design and manufacture, and traditional methods of production were revisited. The results of this research and development went into production in 1994–96:

- for Hooke's jointed driveshafts, weight savings were achieved through new forgings, noise reduction through better balancing and greater durability through improved lubrication.
- for constant velocity joints, one can talk about a "New Generation", employing something more like roller bearing technology, but where the diverging factors are dealt with. The revised Chapters 4 and 5 re-examine the movement patterns and stresses in these joints.

The increased demands for strength and precision led to even intricate shapes being forged or pressed. These processes produce finished parts with tolerances of 0.025 mm. 40 million parts were forged in 1998. These processes were also reviewed.

Finally, advances were made in combined Hooke's and constant velocity jointed driveshafts.

A leading part in these developments was played by the GKN Group in Birmingham and Lohmar, which has supported the creation of this book since 1982. As a leading manufacturer they supply 600,000 Hooke's jointed driveshafts and 500,000 constant velocity driveshafts a year. I would like to thank REINHOLD SCHOEBERL for his project-management and typography to carry out the excellent execution of this book.

I would also like to thank my wife, THERESE, for her elaboration of the indices. Moreover my thanks to these contributors:

GERD FAULBECKER (GWB Essen)
JOACHIM FISCHER (ZF Lenksysteme Gmünd)
WERNER JACOB (Ing. Büro Frankfurt/M)
CHRISTOPH MÜLLER (Ing. Büro Ingolstadt)
PROF. DR. ING. ERNST-GÜNTER PALAND (TU Hannover, IMKT)
JÖRG PAPENDORF (Spicer GWB Essen)
STEFAN SCHIRMER (Freudenberg)
RALF SEDLMEIER (GWB Essen)
ARMIN WEINHOLD (SMS Eumuco Düsseldorf/Leverkusen)

GKN Driveline (Lohmar/GERMANY)
WOLFGANG HILDEBRANDT
WERNER KRUDE
STEPHAN MAUCHER
MICHAEL MIRAU (Offenbach)
CLEMENS NIENHAUS (Walterscheid)
PETER POHL (Walterscheid Trier)
RAINER SCHAEFERDIEK
KARL-ERNST STROBEL (Offenbach)

On 19th July 1989 our triumvirate lost Erich Aucktor. He made a valuable contribution to the development of driveline technology from 1937–1958, as an engine engineer, and from 1958–78 as a designer and inventor of constant velocity joints at Löhr & Bromkamp, Offenbach, where he worked in development, design and testing. It is thanks to him that this book has become a reference work for these engineering components.

COUNT HANS CHRISTOPH SEHERR-THOSS

Contents

Index of Tables XIII

Notation XVII

Chronological Table XXI

Chapter 1 Universal Jointed Driveshafts for Transmitting Rotational Movements 1

1.1	Early Reports on the First Joints	1
1.1.1	Hooke's Universal Joints	1
1.2	Theory of the Transmission of Rotational Movements by Hooke's Joints	5
1.2.1	The Non-univormity of Hooke's Joints According to Poncelet . . .	5
1.2.2	The Double Hooke's Joint to Avoid Non-univormity	8
1.2.3	D'Ocagne's Extension of the Conditions for Constant Velocity . .	10
1.2.4	Simplification of the Double Hooke's Joint	10
1.2.4.1	Fenaille's Tracta Joint	12
1.2.4.2	Various Further Simplifications	14
1.2.4.3	Bouchard's One-and-a-half Times Universal Joint	14
1.3	The Ball Joints	17
1.3.1	Weiss and Rzeppa Ball Joints	19
1.3.2	Developments Towards the Plunging Joint	27
1.4	Development of the Pode-Joints	32
1.5	First Applications of the Science of Strength of Materials to Driveshafts	40
1.5.1	Designing Crosses Against Bending	40
1.5.2	Designing Crosses Against Surface Stress	42
1.5.3	Designing Driveshafts for Durability	47
1.6	Literature to Chapter 1	49

Chapter 2 Theory of Constant Velocity Joints (CVJ) 53

2.1	The Origin of Constant Velocity Joints	54
2.2	First Indirect Method of Proving Constant Velocity According to Metzner	58
2.2.1	Effective Geometry with Straight Tracks	61
2.2.2	Effective Geometry with Circular Tracks	64

2.3	Second, Direct Method of Proving Constant Velocity by Orain	66
2.3.1	Polypode Joints	71
2.3.2	The Free Tripode Joint	75
2.4	Literature to Chapter 2	78

Chapter 3 Hertzian Theory and the Limits of Its Application 81

3.1	Systems of Coordinates	82
3.2	Equations of Body Surfaces	83
3.3	Calculating the Coefficient $\cos \tau$	85
3.4	Calculating the Deformation δ at the Contact Face	88
3.5	Solution of the Elliptical Single Integrals J_1 to J_4	94
3.6	Calculating the Elliptical Integrals K and E	97
3.7	Semiaxes of the Elliptical Contact Face for Point Contact	98
3.8	The Elliptical Coefficients μ and ν	101
3.9	Width of the Rectangular Contact Surface for Line Contact	101
3.10	Deformation and Surface Stress at the Contact Face	104
3.10.1	Point Contact	104
3.10.2	Line Contact	105
3.11	The validity of the Hertzian Theory on ball joints	106
3.12	Literature to Chapter 3	107

Chapter 4 Designing Joints and Driveshafts 109

4.1	Design Principles	109
4.1.1	Comparison of Theory and Practice by Franz Karas 1941	110
4.1.2	Static Stress	111
4.1.3	Dynamic Stress and Durability	112
4.1.4	Universal Torque Equation for Joints	114
4.2	Hooke's Joints and Hooke's Jointed Driveshafts	116
4.2.1	The Static Torque Capacity M_o	117
4.2.2	Dynamic Torque Capacity M_d	118
4.2.3	Mean Equivalent Compressive Force P_m	119
4.2.4	Approximate Calculation of the Equivalent Compressive Force P_m	124
4.2.5	Dynamic Transmission Parameter 2 CR	126
4.2.5.1	Exemple of Specifying Hooke's Jointed Driveshafts in Stationary Applications	128
4.2.6	Motor Vehicle Driveshafts	130
4.2.7	GWB's Design Methodology for Hooke's joints for vehicles	133
4.2.7.1	Example of Specifying Hooke's Jointed Driveshafts for Commercial Vehicles	136
4.2.8	Maximum Values for Speed and Articulation Angle	138
4.2.9	Critical Speed and Shaft Bending Vibration	140
4.2.10	Double Hooke's Joints	144
4.3	Forces on the Support Bearings of Hooke's Jointed Driveshafts	148
4.3.1	Interaction of Forces in Hooke's Joints	148
4.3.2	Forces on the Support Bearings of a Driveshaft in the W-Configuration	150

4.3.3	Forces on the Support Bearings of a Driveshaft in the Z-Configuration	152
4.4	Ball Joints	153
4.4.1	Static and Dynamic Torque Capacity	154
4.4.1.1	Radial bearing connections forces	158
4.4.2	The ball-joint from the perspective of rolling and sliding bearings	158
4.4.3	A common, precise joint centre	159
4.4.3.1	Constant Velocity Ball Joints based on Rzeppa principle	160
4.4.4	Internal centering of the ball-joint	162
4.4.4.1	The axial play s_a	163
4.4.4.2	Three examples for calculating the axial play s_a	164
4.4.4.3	The forced offset of the centre	166
4.4.4.4	Designing of the sherical contact areas	169
4.4.5	The geometry of the tracks	170
4.4.5.1	Longitudinal sections of the tracks	170
4.4.5.2	Shape of the Tracks	174
4.4.5.3	Steering the Balls	177
4.4.5.4	The Motion of the Ball	178
4.4.5.5	The cage in the ball joint	179
4.4.5.6	Supporting surface of the cage in ball joints	183
4.4.5.7	The balls	184
4.4.5.8	Checking for perturbations of motion in ball joints	185
4.4.6	Structural shapes of ball joints	187
4.4.6.1	Configuration and torque capacity of Rzeppa-type fixed joints	187
4.4.6.2	AC Fixed Joints	188
4.4.6.3	UF Fixed joints (undercut free)	193
4.4.6.4	Jacob/Paland's CUF (completely undercut free) Joint for rear wheel drive > 25°	195
4.4.6.5	Calculation example for a CUF joint	197
4.4.7	Plunging Joints	200
4.4.7.1	DO Joints	200
4.4.7.2	VL Joints	202
4.4.8	Service Life of Joints Using the Palmgren/Miner Rule	207
4.5	Pode Joints	209
4.5.1	Bipode Plunging Joints	213
4.5.2	Tripode Joints	214
4.5.2.1	Static Torque Capacity of the Non-articulated Tripode Joint	215
4.5.2.2	Materials and Manufacture	215
4.5.2.3	GI Plunging Tripode Joints	220
4.5.2.4	Torque Capacity of the Articulated Tripode Joint	222
4.5.3	The GI-C Joint	228
4.5.4	The low friction and low vibration plunging tripode joint AAR	229
4.6	Materials, Heat Treatment and Manufacture	231
4.6.1	Stresses	231
4.6.2	Material and hardening	236
4.6.3	Effect of heat treatment on the transmittable static and dynamic torque	238

4.6.4	Forging in manufacturing	239
4.6.5	Manufacturing of joint parts	240
4.7	Basic Procedure for the Applications Engineering of Driveshafts	243
4.8	Literature to Chapter 4	245

Chapter 5 Joint and Driveshaft Configurations 249

5.1	Hooke's Jointed Driveshafts	250
5.1.1	End Connections	251
5.1.2	Cross Trunnions	253
5.1.3	Plunging Elements	258
5.1.4	Friction in the driveline – longitudinal plunges	258
5.1.5	The propshaft	262
5.1.6	Driveshaft tubes made out of composite Fibre materials	263
5.1.7	Designs of Driveshaft	267
5.1.7.1	Driveshafts for Machinery and Motor Vehicles	267
5.1.8	Driveshafts for Steer Drive Axles	268
5.2	The Cardan Compact 2000 series of 1989	269
5.2.1	Multi-part shafts and intermediate bearings	273
5.2.2	American Style Driveshafts	275
5.2.3	Driveshafts for Industrial use	277
5.2.4	Automotive Steering Assemblies	285
5.2.5	Driveshafts to DIN 808	292
5.2.6	Grooved Spherical Ball Jointed Driveshafts	294
5.3	Driveshafts for Agricultural Machinery	296
5.3.1	Types of Driveshaft Design	297
5.3.2	Requirements to meet by Power Take Off Shafts	298
5.3.3	Application of the Driveshafts	300
5.4	Calculation Example for an Agricultural Driveshaft	307
5.5	Ball Jointed Driveshafts	308
5.5.1	Boots for joint protection	310
5.5.2	Ways of connecting constant velocity joints	311
5.5.3	Constant velocity drive shafts in front and rear wheel drive passenger cars	312
5.5.4	Calculation Example of a Driveshaft with Ball Joints	316
5.5.5	Tripode Jointed Driveshaft Designs	321
5.5.6	Calculation for the Tripode Jointed Driveshaft of a Passenger Car	322
5.6	Driveshafts in railway carriages	325
5.6.1	Constant velocity joints	325
5.7	Ball jointed driveshafts in industrial use and special vehicles	327
5.8	Hooke's jointes high speed driveshafts	329
5.9	Design and Configuration Guidelines to Optimise the Drivetrain	332
5.9.1	Exemple of a Calculation for the Driveshafts of a Four Wheel Drive Passenger Car (Section 5.5.4)	333
5.10	Literature to Chapter 5	343

Index of Tables

Independent Tables

Table 1.1	Bipode joints 1902–52	33
Table 1.2	Tripode joints 1935–60	36
Table 1.3	Quattropode-joints 1913–94	39
Table 1.4	Raised demands to the ball joints by the customers	39
Table 3.1	Complete elliptical integrals by A.M. Legendre 1786	96
Table 3.2	Elliptical coefficients according to Hertz	102
Table 3.3	Hertzian Elliptical coefficients	103
Table 4.1	Geometry coefficient f_1 according to INA	120
Table 4.2	Shock or operating factor f_{ST}	127
Table 4.3	Values for the exponents n_1 and n_2 after GWB	138
Table 4.4	Maximum Speeds and maximum permissible values of $n\beta$ arising from the moment of inertia of the connecting parts	139
Table 4.5	Radial bearing connection forces of Constant Velocity Joints (CVJ) with shafts in one plane, from Werner KRUDE	157
Table 4.6	Most favourable track patterns for the two groups of joints	174
Table 4.7	Ball joint family tree	176
Table 4.8	Steering the balls in ball joints	177
Table 4.9	Effect of the ball size on load capacity and service life	185
Table 4.10	Rated torque M_N of AC joints	191
Table 4.11	Dynamic torque capacity M_d of AC joints	191
Table 4.12	Data for UF-constant velocity joints made by GKN Löbro	194
Table 4.13	VL plunging joint applications	206
Table 4.14	Percentage of time in each gear on various types of roads	208
Table 4.15	Percentage of time a_x for passenger cars	208
Table 4.16	Surface stresses in pode joints with roller bearings	212
Table 4.17	Surface stresses in pode joints with plain bearings	212
Table 4.18	Materials and heat treatments of inner and outer races for UF-constant velocity joints	234
Table 4.19	Values for the tri-axial stress state and the required hardness of a ball in the track of the UF-joint	235
Table 4.20	Hardness conditions for the joint parts	238
Table 4.21	Static and dynamic torque capacities for UF constant velocity joints with outer race surface hardness depth (Rht) of 1.1 and 2.4 mm	239

Table 4.22	Applications Engineering procedure for a driveshaft with uniform loading	244
Table 5.1	Maximum articulation angle β_{max} of joints	250
Table 5.2	Examples of longitudinal plunge via balls in drive shafts	260
Table 5.3	Data for composite propshafts for motor vehicles	263
Table 5.4	Comparison of steel and glass fibre reinforced plastic propshafts for a high capacity passenger car	267
Table 5.5	Torque capacity of Hooke's jointed driveshafts	268
Table 5.6	Steering joint data	287
Table 5.7	Comparison of driveshafts	309
Table 5.8	Examples of standard longshaft systems for passenger cars around 1998	312
Table 5.9	Durability values for the selected tripode joint on rear drive	324
Table 5.10	Comparison of a standard and a high speed driveshaft (Fig. 5.83, 5.85)	331
Table 5.11	Durability values for front wheel drive with UF 1300	335
Table 5.12	Calculation of the starting and adhesion torques in the calculation example (Section 5.9.1)	336
Table 5.13	Values for the life of the propshaft for rear wheel drive	339

Tables of principal dimensions, torque capacities and miscellaneous data inside the Figures

Figure 1.13	Pierre Fenaille's Tracta joint 1927	13
Figure 1.16	One-and-a-half times universal joint 1949	16
Figure 1.21	Fixed joint of the Weiss type	21
Figure 4.5	Bearing capacity coefficient f_2 of rolling bearings	121
Figure 4.10	Principal and cross dimensions of a Hooke's jointed driveshaft for light loading	128
Figure 4.45	Tracks with elliptical cross section	175
Figure 4.58	AC fixed joints of the Rzeppa type according to Wm. Cull	189
Figure 4.59	AC fixed joint (improved) 1999	190
Figure 4.61	UF-constant velocity fixed joint (wheel side)	195
Figure 4.64	Six-ball DO plunging joints (Rzeppa type)	201
Figure 4.65	Five-ball DOS plunging joint from Girguis 1971	202
Figure 4.66	VL plunging joints of the Rzeppa type with inclined tracks and 6 balls	203
Figure 4.77	Glaenzer Spicer GE tripode joint with β_{max} = 43° to 45°	217
Figure 4.79	Glaenzer Spicer GI tripode joint	220
Figure 4.87	Plunging tripode joint (AAR)	229
Figure 5.3	Driveshaft flange connections	252
Figure 5.8	Levels of balancing quality Q for driveshafts	257
Figure 5.24	Non-centred double Hooke's jointed driveshaft with stubshaft for steer drive axles	271
Figure 5.27	Specifications and physical data of Hooke's joints for commercial vehicles with length compensation	274

Index of Tables

Figure 5.29	Hooke's jointed driveshaft from Mechanics/USA	277
Figure 5.30	Centred double-joint for high and variable articulation angles	278
Figure 5.31	Driving dog and connection kit of a wing bearing driveshaft	279
Figure 5.32	Specifications, physical data and dimensions of the cross trunnion of a Hooke's jointed driveshaft for stationary (industrial) use	280
Figure 5.33	Universal joints crosses for medium and heavy industrial shafts	281
Figure 5.34	Specifications, physical data and dimensions of the cross trunnion of Hooke's jointed driveshafts for industrial use, heavy type with length compensation and dismantable bearing cocer	282
Figure 5.35	Specifications of Hooke's heavy driveshafts with length compensation and flange connections for rolling-mills and other big machinery	282
Figure 5.40	Full complement roller bearing. Thin wall sheet metal bush	287
Figure 5.42	Steering joints with sliding serration ca. 1970	289
Figure 5.46	Single and double joint with plain and needle bearings	292
Figure 5.51	Connection measurement for three tractor sizes from ISO standards	296
Figure 5.55	Joint sizes for agricultural applications	299
Figure 5.56	Sizes for sliding profiles of driveshafts on agricultural machinery	301
Figure 5.58	Hooke's joints in agricultural use with protection covers	302
Figure 5.61	Double joints with misaligned trunnions	305
Figure 5.70	Drive shafts with CV joints for Commercial and Special vehicles	315

Notation

Symbol	Unit	Meaning
1. Coordinate Systems		
$0, x, y, z$	mm	orthogonal, right-handed body system
$0, x', y', z'$	mm	spatial system, arising from the transformation with the rotation matrix $D_\varphi = \begin{pmatrix} \sin\varphi & -\cos\varphi & 0 \\ \cos\varphi & \sin\varphi & 0 \\ 0 & 0 & 1 \end{pmatrix}$
$0, x'', y'', z''$	mm	spatial system, arising from the transformation with the translation matrix $D_\beta = \begin{pmatrix} 1 & 0 & 0 \\ 0 & \cos\beta & -\sin\beta \\ 0 & \sin\beta & \cos\beta \end{pmatrix}$
$0, r, \varphi$	mm, degrees	system of polar coordinates in Boussinesq half space
$r = a + tu$		vectorial representation of a straight track
$a = \begin{pmatrix} k \\ l \\ m \end{pmatrix}$		location vector and its components
$u = \begin{pmatrix} n \\ p \\ q \end{pmatrix}$		direction vector and its components
Indices 1		body 1, driving unit
Indices 2		body 2, driven unit or intermediate body
Indices 3		body 3, driven unit for three bodies
$i = 1, 2, 3, \ldots n$		sequential numbering

Symbol	Unit	Meaning
2. Angles		
α	degrees	pressure angle
β	degrees	articulation angle
γ	degrees	skew angle of the track
δ	degrees	divergence of opening angle ($\delta = 2\varepsilon$), offset angle
ε	degrees	tilt or inclination angle of the track (Stuber pledge angle)
ϑ	degrees	complementary angle to α ($\alpha = 90° - \vartheta$)
τ	degrees	Hertzian auxiliary angle
$\cos \tau$	–	Hertzian coefficient
φ	degrees	angle of rotation through the joint body
χ	degrees	angle of intersection of the tracks
ψ	degrees	angle of the straight generators r
3. Rolling body data		
A	mm²	contact area
$2a$	mm	major axis of contact ellipse
$2b$	mm	minor axis of contact ellipse
$2c$	mm	separation of cross axes of double Hooke's joints
c	mm	offset (displacement of generating centres)
d	mm	roller diameter
D	mm	trunnion diameter
D_m	mm	diameter of the pitch circle of the rolling bodies
k	N/mm²	specific loading
l	mm	roller length (from catalogue)
l_w	mm	effective length of roller
r	mm	radius of curvature of rolling surfaces
R	m	effective joint radius
s_w	mm	joint plunge
ψ		reciprocal of the conformity in the track cross section
κ_Q		conformity in the track cross section
κ_L		conformity in the longitudinal section of the track
$\varrho = 1/r$	1/mm	curvature of the rolling surfaces
c_p	N/mm²	coefficient of conformity
p_0	N/mm²	Hertzian pressure
δ_0	mm	total elastic deformation at a contact point
δ_b	mm	plastic deformation
μ, ν		Hertzian elliptical coefficients
E	N/mm²	Young's modulus ($2.08 \cdot 10^{11}$ for steel)
m		Poisson's ratio = 3/10
$\vartheta = \dfrac{4}{E}(1 - m^2)$	mm²/N	abbreviation used by Hertz
z	–	number of balls

Symbol	Unit	Meaning

4. Forces

P	N	equivalent dynamic compressive force
Q	N	Hertzian compressive force
F	N	radial component of equivalent compressive force P
A	N	axial component of equivalent compressive force P
Q_{total}	N	total radial force on a roller bearing used by Hertz/Stribeck

5. Moments

M	Nm	moments in general
M_0	Nm	static moment
M_d	Nm	dynamic moment
M_N	Nm	nominal moment (from catalogue)
M_b	Nm	bending moment
M_B	Nm	design moment

6. Mathematical constants and coefficients

ε		ratio of the front and rear axle loads A_F/A_R
ε_F		fraction of torque to the front axle
ε_R		fraction of torque to the rear axle
f_β		articulation coefficient
k_ω		equivalence factor for cyclic compressive forces
s		Stribeck's distribution factor = 5
s_0		static safety factor for oscillating bearings of Hooke's joints = 0.8 to 1.0
u		number of driveshafts
μ		friction coefficient of road

7. Other designations

P_{eff}	kW	output power
V	kph	driving speed
ω	s^{-1}	angular velocity
m		effective number of transmitting elements
n	rpm	rotational speed (revolutions per minute)
π		plane of symmetry

Designations which have not been mentioned are explained in the text or shown in the figures.

Chronological Table

1352–54	Universal jointed driveshaft in the clock mechanism of Strasbourg Cathedral.
1550	*Geronimo Cardano's* gimbal suspension.
1663	*Robert Hooke's* universal joint. 1683 double Hooke's joint.
1824	Analysis of the motion of Hooke's joints with the aid of spherical trigonometry and differential calculus, and the calculation of the forces on the cross by *Jean Victor Poncelet*.
1841	Kinematic treatment of the Hooke's joint by *Robert Willis*.
1894	Calculation of surface stresses for crosses by *Carl Bach*.
1901/02	Patents for automotive joints by *Arthur Hardt* and *Robert Schwenke*.
1904	Series production of Hooke's joints and driveshafts by *Clarence Winfred Spicer*.
1908	First ball joint by *William A. Whitney*. Plunging + articulation separated.
1918	Special conditions for the uniform transmission of motion by *Maurice d'Ocagne*. 1930 geometrical evidence for the constant velocity characteristics of the Tracta joint.
1923	Fixed ball joint steered by generating centres widely separated from the joint mid-point, by *Carl William Weiss*. Licence granted to the Bendix Corp.
1926	*Pierre Fenaille's* "homokinetic" joint.
1927	Six-ball fixed joint with 45° articulation angle by *Alfred H. Rzeppa*. 1934 with offset steering of the balls. First joint with concentric meridian tracks.
1928	First Hooke's joint with needle bearings for the crosses by *Clarence Winfred Spicer*. Bipode joint by *Richard Bussien*.
1933	Ball joint with track-offset by *Bernard K. Stuber*.
1935	Tripode joint by *J. W. Kittredge*, 1937 by *Edmund B. Anderson*.
1938	Plunging ball joint according to the offset principle by *Robert Suczek*.
1946	Birfield-fixed joint with elliptical tracks, 1955 plunging joint, both by *William Cull*.
1951	Driveshaft with separated Hooke's joint and middle sections by Borg-Warner.
1953	Wide angle fixed joint ($\beta = 45°$) by *Kurt Schroeter*, 1971 by *H. Geisthoff, Heinrich Welschof* and *H. Grosse-Entrop*.
1959	AC fixed joint by *William Cull* for British Motor Corp., produced by Hardy-Spicer.
1960	Löbro-fixed joint with semicircular tracks by *Erich Aucktor/Walter Willimek*. Tripode plunging joint, 1963 fixed joint, both by *Michel Orain*.

1961	Four-ball plunging joint with a pair of crossed tracks by *Henri Faure*. DANA-plunging joint by *Phil. J. Mazziotti, E. H. Sharp, Zech*.
1962	VL-plunging joint with crossed tracks, six balls and spheric cage by *Erich Aucktor*.
1965	DO-plunging joint by *Gaston Devos*, completed with parallel tracks and cage offset by Birfield. 1966 series for Renault R 4.
1970	GI-tripode plunging joint by Glaenzer-Spicer. UF-fixed joint by *Heinr. Welschof/Erich Aucktor*. Series production 1972.
1985	Cage-guided balls for plunging in the Triplan joint by *Michel Orain*.
1989	AAR-joint by Löbro in series production.

1 Universal Jointed Driveshafts for Transmitting Rotational Movements

The earliest information about joints came from Philon of Byzantium around 230 BC in his description of censers and inkpots with articulated suspension. In 1245 AD the French church architect, Villard de Honnecourt, sketched a small, spherical oven which was suspended on circular rings. Around 1500, Leonardo da Vinci drew a compass and a pail which were mounted in rings [1.1].

1.1 Early Reports on the First Joints

Swiveling gimbals were generally known in Europe through the report of the mathematician, doctor and philosopher, Geronimo Cardano. He also worked in the field of engineering and in 1550 he mentioned in his book "De subtilitate libri XXI" a sedan chair of Emperor Charles V "which was mounted in a gimbal" [1.2]. In 1557 he described a ring joint in "De armillarum instrumento". Pivots staggered by 90° connected the three rings to one another giving rise to three degrees of freedom (Fig. 1.1). This suspension and the joint formed from it were named the "cardan suspension" or the "cardan joint" after the author.

The need to transmit a rotary movement via an angled shaft arose as early as 1300 in the construction of clocktowers. Here, because of the architecture of the tower, the clockwork and the clockface did not always lie on the same axis so that the transmission of the rotation to the hands had to be displaced upwards, downwards or sideways. An example described in 1664 by the Jesuit, Caspar Schott, was the clock in Strasbourg Cathedral of 1354 [1.3]. He wrote that the inclined drive could best be executed through a cross with four pivots which connected two shafts with forks (fuscinula) fitted to their ends (Fig. 1.2). The universal joint was therefore known long before Schott. He took his description from the unpublished manuscript "Chronometria Mechanica Nova" by a certain Amicus who was no longer alive. If one analyses Amicus's joint, the close relationship between the gimbal and the universal joint can be seen clearly (Figs. 1.3a and b). The mathematics of the transmission of movement were however not clear to Schott, because he believed that one fork must rotate as quickly as the other.

1.1.1 Hooke's Universal Joints

In 1663 the English physicist, Robert Hooke, built a piece of apparatus which incorporated an articulated transmission not quite in the form of Amicus's joint. In 1674 he described in his "Animadversions" [1.4] the helioscope of the Danzig astronomer,

CAPVT VII.

De Armillarum instrumento.

COnstat ex circulis tribus instrumentum armillis simile, quorum superiores sunt duplicati, & polis secundus primo fixis infigitur. Vt sit A B circulus primus, cui infixi sint ad rectos polorum CD

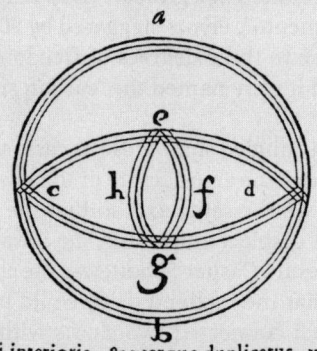

poli interioris, & vterque duplicatus, vel vt media pars circumagi possit sub eisdem polis, quia inferior; vel quia ex dimidio stabilis praeter polos, ex dimidio mobilis. Tertius autem in medio secundi, ita vt

vt circumagi possint poli eius ex E in C, & C in G, & G in D. Et rursus polis quasi nullis ex E in F, & F in G, & G in H. Et vt lateat prorsus coniunctio, adeò vt annexus alter alteri videatur. Tale instrumentum vidi apud virum Maximiliani Caesaris, Mathematicum Medicum & Philosophum insignem, Ioannem Sagerum Gisenhaigen Vratisleuiensem, quanquam neque ipse docuerit, quomodo infertus esset; neque ego interrogauerim. Ergo fieri potest, vt circulus inferior D E C G circumuettatur, superiore immobili: atque ita poli ferentur per E C G D: sed tunc necessaria erit cauitas infra circulum secundum, per quam feratur. Sed si circulus E C G D integer sit, fieri non potest vt ferantur poli nisi cum circulo, cui infixi sint: hic autem est pars circuli praedicti media, aut etiam inferior. Erunt ergo tres modi. Inferior autem circulus, cum in seipso reuoluitur, poterit manentibus quidem polis circumduci à lateribus, fixo manente medio: nam si medius transferatur cui polus infixus est, exibit polus circumductus circulum E C G D secundum latitudinem: aut transferetur polus per cauitatem. At tunc non erit circulus F G H E solidus. Cum ergo voluerimus circulos ambos esse solidos, relinquentur duo modi tantum, vt pars media circuli secundi, cui infixi sint circuli, circumuoluatur; & extremae inferioris partes, seu latera: aut vt pars inferior secundi circuli intrusa superiori, & in qua sint poli fixi edem modo sub superiore circumducatur, atque eo modo totus circulus E F G H circumagatur per E D G C. Ipse verò circulus E D G C in seipso vt prius manente fixa parte media, in qua sunt poli infixi, circumducatur lateribus suis. Commune autem est ambobus, vt poli sint infixi vtrisque circulis secundo & tertio, & quòd latera iuferioris manente medio circumagantur. In secundo tantum differunt; cum vel media pars manentibus lateribus, vel inferior superiore fixa circumduci possit.

Fig. 1.1. Ring joint of Hieronymus Cardanus 1557. In his "Mediolanensis philosophi ac medici", p. 163, he wrote the following about it: "I saw such a device at the house of Emperor Maximilian's adviser, Joannem Sagerum (Johann Sager) from Gisenhaigen (Gießenhagen, Geißenhain), an important mathematician, doctor and philosopher in the Pressburg diocese. He did not explain how he had arrived at this, on the other hand I did not ask him about this." This is proof that the suspension was already known before Cardanus and was only called the "cardan suspension" after this description [1.2]. Translated from the Latin by Theodor Straub, Ingolstadt. Photograph: Deutsches Museum, Munich

1.1 Early Reports on the First Joints

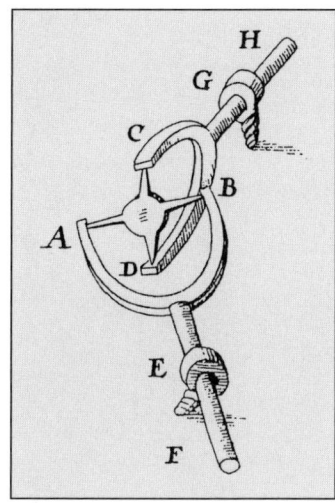

Fig. 1.2. Universal joint of Amicus (16th century). *ABCD* is a cross, the opposing arms *AB* of which are fitted into holes on the ends of a fork *ABF* (fuscinula). The other pair of arms *CD* is received in the same way by the fork *CDH* and the forks themselves are mounted in fixed rings *G* and *E* [1.3, 1.7]. Photograph: Deutsches Museum, Munich

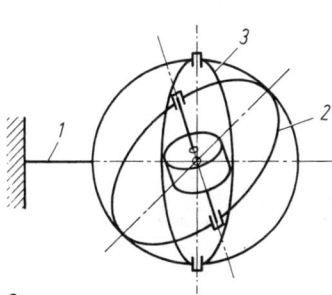 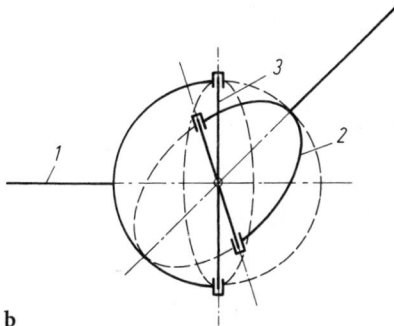

Fig. 1.3a, b. Relationship between gimbal and universal joint. a swiveling gimble, the so-called "cardan suspension" (16th century). *1* attachment, *2* revolving ring, *3* pivot ring; b universal joint of Amicus (17th century), extended to make the cardan suspension. *1* drive fork, *2* driven fork, *3* cross

Johannes Hevelius, which comprised a universal joint similar to that of Amicus (Fig. 1.4). In 1676 he spoke of a "joynt" and a "universal joynt" because it is capable of many kinds of movements [1.5–1.7].

Hooke was fully conversant with the mathematics of the time and was also skilled in practical kinematics. In contrast to Schott he knew that the universal joint does not transmit the rotary movement evenly. Although Hooke did not publish any theory about this we must assume that he knew of the principle of the non-uniformity. He applied his joint for the first time to a machine with which he graduated the faces of sun dials (Fig. 1.5).

Cardano and Hooke can therefore be considered as having prepared the way for universal joints and driveshafts. The specialised terms "cardan joint" in Continental Europe and "Hooke's joint" in Anglo-Saxon based languages still remind us of the two early scholars.

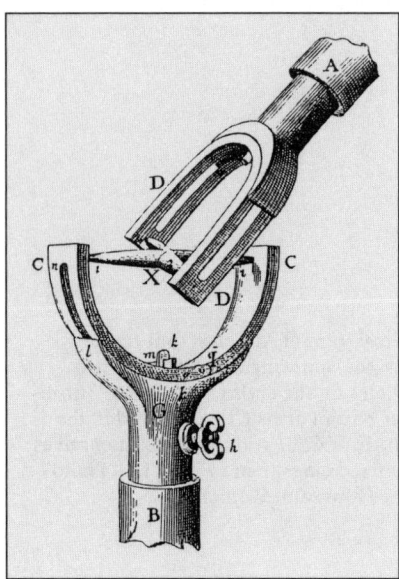

Fig. 1.4. Universal joint from the helioscope by Johannes Hevelius [1.5, 1.7]. Photograph: Deutsches Museum, Munich

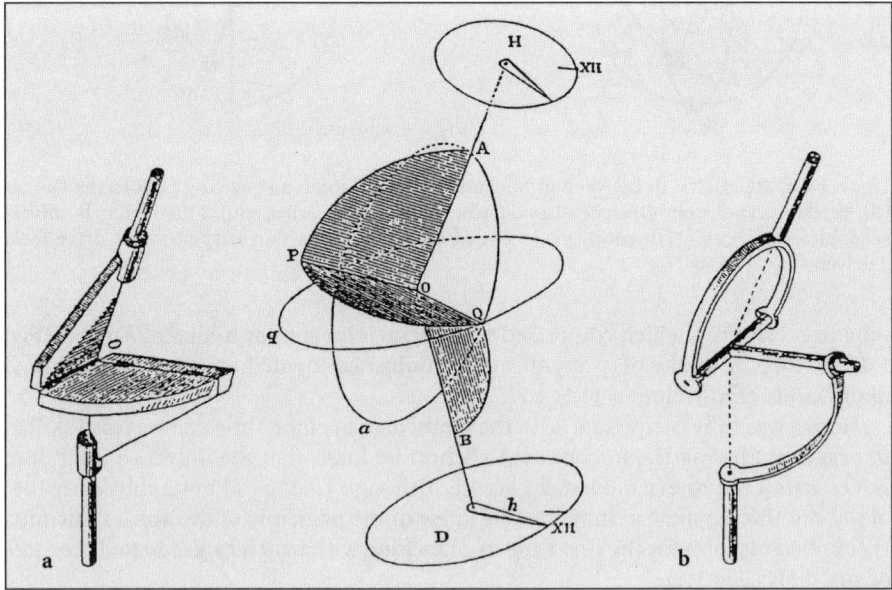

Fig. 1.5a, b. Principle of the apparatus for graduating the faces of sun dials, according to Robert Hooke 1674 [1.11]. **a** Double hinge of the dividing apparatus in a shaft system, as in **b**; **b** relationship between the universal joint and this element. Photograph: Deutsches Museum, Munich

1.2 Theory of the Transmission of Rotational Movements by Hooke's Joints

Gaspard Monge established the fundamental principles of his Descriptive Geometry in 1794 at the École Polytechnique for the study of machine parts in Paris. The most important advances then came in the 19th century from the mathematician and engineer officer Jean-Victor Poncelet, who had taught applied mathematics and engineering since 1824 in Metz at the Ecole d'Application, at a time when mechanical, civiland industrial engineering were coming decisively to the fore. As the creator of projective geometry in 1822 he perceived the spatial relationships of machine parts so well that he was also able to derive the movement of the Hooke's joint [1.8, 1.9]. It was used a great deal at that time in the windmills of Holland to drive Archimedes screws for pumping water.

1.2.1 The Non-Uniformity of Hooke's Joints According to Poncelet

In 1824 Poncelet proved, with the aid of spherical trigonometry, that the rotational movement of Hooke's joints is non-uniform (Fig. 1.6a and b).

Let the plane, given by the two shafts CL and CM, be the horizontal plane. The starting position of the cross axis AA' is perpendicular to it[1]. The points of reference $ABA'B'$ of the cross and the yokes move along circles on the surface of the sphere K with the radius $AC = BC = A'C = B'C$. The circular surface $EBE'B'$ stands perpendicular to the shaft CL *and the circular surface $DAD'A'$ to the shaft CM. The angle of inclination DCE of these surfaces toone another is the angle of articulation β of the two shafts CL and CM.

The relationships between the movements were derived by Poncelet from the spherical triangle ABE in Fig. 1.6b. If the axis CM has rotated about the arc $EA = \varphi_1$

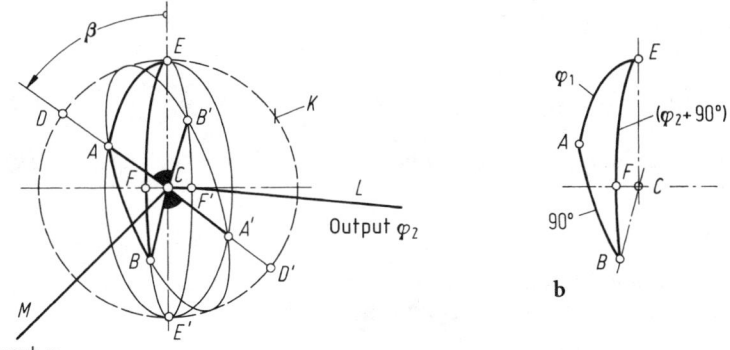

Fig. 1.6a, b. Proof of the non-uniform rotational movement of a Hooke's joint by J. V. Poncelet 1824. **a** Original figure [1.8, 1.9]; **b** spherical triangle from **a**

---

[1] Following the VDI 2722 directive of 1978 this is called "orthogonal". If AA' lies in the plane of the shaft then the starting position is "in-phase" [2.14].

while the axis CL turns about the arc $FB = \varphi_2$ then according to the cosine theorem [1.10, Sect. 3.3.12, p. 92]

$$\cos 90° = \cos \varphi_1 \cos(90 + \varphi_2) + \sin \varphi_1 \sin(90 + \varphi_2) \cos \beta.$$

Since $\cos 90° = 0$ it follows that

$$0 = \cos \varphi_1 (-\sin \varphi_2) + \sin \varphi_1 \cos \varphi_2 \cos \beta.$$

After dividing by $\cos \varphi_1 \sin \varphi_2$ Poncelet obtained

$$\tan \varphi_2 = \cos \beta \tan \varphi_1 \tag{1.1a}$$

or

$$\varphi_2 = \arctan(\cos \beta \tan \varphi_1). \tag{1.1b}$$

For the in-phase starting position, with $\varphi_1 + 90°$ and $\varphi_2 + 90°$ (1.1a). The following applies

$$\tan(\varphi_2 + 90°) = \cos \beta \tan(\varphi_1 + 90°) \Rightarrow \cot \varphi_2 = \cos \beta \cot \varphi_1$$

or

$$\frac{1}{\tan \varphi_2} = \cos \beta \frac{1}{\tan \varphi_1} \Rightarrow \tan \varphi_2 = \frac{\tan \varphi_1}{\cos \beta}. \tag{1.1c}$$

The *first derivative* of (1.1b) with respect to time gives the angular velocity

$$\frac{d\varphi_2}{dt} = \frac{1}{1 + \tan^2 \varphi_1 \cos^2 \beta} \frac{\cos \beta}{\cos^2 \varphi_1} \frac{d\varphi_1}{dt}$$

$$= \frac{\cos \beta}{\cos^2 \varphi_1 + \sin^2 \varphi_1 (1 - \sin^2 \beta)} \frac{d\varphi_1}{dt}$$

$$\frac{d\varphi_2}{dt} = \frac{\cos \beta}{1 - \sin^2 \varphi_1 \sin^2 \beta} \frac{d\varphi_1}{dt} \quad [1.10, \text{Sect. } 4.3.3.13, \text{p. } 107]$$

or because $d\varphi_2/dt = \omega_2$ and $d\varphi_1/dt = \omega_1$

$$\frac{\omega_2}{\omega_1} = \frac{\cos \beta}{1 - \sin^2 \beta \sin^2 \varphi_1}. \tag{1.2}$$

Equations (1.1a–c) and (1.2) form the basis for calculating the angular difference shown in Fig. 1.7a:

$$\Delta \varphi = \varphi_2 - \varphi_1$$

and the ratio of angular velocities ω_2/ω_1 shown in Fig. 1.7b.

For the two boundary conditions, (1.2) gives

$$\varphi_2 = 0°, \quad \varphi_1 = 90° \Rightarrow \omega_2 = \omega_{min} = \omega_1 \cos \beta,$$

$$\varphi_2 = 90°, \quad \varphi_1 = 180° \Rightarrow \omega_2 = \omega_{max} = \omega_1/\cos \beta.$$

Figure 1.7a shows that an articulation of 45° gives rise to a lead and lag $\Delta \varphi$ about ±10°, and very unpleasant vibrations ensue. The reduction of these vibrations has been very important generally for the mechanical engineering and the motor

1.2 Theory of the Transmission of Rotational Movements by Hooke's Joints

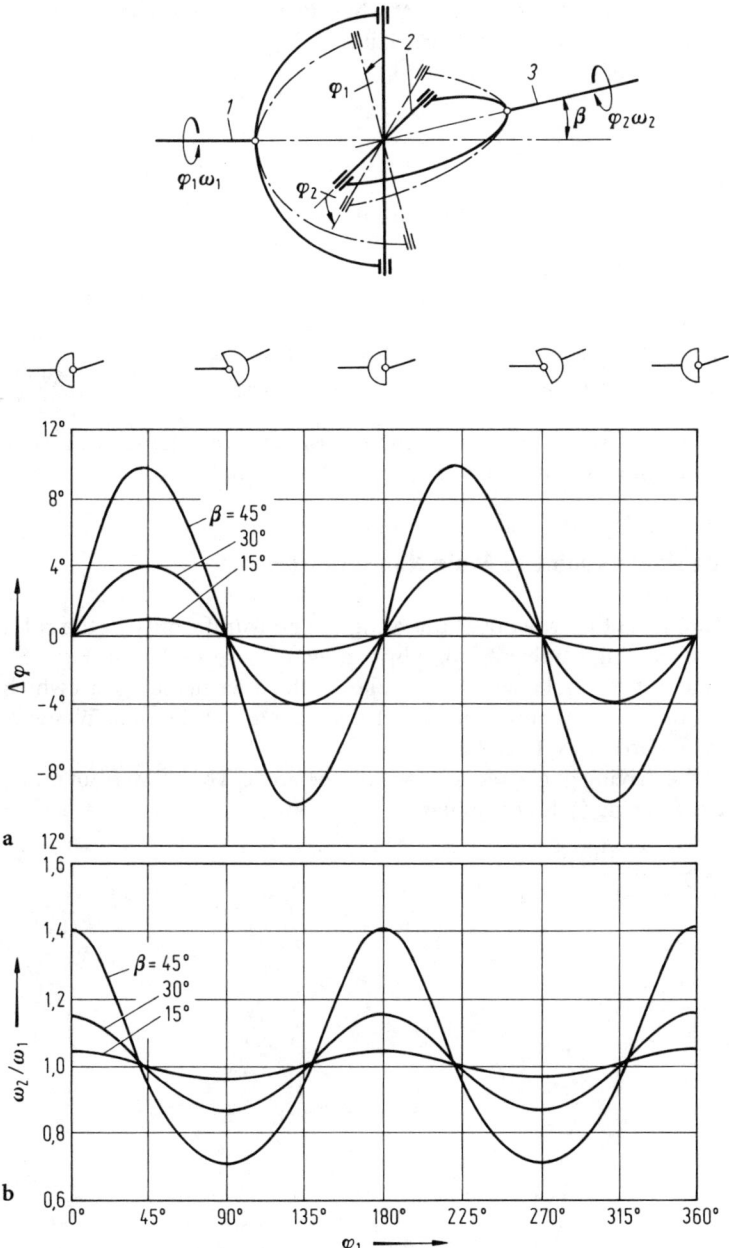

Fig. 1.7 a, b. Angular difference and angular velocities of Hooke's joints for articulation angles of 15°, 30° and 45°. Single Hooke's joint (schematically) [1.36]. *1* Input shaft; *2* intermediate member (cross); *3* output shaft. ω_1, ω_2 angular velocities of the input and output shafts, β angle of articulation, φ_1, φ_2 angle of rotation of the input and output shafts. **a** Angular difference $\Delta\varphi$ (cardan error); **b** angular velocities. Graph of the angular velocity ω_2 for articulation angles of 15°, 30° and 45°

vehicle industries: for the development of front wheel drive cars it has been imperative. $\Delta\varphi$ is also called the angular error or "cardan error".

The *second derivative* of (1.1b)

$$\frac{d\varphi_2}{dt} = \frac{d\varphi_1}{dt} \frac{\cos\beta}{1 - \sin^2\beta \sin^2\varphi_1} = \frac{k'}{1 - k^2\sin^2\varphi_1}$$

with respect to time gives the angular acceleration

$$\alpha_2 = \frac{d^2\varphi_2}{dt^2} = \frac{d\varphi_1}{dt} k' \frac{0(1 - k^2\sin^2\varphi_1) - 1(-k^2 2\sin\varphi_1\cos\varphi_1)}{(1 - k^2\sin^2\varphi_1)^2} \frac{d\varphi_1}{dt}$$

$$= \left(\frac{d\varphi_1}{dt}\right)^2 \frac{k'k^2 2\sin\varphi_1\cos\varphi_1}{(1 - k^2\sin^2\varphi_1)^2} = \omega_1^2 \frac{\cos\beta \sin^2\beta \sin 2\varphi_1}{(1 - \sin^2\beta \sin^2\varphi_1)^2}. \quad (1.3)$$

This angular acceleration α_2 will be required in Sect. 4.2.7, for calculating the resulting torque, and for the maximum values of speed and articulation $n\beta$.

1.2.2 The Double Hooke's Joint to Avoid Non-uniformity

In 1683 Robert Hooke had the idea of eliminating the non-uniformity in the rotational movement of the single universal joint by connecting a second joint (Fig. 1.8). In this arrangement of two Hooke's joints the yokes of the intermediate shaft which rotates at ω_2 lie in-phase, i.e. as shown in Figs. 1.8, 1.9a and b in the same plane, as Robert Willis put forward in 1841 [1.11].

For joint 2 φ_2 is the driving angle and φ_3 is the driven angle. Therefore from (1.1c) tan φ_3 = tan $\varphi_2/\cos\beta_2$. Using (1.1c) one obtains

$$\tan\varphi_3 = \frac{\cos\beta_1 \tan\varphi_1}{\cos\beta_2}.$$

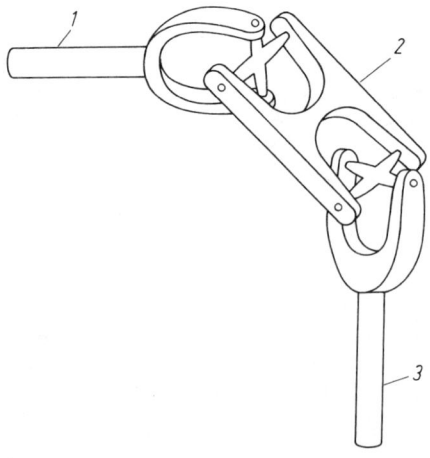

Fig. 1.8. Robert Hooke's double universal joint of 1683 [1.6]. *1* Input, *2* intermediate part, *3* output

1.2 Theory of the Transmission of Rotational Movements by Hooke's Joints

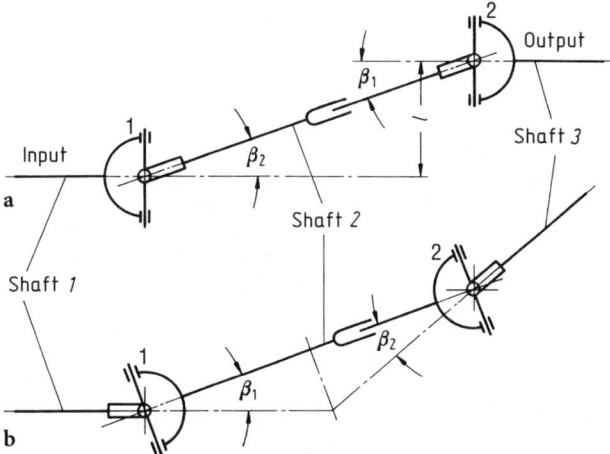

Fig. 1.9a, b. Universal joints with in-phase yokes and $\beta_1 = \beta_2$ as a precondition for constant velocity $\omega_3 = \omega_1$ [1.37]. **a** Z-configuration, **b** W-configuration

From this uniformity of the driving and driven angles then follows for the condition $\beta_2 + \pm \beta_1$, because $\cos(\pm \beta)$ gives the same positive value in both cases. Therefore $\tan \varphi_3 = \tan \varphi_1$ both in the Z-configuration in Fig. 1.9a and in the W-configuration in Fig. 1.9b, and also in the double Hooke's joint in Fig. 1.8. It follows that if $\varphi_3 = \varphi_1$

$$d\varphi_3/dt = d\varphi_1/dt \quad \text{or} \quad \omega_3 = \omega_1 = \text{const.} \tag{1.4}$$

Driveshafts in the Z-configuration are well suited for transmitting uniform angular velocities $\omega_3 = \omega_1$ to parallel axes if the intermediate shaft 2 consists of two prismatic parts which can slide relative to one another, e.g. splined shafts. The intermediate shaft 2 then allows the distance l to be altered so that the driven shaft 3 which revolves with $\omega_3 = \omega_1$ can move in any way required, which happens for example with the table movements of machine tools.

Arthur Hardt suggested in 1901 a double Hooke's joint in the W-configuration for the steering axle of cars (DRP 136605), "…in order to make possible a greater wheel lock…". He correctly believed that a single Hooke's joint is insufficient to accommodate on the one hand variations in the height of the drive relative to the wheel, and on the other hand sharp steering, up to 45°, with respect to the axle. For this reason he divided the steer angle over two Hooke's joints, the centre of which was on the axis of rotation of the steering knuckle, "… as symmetrical as possible to this… in order to achieve a uniform transmission…". Hardt saw here the possibility of turning the inner wheel twice as much as with a single Hooke's joint.

If, however, the yokes of the intermediate shaft 2 which revolves at ω_3 are out of phase by 90° then the non-uniformity in the case of shaft 3 increases to

$$\omega_{max} = \frac{\omega_{min}}{\cos\beta_1 \cos\beta_2},$$

which can be deduced in a similar way to (1.4).

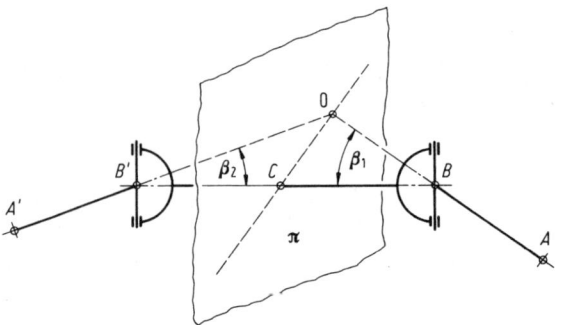

Fig. 1.10. Showing the constant velocity conditions for double Hooke's joints, according to d'Ocagne 1930 [1.14]

1.2.3 D'Ocagne's Extension of the Conditions for Constant Velocity

Is the condition $\beta_2 = \pm \beta_1$ sufficient for constant velocity? In 1841 Robert Willis deduced that the double Hooke's joint possesses constant velocity properties, even when the input and output shafts are neither parallel nor intersecting [1.11]. On these spatially oriented shafts constant velocity can only be achieved if the drive and driven axes are fixed. The yokes of the intermediate shaft 2 must be oppositely phased and disposed about the articulation angle φ according to the equation $\tan \varphi/2 = \tan \beta_2/\tan \beta_1$. In the case of planar axes $\beta_2 = \pm \beta_1$ can only be attained if the input and output shafts intersect one another. This was shown by Maurice d'Ocagne in 1918 [1.12]. According to his theory a double Hooke's joint only transmits in a uniform manner if, in accordance with Fig. 1.10, the following two conditions are fulfilled:

- the axes of the input and output shafts meet at the point O,
- the two Hooke's joints are arranged symmetrically to a plane π which goes through points O and C.

These conditions are not so simple to fulfil in practice. Without special devices it is not possible to guarantee that the axes will always intersect and that the angle β_1 and β_2 are always the same. Figure 1.11a shows a modern design of double Hooke's joint without centring. There are however joint designs (Fig. 1.11b) in which the conditions are approximated (quasi-homokinetic joint) and some (Fig. 1.11c) where they are strictly fulfilled (homokinetic joint)[2]. In 1971 Florea Duditza dealt with these three cases with the methods of kinematics [1.13].

1.2.4 Simplification of the Double Hooke's Joint

The quasi-homokinetic and also the strictly homokinetic double Hooke's joint as shown in Fig. 1.11b and c are complicated engineering systems made up of many dif-

---

[2] The term "homokinetic" (homo = same, kine = to move) was originated by two Frenchman Charles Nugue and Andre Planiol in the 1920s.

1.2 Theory of the Transmission of Rotational Movements by Hooke's Joints

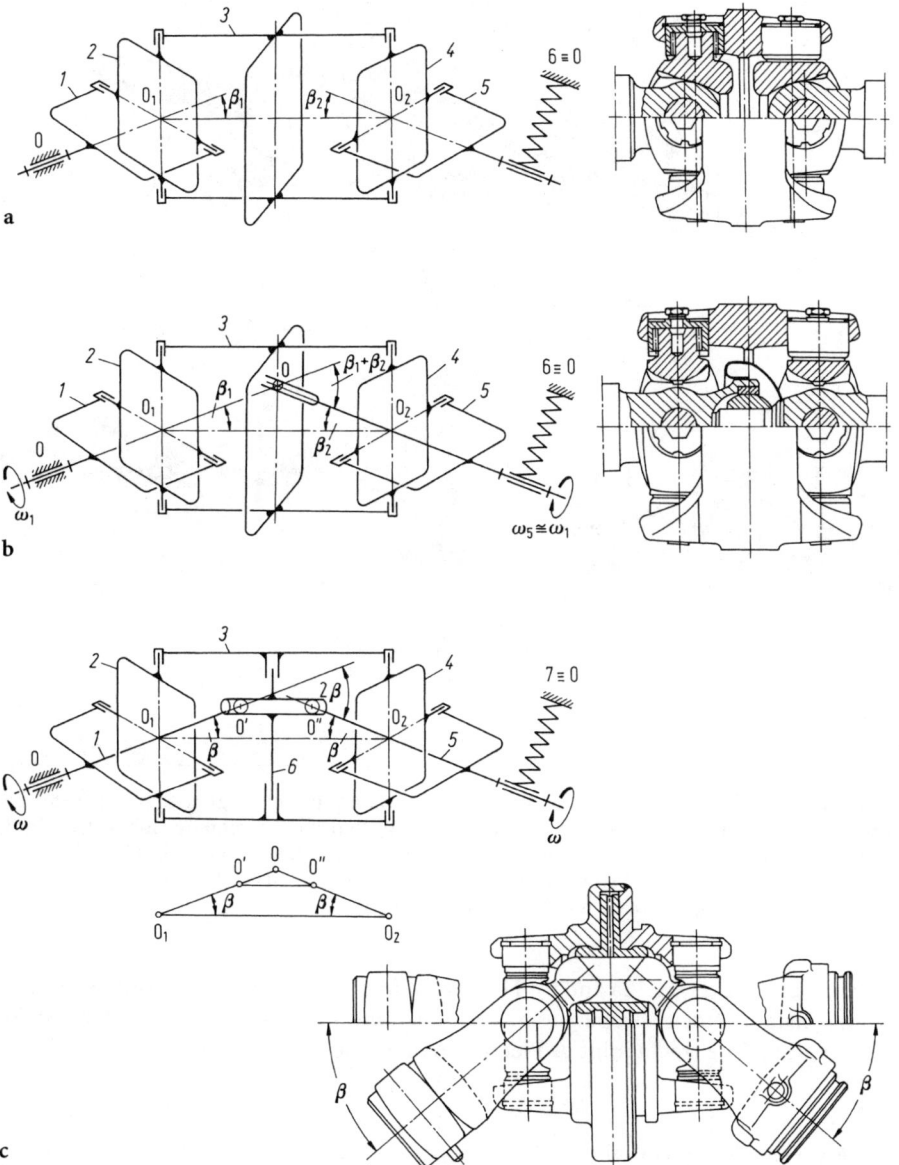

Fig. 1.11a–c. Various designs of double Hooke's joints. **a** Double Hooke's joint without centring. GWB design; **b** double Hooke's joint with centring (quasi-homokinetic), GWB design; **c** double Hooke's joint with steereing attachment (strictly homokinetic) according to Paul Herchenbach (German patent 2802 572/1978), made by J. Walterscheid GmbH [2.14]

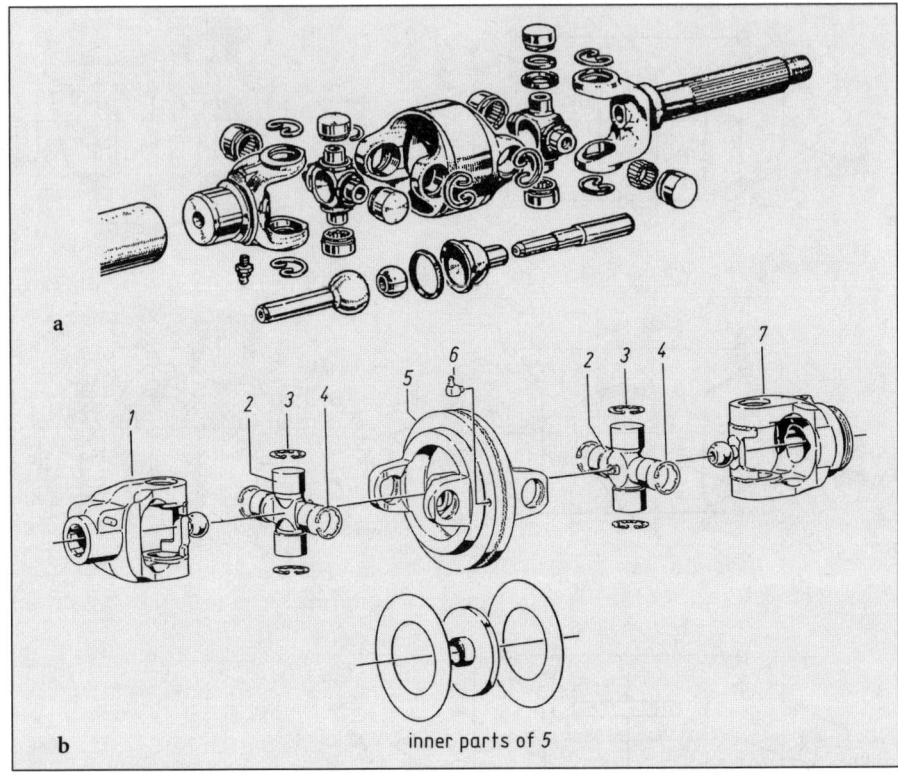

Fig. 1.12a, b. Exploded views of double Hooke's joints. **a** Citroën design with needle bearings and centring, 1934. Quasi-homokinetic, $\beta = 40°$; **b** Walterscheid design by Hubert Geisthoff, Heinrich Welschof and Paul Herchenbach 1966–78 (German patent 1302735, 2802572). Fully-homokinetic, $\beta = 80°$. *1* ball stud yoke; *2* unit pack; *3* circlip; *4* circlip for double yoke; *5* double yoke; *6* right angled grease nipple; *7* inboard yoke/guide hub

ferent elements (Fig. 1.12a and b). The task of finding simpler and cheaper solutions with constant velocity properties has occupied inventors since the second decade of this century. The tracta joint of Pierre Fenaille 1926 has had the best success [1.14, 1.15].

1.2.4.1 Fenaille's Tracta Joint

The Tracta joint works on the principle of the double tongue and groove joint (Fig. 1.13a–c). It comprises only four individual parts: the two forks F and F' and the two sliding pieces T and M (centring spheres) which interlock. These two sliding pieces, guided by spherical grooves, have their centre points C always the same symmetrical distance from the centre of the joint O. Constant velocity is thus ensured; independently of the angle of articulation the plane of symmetry π passes at $\beta/2$ through O. In 1930 Maurice d'Ocagne proved the constant velocity properties of the Tracta joint before the Academie des Sciences in Paris (Fig. 1.13b) [1.14].

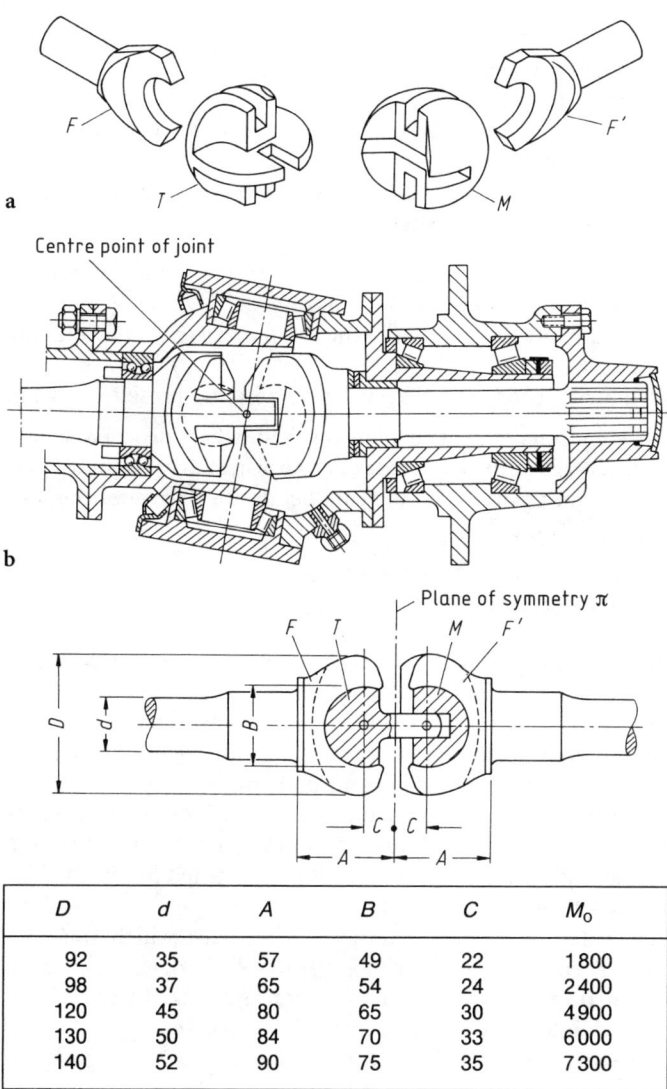

Fig. 1.13a–c. Pierre Fenaille's 1926 Tracta joint [German Patent 617356 of 1927]. **a** Principle of the double tonque and groove joint; **b** as fitted into the king pin bearing of a commercial vehicle; **c** principal dimensions (mm) and torque capacity (Nm) for $\beta = 32°$. Double the torque is permitted for $\beta = 10°$

The *advantages* of the Tracta joint are:
- it transmits movement at articulation angles up to about 32°
- it appears simpler than the double Hooke's joint
- it can be lubricated well because of the large sliding surfaces; the contact faces of the joint spheres slide on each other
- it is easy to fit into the swivel of a rigid front axle with the joint being located axially by thrust faces on the bearing bushes.

The *disadvantages* of the Tracta joint are:
- it gets very hot at high speeds due to the sliding movement
- the centre of the joint and intersection point of the rotational axes of the joint are hard to adjust
- an elaborate bearing housing is necessary because the joint has no self-centring.

The Tracta joint was fitted in independent suspension, front wheel drive passenger cars and in the steering knuckles of the front axles of four wheel drive military personnel cars. It held a dominant position for 40 years before the arrival of the self-centring constant velocity ball joint.

1.2.4.2 Various Further Simplifications

Approximate solutions to constant velocity joints were attempted in 1922 by Mechanics and Julie-Marie-René Retel (French patent 550880), in 1930 by Chenard & Walcker, and in 1931 by Hanns Jung for the DKW front wheel drive small car. With Jung's solution the displacements occurred in the steering yoke between the input and the shell (Fig. 1.14); the ball pivot formed the mid-point of the king pin axis. This DKW joint worked like a hinge in one direction and as a double joint in the other. In later improved versions the travel of the ball pivot was increased through a spherical joint which was open at both sides. Jung's hinged joint was fitted in several Auto Union passenger car models from 1931 to 1963.

In 1949 Marcel Villard designed a five-part joint (Fig. 1.15) which was subsequently fitted in the Citroën 2 CV small car with front wheel drive. Four parts (3 – 6), sliding in semicircular recesses allowed articulation and rocked on machined serrations, the axes of which intersected in the centre of the recesses.

This joint was not any simpler but the durability equalled that of the engine.

1.2.4.3 Bouchard's One-and-a-half Times Universal Joint

Robert Bouchard developed the double Hooke's joint for the front wheel drive Citroën 15 CV car in 1934 (Fig. 1.12a). His experience led him 1949 to the idea of a $1^1/_2$ times quasi-homokinetic universal joint [1.16, 1.17].

The wide-angle *fixed joint* is built up symmetrically and consists of four main parts (Fig. 1.16a): the two identical fork-shaped, parts *1*, the two identical joint crosses *2* which interlock and which have only three bearing journals each. Theses are guided in standard bearing cups *3* and are thus able to support the same loads and have the same durability as Hooke's joints with needle bearings. The ball-shaped design of the joint reduces the amount of axial space needed (Fig. 1.16b). The joint

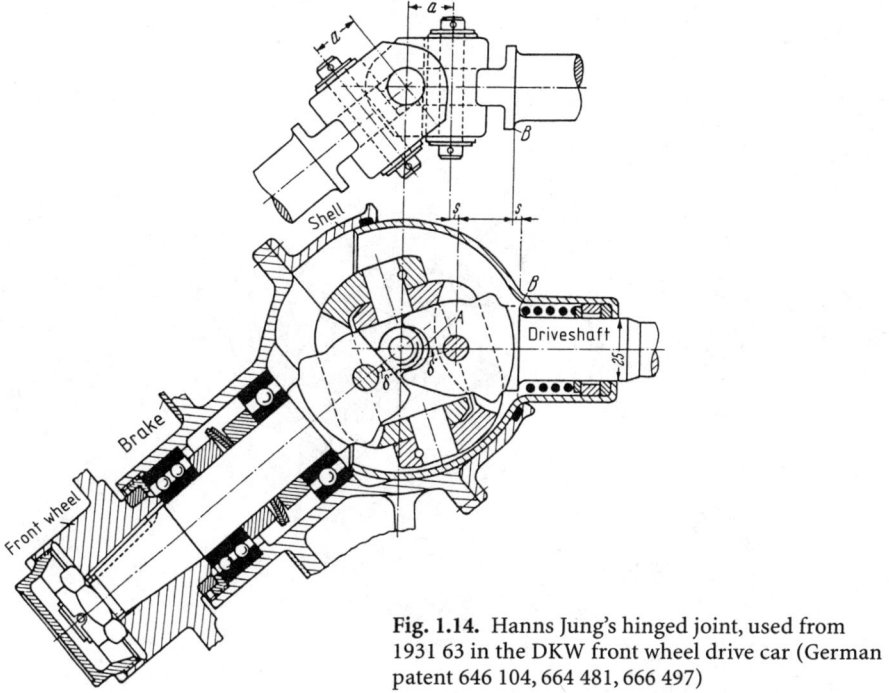

Fig. 1.14. Hanns Jung's hinged joint, used from 1931 63 in the DKW front wheel drive car (German patent 646 104, 664 481, 666 497)

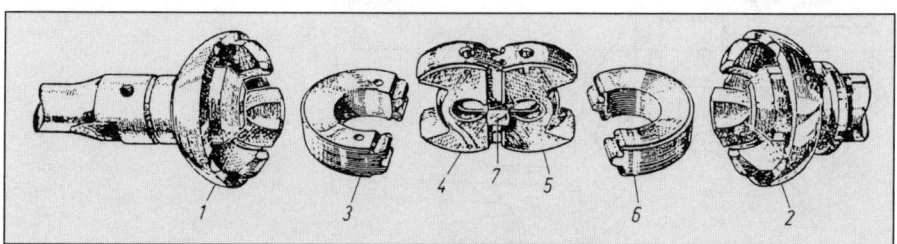

Fig. 1.15. Marcel Villard's joint of 1949 for the Citroën 2CV [1.33, 1.38]. *1* Input, *2* output: *3–6* sliding parts; *7* coupling piece. Photograph: M. Kunath

can be fitted in open steer drive axles which means that the full articulation angle $\beta = 45°$ can be utilised. Any axial displacement of the shaft on articulation is avoided because the mounting is fixed radially and axially in the housing. Since the holes of the joint yoke are positioned eccentrically to the axis of rotation the centre plane of the cross adjusts itself automatically in the plane of symmetry π. It moves with each revolution of the shaft, with the amplitude $\pm a$ as a function of the articulation angle, parallel to the homokinetic plane of the joint. The amplitude a oscillates between 0 (in-line joint) and the highest $\pm$ value (joint at β_{max}), which impairs the smooth running of the joint at large articulation angles and high speeds. The joint was made in France only from 1952 to 1970 in five sizes for the steer drive axles of slow speed commercial vehicles (see table in Fig. 1.16b).

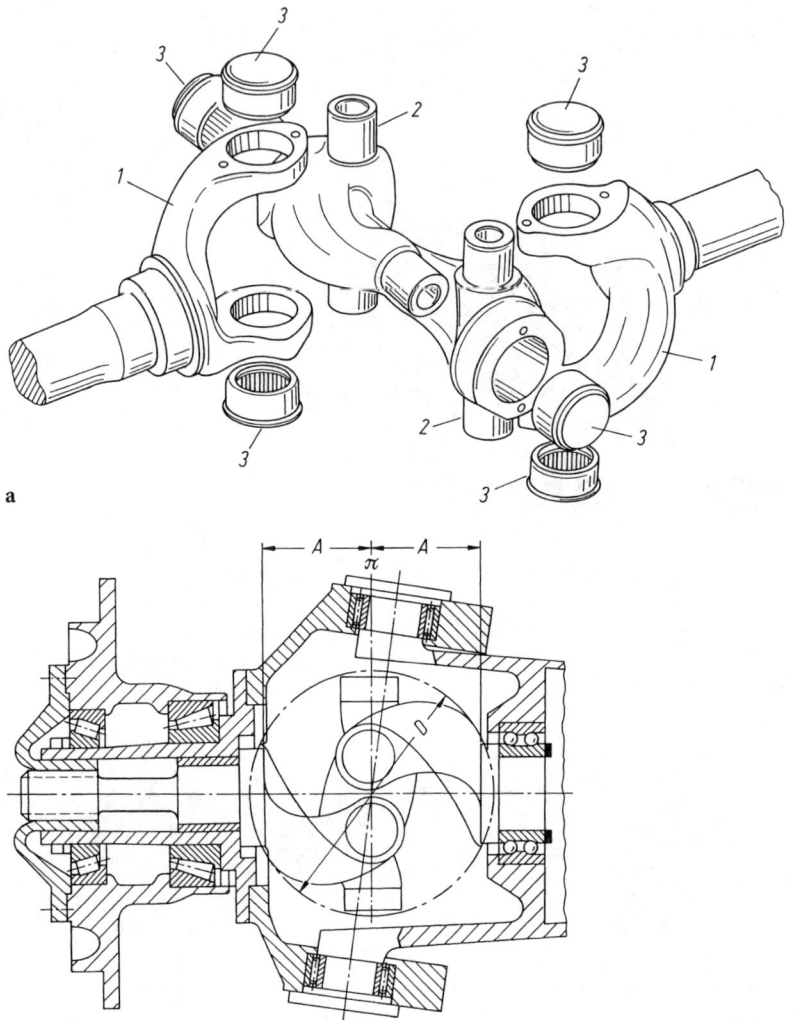

Joint size	D	2A	M_d	M_0
1000	140	130	60	320
2000	185	172	130	715
3000	220	205	240	1320
4000	264	235	340	1870
5000	315	280	570	3130

Fig. 1.16a, b. One-and-a-half times universal joint; Robert Bouchard's fixed joint of 1949 (German patent 838 552). **a:** *1* Drive and driven yokes, *2* crosses, *3* needle bearings; **b** fixed joint in a steer drive axle. Principal dimensions in mm and torque capacity in Nm

1.3 The Ball Joints

The transmission of rotary movement between two angled shafts by means of two bevel gears (Fig. 1.17a) is uniform, apart from small discrepancies due to play and manufacturing variations. It was an imaginative concept to replace the gear teeth by balls and spherical pockets to provide the transmission (Fig. 1.17b). The balls are held as if on a ring and circulate between the input and output members of the joint. However, in order to give the ball joint the ability to articulate an important step remains. The spherical pockets have to be extended into ball tracks in a bell-shaped outer race and a spherical inner race (Fig. 1.17c). The ball-track principle allowed the development of universal joints to begin anew and led to the compact constant velocity joint (Fig. 1.17d). This principle goes back to the 1908 work of the American William Whitney (Fig. 1.18); USA-pat. 1022 909.

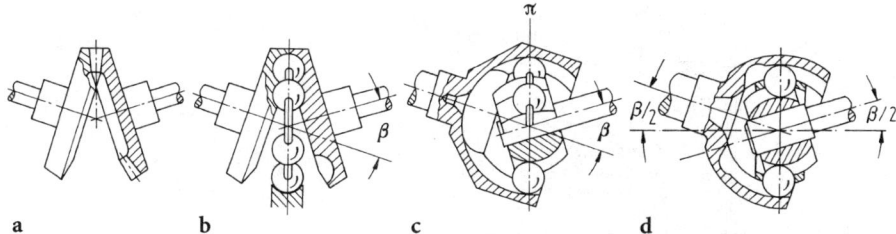

Fig. 1.17 a–d. Analogy of a pair of bevel gears at a fixed angle developing into a ball joint with a variable articulation angle β. **a** Pair of bevel gears; **b** ball joint with fixed articulation angle β; **c** ball joint with variable articulation angle β, ball tracks and spherical inner race; **d** ball joint with cage

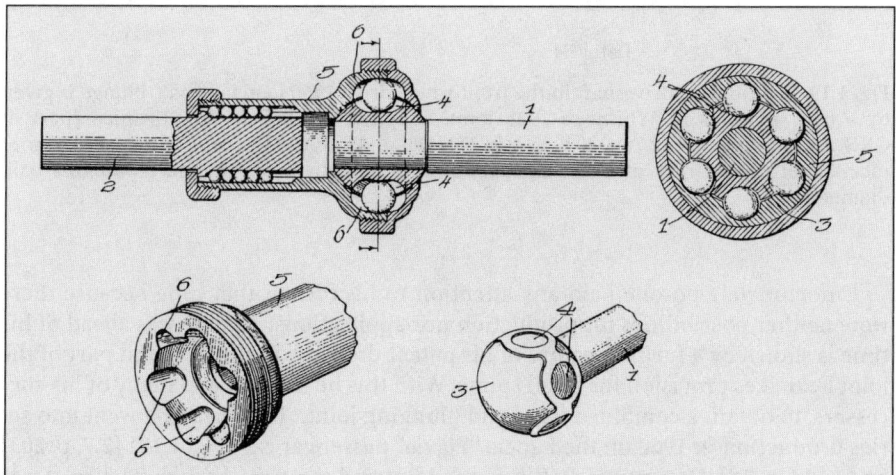

Fig. 1.18. William Whitney's ball joint of 1908 with cage and concentric tracks (US patent 1022909). *1* and *2* Shafts, *3* inner race, *4* inner track, *5* outer race, *6* outer track

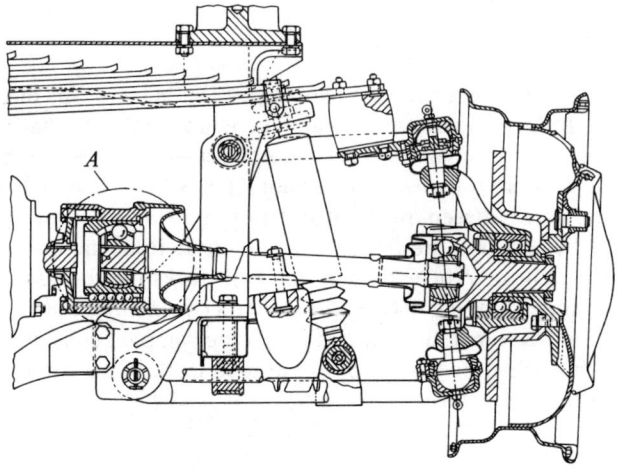

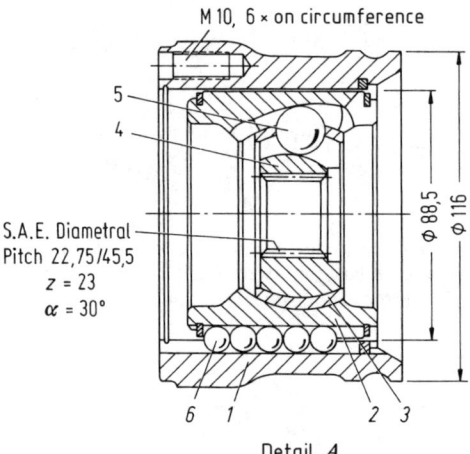

Fig. 1.19. Ball jointed driveshaft in the front wheel drive 1960 Lancia Flavia. Plunge is given by a ball spline as in Whitney's 1908 joint. Supplied by Birfield SpA Brunico (part of GKN Automotive AG) [2.7, p. 203; 2.14, p. 18]. Detail A: inboard joint. *1* outer race; *2* inner race; *3* ball cage; *4* rack of one main ball; *5* main ball, diameter 18 mm; *6* plunging ball, diameter 3/8″

Unfortunately no-one paid any attention to his idea at that time because there were neither possibilities for production nor applications. That he was ahead of his time is shown by a further feature of his patent drawing: in the left hand part of the joint he makes provision for a ball spline. With this he tried, like so many of his successors, to obtain a combined fixed and plunging joint. This idea first went into series production in 1960 on the Lancia "Flavia" passenger car (Fig. 1.19) [2.7, p. 203] and [2.14, p. 18]. Unfortunately Whitney's joint had the deficiency that when it was straight no forces acted on the balls to guide them into the plane of symmetry π (Fig. 1.17c) so that their position in the concentric tracks was indeterminate.

1.3 The Ball Joints

1.3.1 Weiss and Rzeppa Ball Joints

Carl W. Weiss overcame this deficiency in 1923 by offsetting the generating centres of the ball tracks in the input and output bodies from the centre 0 of the joint by an amount c (Fig. 1.20). This created intersecting tracks which steer the balls automatically – even with a straight joint – into the angle-bisecting plane π [1.47].

Of the various designs made using the Weiss principle (F5 to F8), the F7 joint shown in Figs. 1.21 and 1.20 has proved most successful in practice. The large offset c of the generating centres O_1 and O_2 from the joint centre O (Fig. 1.21) produces tracks which, even when the joint is straight, intersect with the large angle $\delta = 2\varepsilon = 50°$ to $60°$ and steer the balls into the plane of symmetry π.

The joint consists of two identical ball yokes which are positively located by four balls. Two balls in circular tracks transmit the torque while the other two preload the joint and ensure that there is no backlash when the direction of loading changes. The two joints are centred by means of a ball with a hole in the middle. This is prevented from falling out, when the joint is at an angle, by a cylindrical pin to DIN 7 or with a conical pin to DIN 1473. The largest articulation angle of the Weiss joint is $32°$. The static torque capacity of five joint sizes is given in the table in Fig. 1.21.

Its offset c was big enough to allow for a cage-less ball joint. The ball tracks run parallel to the axes of the input and output shafts, which is why the skew angle γ the pressure angle $\alpha = 90°$ and the conformity in the circular cross section of the $\varkappa_Q = -0.92$ to -0.96 ($\varphi_Q = 1.04$ bis 1.08).

Weiss however lost Whitney's advantageous division of the joint into a bell-shaped outer race and a spherical inner race with internal axial location. He chose instead a radial division with a complicated external method for axial location. A disadvantage arising from this is an articulation angle of only about $30°$.

In the unarticulated Bendix-Weiss joint the balls transmit the torque with a constant pressure angle α and skew angle γ. In the articulated joint on the other hand, the values of both α and γ depending on the articulation angle β. With the spread of front wheel drive vehicles with independent suspension, driveshafts were needed with inboard joints able to articulate up to $20°$, which was achieved at full wheel travel. In addition a plunge capacity similar to the 10 to 16 mm provided by the V5 to V8 joints was needed. Initially this requirement was met with Hooke's joints, with splines providing the plunge. After a while however, this solution was recognised to be unsatisfactory because of excessive plunging forces and the generation of too much noise by the nonuniform movement of the Hooke's joint. The fixed Bendix-Weiss ball joint was therefore redesigned into a plunging joint.

The most advanced plunging joint which works on the Weiss principle is the six-ball star joint (V7). It was developed for independent suspension rear wheel drive passenger cars and allows an articulation angle of $25°$ and a plunge of 20 mm (Fig. 1.22) [5.10]. It has a greater torque capacity than the four-ball plunging joint (Fig. 1.21).

The 1963 six-ball star joint of Kurt Enke (French pat. 1353 407) consists of an outer three-pointed star and an inner three armed inner race (Fig. 1.25) [5.10]. There are three pairs of parallel tracks in these two parts, which run on the surface of a $50°$ cone, symmetrically about the axis of rotation of the joint. The six balls, 18 mm in diameter, are steered firmly by the track intersections of both ball yokes: this means

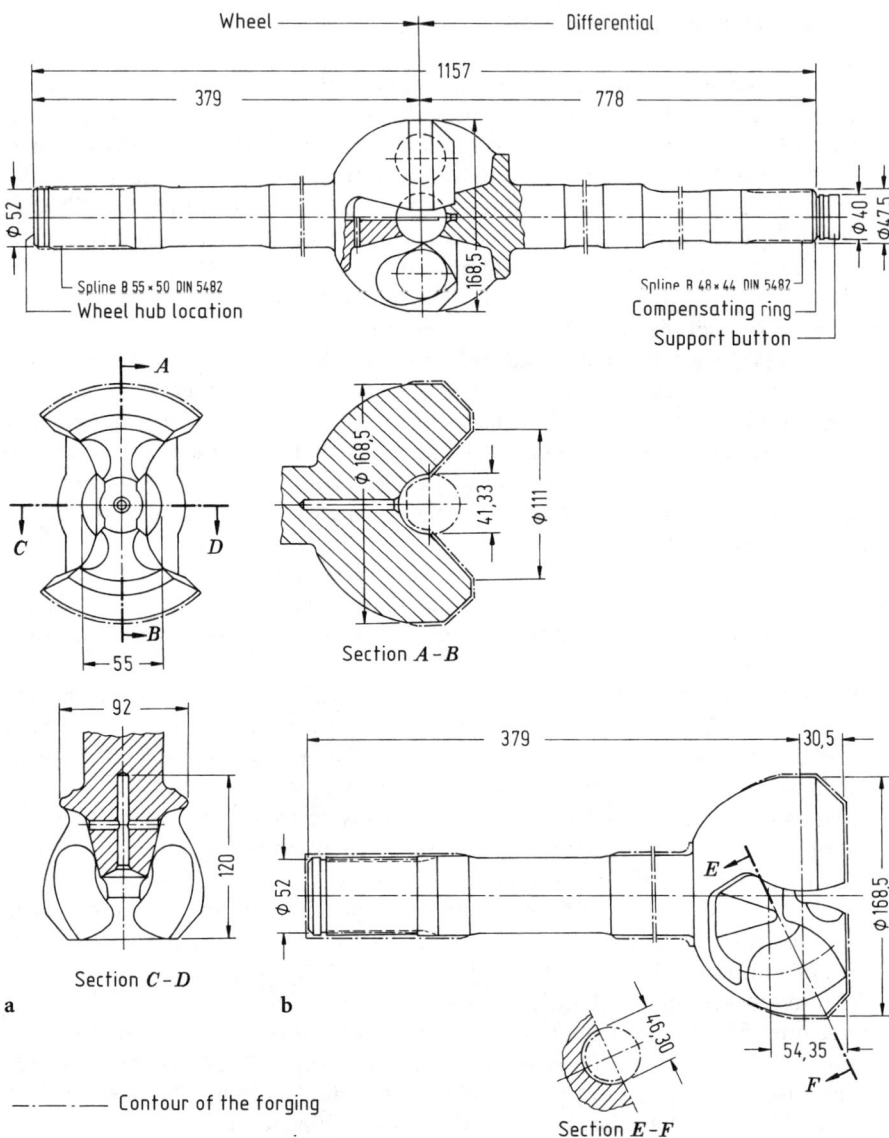

Fig. 1.20a,b. Weiss joint in the steer drive axle of a 6 tonne lorry built from 1953–60. **a** Ball yoke, **b** ball yoke with stub shaft for driving the sun wheel of the planetary hub reduction gear [1.22]

that they lie in the plane of symmetry π. The torque is transmitted by only three balls, depending on the direction of rotation, while the remaining three centre and hold it together. The balls are preloaded because this cageless joint only works fully without play. The joint is completely encapsulated and thus protected from falling apart. It is lubricated for life with SAE 90 hypoid oil and is sealed against water by a very strong boot.

1.3 The Ball Joints

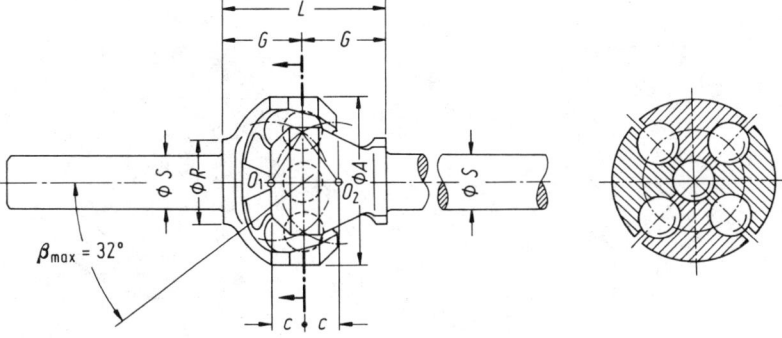

Principal dimensions (mm) and torque (Nm)

Joint size	G	L	S	A	R	M_N
253	52	104	35	112	72	2200
203	55	110	40	120	75	3000
201	62	124	43	130	82	3400
10.5	78	156	53	169	110	7500
10.6	100	200	65	200	130	14500

Fig. 1.21. Fixed joint of the Weiss type

In the eighties, this joint was replaced by the VL joint because it is lighter, cheaper, easier to repair and less likely to fail.

Fixed joints of the Weiss type became known worldwide after they were used as outboard joints in the part-time front wheel drives of American military trucks in the Second World War. They were produced in Germany from 1947 to 1965 for spares, but were discontinued because the permissible speed, articulation angle and durability were too low.

A two-ball joint following the Weiss principle was developed in 1956 by Gaston Devos (Fig. 1.30). His two balls run in straight tracks inclined at an angle of approximately 12° to the axes of the ball yokes compared with 25 to 30° for the four-ball Weiss joints. Devos obtained very smooth running and high torque capacity, a large articulation angle of 42° and high mechanical efficiency. One ball transmits the torque while the other grips the joint and preloads for the reverse travel. This joint was fitted in the small Citroën 2CV instead of the production Hooke's joints which demonstrated their non-uniformity under hard cornering.

In 1927 Alfred Rzeppa filed an application (US-Pat. 1665 280) for a joint which still included the deficiency of Whitney's joint. He therefore introduced in 1933 positive steering of the balls through a cage with a pilot (guide) lever (Fig. 1.23b) as an auxiliary control device. The cage is enlarged by means of a hemispherical cup. The lever with two outer and a middle ball is mounted in three places: in the joint shank, in the hemispherical part and in the inner part of the joint. The position of the three

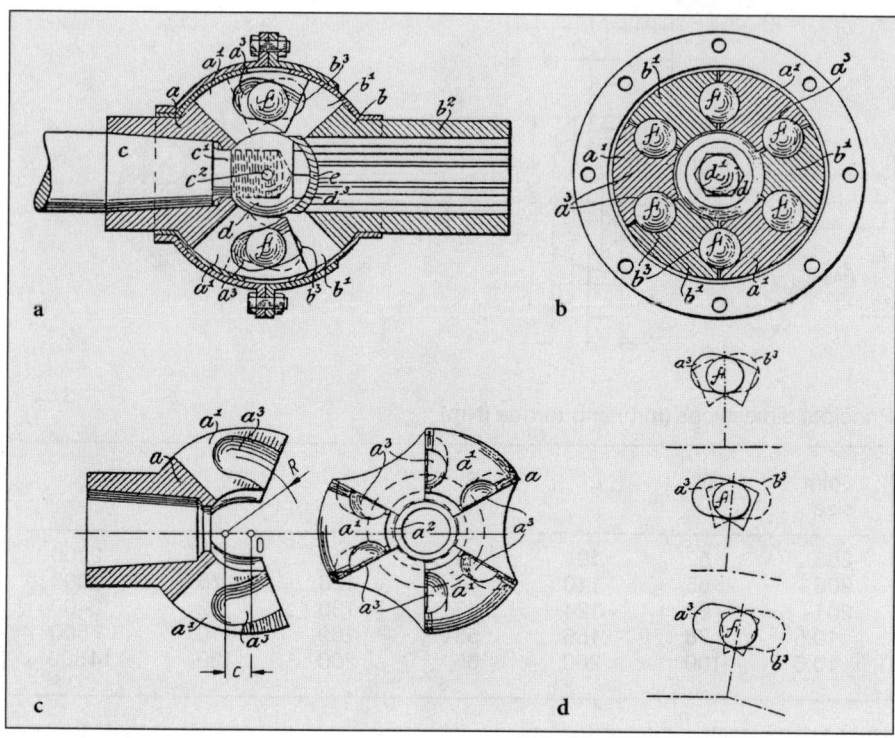

Fig. 1.22. a 1923 ball joint by Carl Weiss with generating centres of the tracks offset by c from the mid-point O of the joint (US Patent 1522351); **b** section through the joint centre O; **c** ball tracks in transverse section; offset c from the mid-point θ of the joint; **d** crossing of the tracks

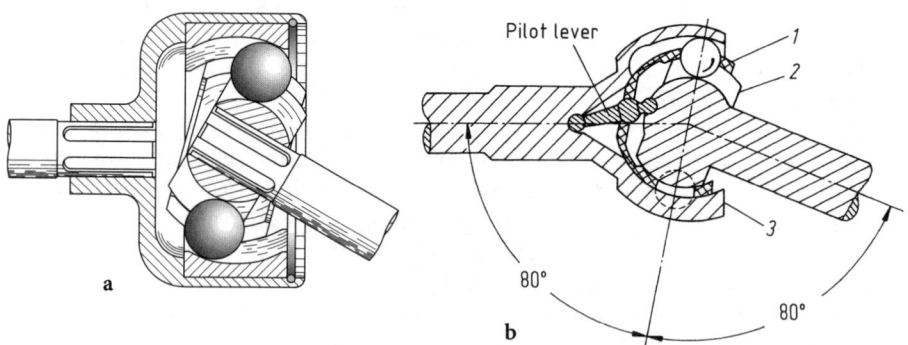

Fig. 1.23. a fundamental patents by Alfred H. Rzeppa (US-PS 1665280, 1916442, 2010899); **b** with forced steering of the balls by a cage and pilot lever. *1* cage, *2* inner body of the joint, *3* outer body of the joint

1.3 The Ball Joints

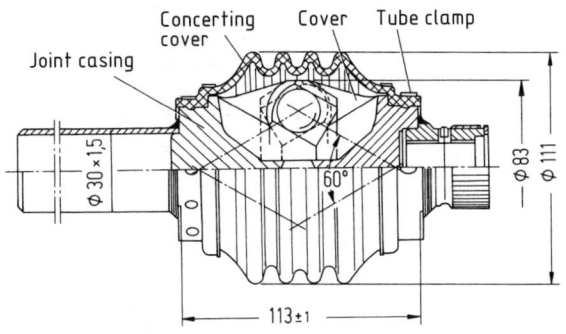

Fig. 1.24. Four ball plunging joint, in the halfshaft of the front wheel drive Renault R16. Bendix-Weiss-Spicer design. Largest articulation angle $\beta_{max} = 20°$; ball diameter = 22.2 mm, axial plunge $s = 20$ mm

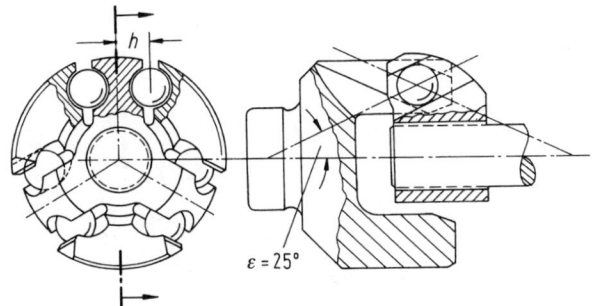

Fig. 1.25. Kurt Enke's 1963 six ball conical ruled or star joint. Ball tracks in pairs offset by h and inclined at 25° to the joint axis [5.9]

steering balls is chosen so that, when the joint articulates, the cage locates the balls in the plane of symmetry π. With this method of steering the whole depth of the track is retained which gives a high torque capacity. The pilot lever was however the weak point [1.23][3].

Rzeppa's principle did not achieve a breakthrough until Bernard Stuber, also in 1933, again with a "ball and socket" joint (US Patent 1975 758), offset the generating centres of the tracks by the amount c from the centre of the joint (Fig. 1.26a–d, 1.28). In so doing he created tracks which intersected (Fig. 1.26b) and even with a non-articulated joint, automatic positioning into the bisecting plane π of the two joint bodies. The big advantages of the Rzeppa joint of 1934 with offset steering (US-Pat. 2046 584) are an articulation angle of up to 45° and simple assembly: it is not necessary to separate and then bolt together the bell-shaped outer part.

Having shown the principle of constant velocity and the steering of the balls, the conditions for constant velocity will now be analysed. Two different methods can be used:

– the indirect method, where to output angle φ_2 must be equal to the input angle φ_1. The design of the joint must then be determined.
– the direct method, in which for a given joint, the relationship $\varphi_2 = f(r, \varphi_1, \beta)$ is determined mathematically and a check is then made to see whether the condition $\varphi_2 = \varphi_1$ is fulfilled.

---

[3] Further pilot levers as steering aid to the cage even with small angles: Rzeppa (US-PS 2010899/1933 and 2046584/1934), John Wooler (GB-PS 760681/1952), E. Aucktor (DE-PS 1232411/1963).

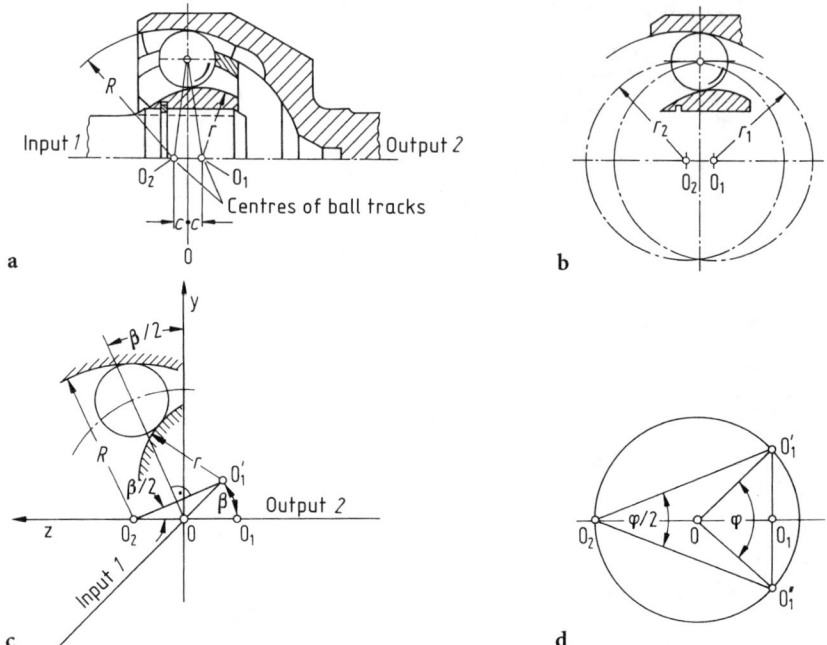

Fig. 1.26a–d. Bernard Stuber's offset steering 1933 (US patent 1975758). **a, b** The offset (by c) of the generating centres O_1 and O_2 means that the tracks R and r intersect even in the non-articulated joint and position the balls automatically in the plane of symmetry. **c** The generating centres of the tracks O_1 and O_2 lie on the joint axis $O_1 O\ O_2$ when the joint is not at an angle. When it articulates through β, O_1 is lifted on an arc of radius c about O towards O_1'. Point O_2 does not move. Hence line $O_1'O_2$ forms the semi-articulation angle $\beta/2$ with the axis of the output member 2. The balls are thus steered into the plane of symmetry π which, as shown in **a–d**, is the condition for the *uniform* transmission of rotation by intersecting shafts. **d** The relationship of **c–d** is based on Euclid's subtended angle theorem. *

Intersecting Angles $\chi = 18°$ to $40°$

These angles still steer the balls properly but need auxiliary elements (HE) to guide them. For VL-plunging joints, the tracks intersect at 32° (SR). They need a ball cage so that when the joint articulates, the balls do not fall out of the tracks. On the Devos joint with straight tracks (Fig. 1.30) and intersecting angles of about 24°, a pin guides the rotatable and axially free outer balls of its three-ball unit into the plane of symmetry π.

The ball tracks of RF fixed joints intersect at 18° to 20° due to the track offset (RO). The opening angle δ of the tracks must be larger than twice the friction

---

* "The subtended angle is half as big as the centre angle over the same arec." (Fig. 1.26d). Euclid/Clemens Thaer, Die Elemente (The Elements), Book 3, Section 21, Darmstadt: Wissenschaftl. Buchges. 1980.

1.3 The Ball Joints

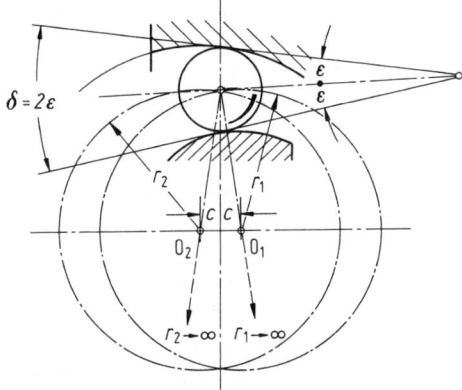

Fig. 1.27. The divergence or opening angle δ of the fixed ball joint is equal to 2ε, being the sum of the angles made by the tangents to the tracks. If the radius $r \to \infty$, inclined, straight tracks occur. If the balls are to move in the tracks, the tilt angle ε at the point of contact between the ball and the track flank must be larger than the friction angle of friction p. This is achieved by using a sufficiently large offset c

angle ϱ (Fig. 1.27). The balls are steered into the plane of symmetry π by the offset (Figs. 1.28a and b). The condition for constant velocity is

$$y'_2 = - z'_2 \cot \beta/2.$$

The offset c must not be too large (Fig. 4.55), or else the track flanks 9 and 10 become too short and are no longer able to transmit the compressive forces. Moreover, since the tracks open a cage must be used to prevent the balls from falling out.

Intersection Angle $\chi = 0°$

In the initial ideas of Whitney 1908 and Rzeppa 1927, the tracks intersect at 0° when the joint is straight, i.e. they are parallel. This means that the position of the balls in

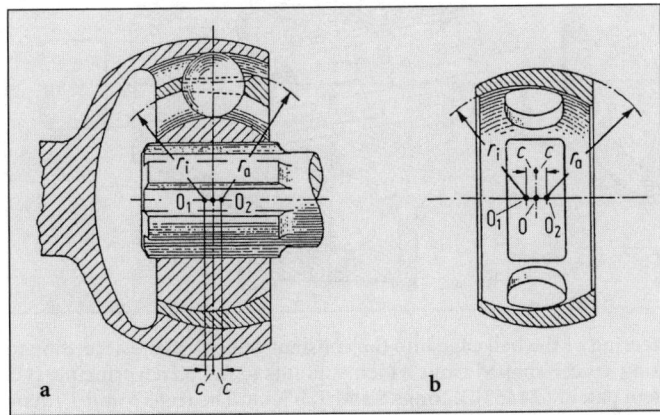

Fig. 1.28a,b. Steering the balls into the angle-bisecting plane by offsetting the generating centres of the cage spheres, according to Bernard K. Stuber 1933 (US-Patent 1975758). **a** Effect of the offset c on the ball joint; **b** the centres O_1 and O_2 of the cage spheres r_i and r_a are symmetrical with respect to the joint centre 0 on the joint axis

the tracks is undefined [2.9, p. 47] and the joints cannot function at articulation angles less than about 18°. Hence in 1933 Rzeppa introduced the "pilot lever" as an auxiliary control device (Fig. 1.23b). However with wear the auxiliary steering fails and the balls jam and so in about 1960 these joints were superseded by joints with track steering.

The cage is extended spherically. The lever with two outer balls and a centre ball is located in three places: in the joint shaft, in the spherical extension and in the inside part of the joint. The spacing of the three balls is chosen so that, when the joint articulates, the cage puts the balls into the plane of symmetry π. With this steering system the full track depth is maintained which gives a high transmission capacity. With increasing wear however, this auxiliary steering fails and the balls jam.

In 1933 Rzeppa also suggested the use of saucer-shaped rings as a mean of auxiliary steering (Fig. 1.29). The spherical surfaces S and T with generating radii r_a and r_i steer the cage, together with balls, into the plane of symmetry π, implementing the offset principle once more.

Further steering devices or combinations of the various types of steering are feasible.

From 1934/39 Rzeppa provided for the tracks of the outer joint body to be undercut free; they followed straight lines inclined to the axis and they diverged from one another towards one side and could only be incorporated from there into the joint body (US Patent 20 46 584). With this design, however, it was no longer possible to achieve an adequate track depth at large articulation angles so that there

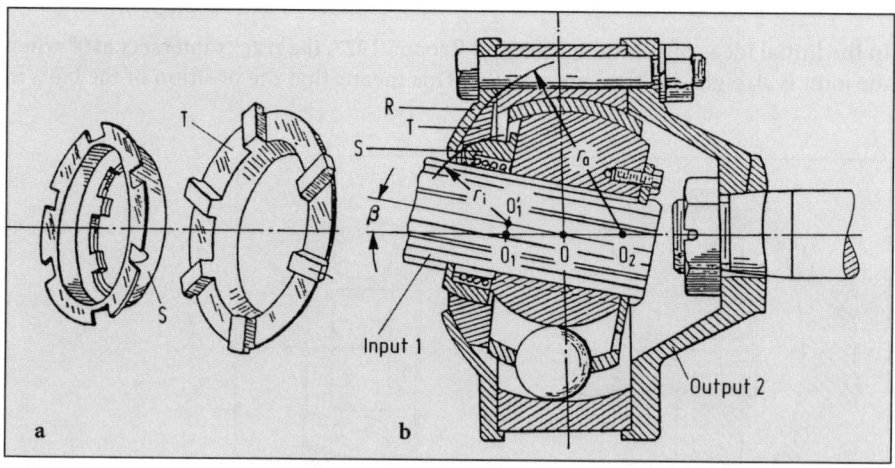

Fig. 1.29a, b. Auxiliary steering of the ball cage into the constant velocity plane according to Alfred H. Rzeppa 1933 using saucer-shaped rings which amounts to the offset-principle (US patent 2010899 and German patent 624463). **a** Rings S and T, **b** joint. The rings S and T rotate about their centres O_1 and O_2 which are displaced by the same amount c from the joint centre O. Ring T slides on the spherical face R of the outer body and on the spherical face S which is joined to the inner race. Thus the centre O_1 of the hemisphere R is raised during articulation, while O_2 remains stationary. The line $O'_1 O_2$ steers the cage, with the balls, into the plane of symmetry π again according to the offset-principle

1.3 The Ball Joints

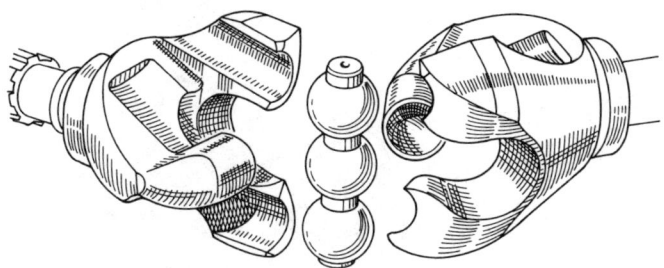

Fig. 1.30. Two-ball fixed joint with straight tracks, developed by Gaston Devos in 1956 (French Patent 1 157 482), with a spherical cup in the centre locating the through-bored centring ball. This ball is pressed firmly onto the middle of a pin. Two other through-bored balls are mounted on the ends of the pin so that they can rotate and can move longitudinally. If the bearing pin with its three balls is preloaded between the two yoke heads, it holds the whole joint together so that it is self-centred (Citroën 2 CV)

was too much pressure between the balls and the track walls. Rzeppa's inventions of "wheel-side" fixed joints helped to promote front wheel drive vehicles and the automotive industry still maintains an interest in ball joints of this kind. The first firm to manufacture the Rzeppa joints was the Gear Grinding Machine Co. of Detroit in 1929, for shaft diameters of $1^1/_4''$ (passenger cars) and $1^1/_3''-2^1/_2''$ (heavy goods vehicles). As the automobile industry has demanded larger angles of steering lock, up to 50°, improvements have had to be made in design and production methods.

1.3.2 Developments Towards the Plunging Joint

In many cases where torques are being transmitted at an angle, an alteration in length or displacement must also take place. With a Z-drive either a parallelogram linkage, such as that designed by Herbert Vanderbeek in 1908 (Fig. 1.31), or a plunging device is required for trouble-free transmission.

As early as 1842 Wilhelm Salzenberg [1.18] illustrated the idea of plunge in the drive of a rack for moving a carriage in a sawmill or for a planing machine (Fig. 1.32).

A machine part consisting of the drive element *d* and the driven element *b* is now called a bipode (literally "two-footed") joint. Joints allowing plunge are particularly important in motor vehicles, where there are many applications for driveshafts. In the case of driven steer axles with a front wheel steer angle of at least 30°, substantial movement in the driveshaft has to be considered. Robert Schwenke recognised this in 1902. In his Patent Specification he drew a complete driveshaft from the axle to the front wheel of the car, with a fixed joint outboard but a plunging joint inboard. The inner bipode was guided by two rectangular blocks with spherically shaped bearing surfaces (Fig. 1.33).

The splined shaft with two, four or more splines was the most popular choice for permitting plunging movements in shafts from the 19th until the middle of the 20th century. August Horch in 1904 chose this type of design (German patent 158 897) for

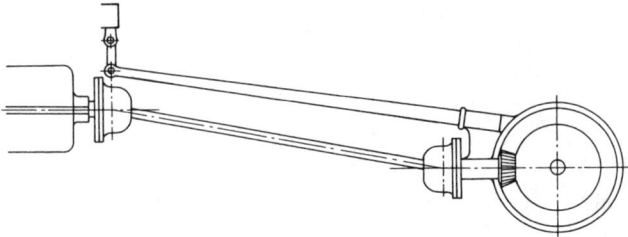

Fig. 1.31. Rear axle drive on automobiles with approximated parallel guiding by H. Vanderbeek 1908 [1.34]

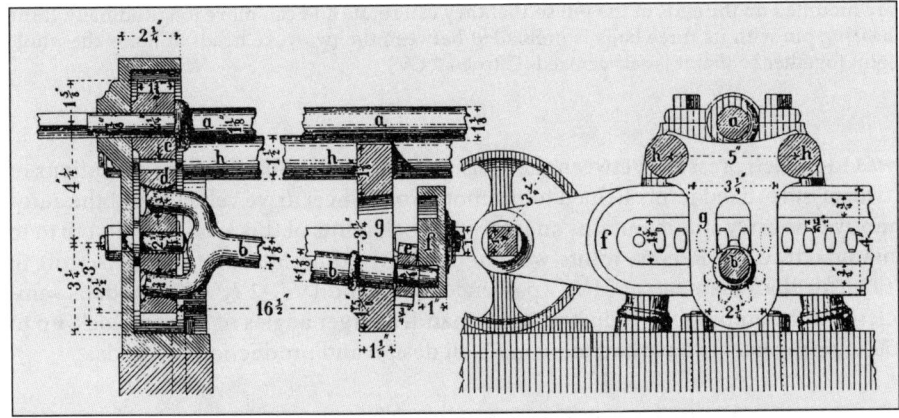

Fig. 1.32. Plunging fork as bipode joint on a machine from the 19th century [1.18]. Repro: M. Kunath

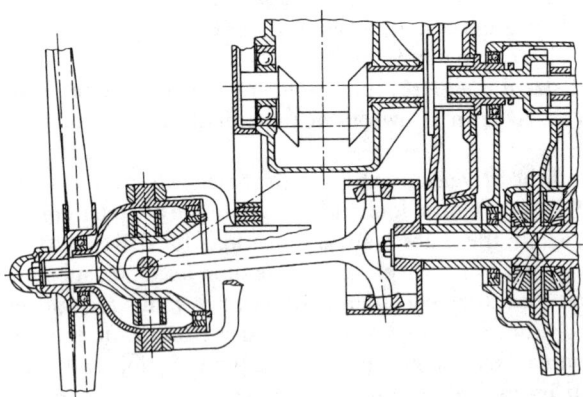

Fig. 1.33. Driveshaft for a front wheel drive motor car by Robert Schwenke 1902, comprising a fixed and a plunging joint (German patent 155834)

1.3 The Ball Joints

motor vehicles, while in 1905 Daimler used a round muff to achieve considerable plunge relative to the back axle (Fig. 1.34).

Centric plunging joints go back to Robert Suczek 1938 and Henri Faure 1961. In general the plunging joint was used where the gap between the two joints varies. As with the RF fixed joints, Rzeppa type plunging joints find applications in cars, this time as inner "differential-side" joints.

In 1955 Robert Bouchard invented a plunging pivoting joint, in addition to his fixed joint (Fig. 1.16, 1.36). The real advantage of this joint was its easy plunging action without axial force. Against this however, when the joint was at an angle, undesirable vibrations occurred due to unbalance. Hence it could not compete with the ball plunging joint and the tripode joint.

Robert Suczek's joint with straight tracks with parallel axes needs a cage to steer the balls- and auxiliary steering. The balls are steered on the offset principle into the plane of symmetry π using auxiliary cup-shaped members which act on the

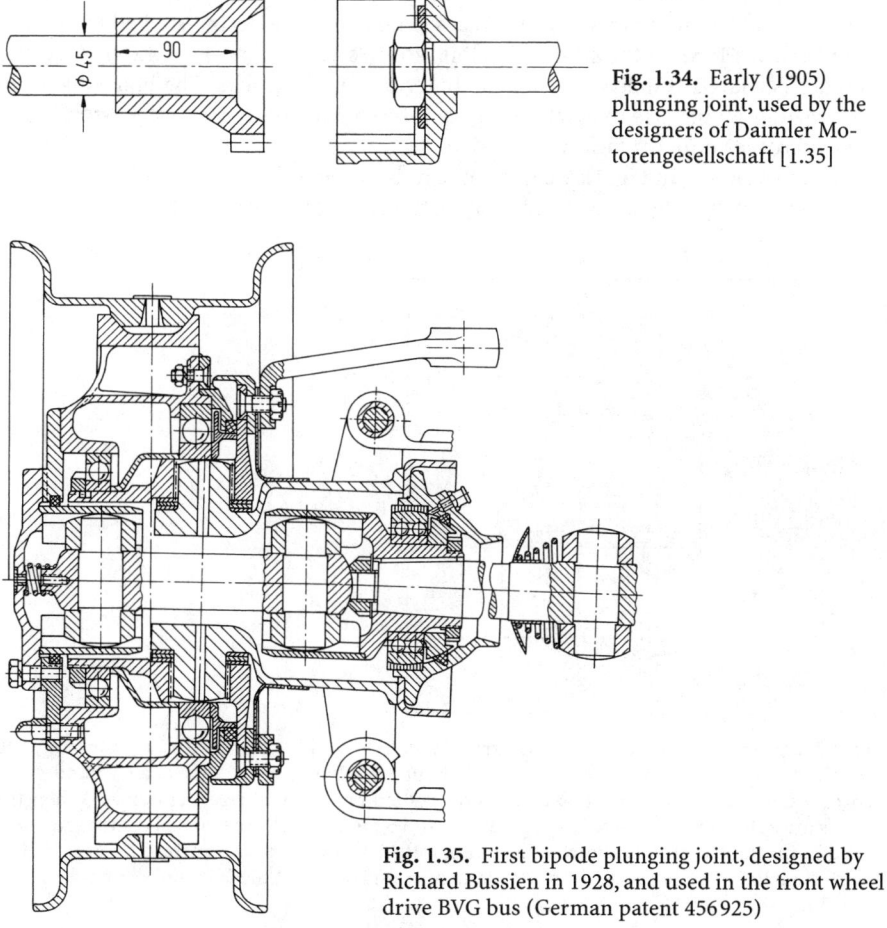

Fig. 1.34. Early (1905) plunging joint, used by the designers of Daimler Motorengesellschaft [1.35]

Fig. 1.35. First bipode plunging joint, designed by Richard Bussien in 1928, and used in the front wheel drive BVG bus (German patent 456925)

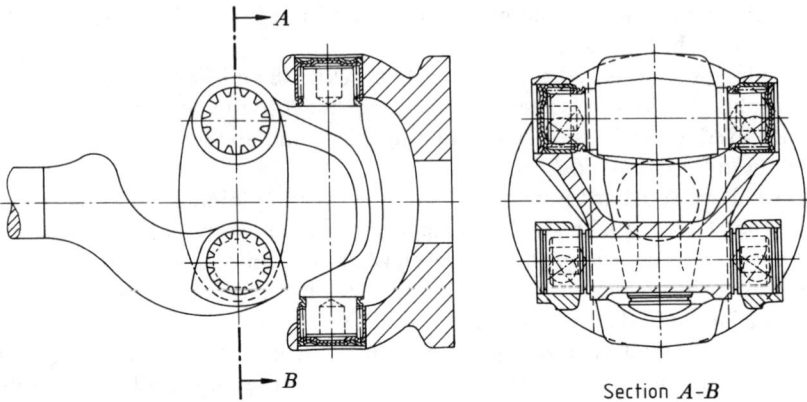

Fig. 1.36. Plunging pivoting joint by Robert Bouchard 1955 (German patent 1072 108 of 1956)

cage. In 1961, Henri Faure steered the four balls of his (V2) joint by pairs of intersecting tracks in the inner and outer races (Fig. 1.38). The design started to achieve importance in 1962, as the six-ball VL joint for passenger cars (Figs. 4.69). Here, the plunge s is shared equally between the outer and inner tracks. The ball rolls more than it slides giving rise to small plunging forces which decrease as the articulation angle increases under rotation.

The joint shown in Fig. 1.39 can plunge in both directions until the sphere of the inner race touches the cage at radius r, or the cage touches the sphere of the outer

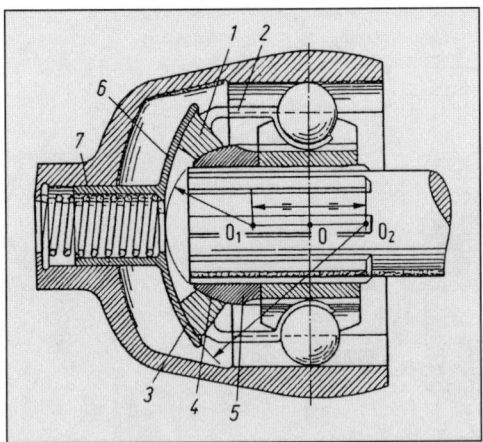

Fig. 1.37. First patent granted for a centrally divided plunging joint with cage steering and straight tracks by Robert Suczek, 1938 (US Patent 2 313 279). *1* and *2* offset cage; *3* and *4* ball centring device; *5* inner race; *6* and *7* axially guided cup-shaped steering device. When the joint is articulated, the three rotating parts *1*, *5* and *6* move about the two ball centring devices *3* and *4*, the mid points of which are equally offset from the joint centre. The cage *1* thus steers the balls with its arms *2* into the plane of symmetry π of the joint, following the offset principle

1.3 The Ball Joints

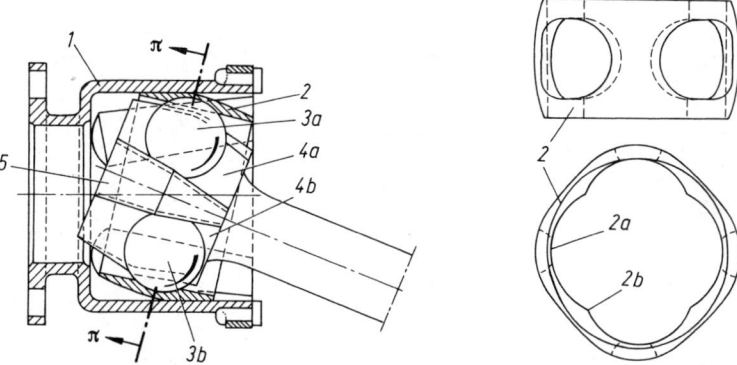

Fig. 1.38. Four-ball plunging joint with inclined track steering, by Henri Faure 1961 (French Patent 1 287 546). The four intersecting tracks, arranged in pairs inclined to the axis, steer the balls which are guided in the cage into the plane of symmetry π of the joint. *1* outer race; *2* cage; *3 a* and *b* = balls; *4 a* and *b* = alterately disposed ball tracks; *5* inner race

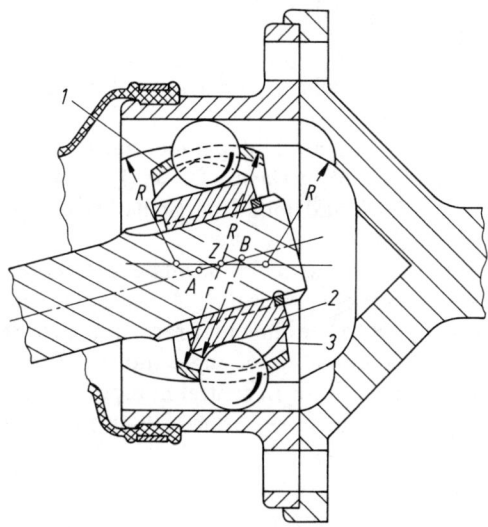

Fig. 1.39. Erich Aucktor's 1963 six-ball, track steered plunging joint, with stops to limit the plunge. Löbro design (German Patent 1 232 411). Used in DKW F 102 passenger cars 1963–1966, Audi 72 1965–1968 and Audi 80 1966–1968. *r* Locating radius on the cage; *R* centering radius; *1* cage; *2* inner race; *3* outer contour of inner race; *Z* joint centre. The spherical location between the contour 3 and the cage inner sphere, both with radius *r*, limits the plunge of the joint to twice the distance AB, giving the same effect as an RF joint. It can therefore transmit an axial load without changing the function of the joint or generating noise

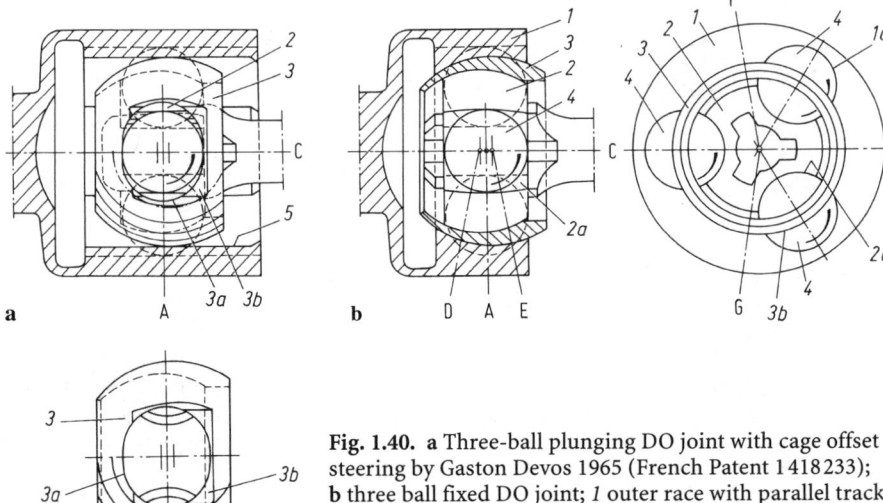

Fig. 1.40. a Three-ball plunging DO joint with cage offset steering by Gaston Devos 1965 (French Patent 1418233); b three ball fixed DO joint; *1* outer race with parallel tracks *1a*; *2* inner race with parallel tracks *2a*; *3* offset cage; *3a* cage window; *3b* outer sphere; *4* ball; *5* cylindrical track

race at radius R. This valuable property can be sed where plunge limited in one or both directions is needed and there are axial loads.

A three-ball plunging or fixed joint (V1/F1) with straight, parallel tracks, using cage and Stuber's offset steering, was invented in 1965 by Gaston Devos (Fig. 1.40). It was further developed into the six-ball DO joint and found particular favour in Japanese front wheel drive passenger cars (Fig. 4.66).

Due to the dynamic torque capacity of these joints, articulation angles are limited to $\beta = 10°$ for long term operation; for short term operation $\beta = 18°$ is allowed. The working temperature should not be more than 80 °C on the outer race; in the short term in should not be more than 120 °C. The combination of angle and speed provides further restrictions. For air-cooled joints operating in central European ambient temperatures it is recommended that $n\beta$ does not exceed 16 000 (30 000 for short term operation); Table 4.4.

1.4 Development of the Pode-Joints

The idea of the pode joint was:

- greater plunging distance inside the joint allowing articulation angles (β) up to 60° without special measures
- very little friction from sliding, and since 1921, the incorporation of rolling members in guides parallel to the joint axis.

For its diameter the pode joint can absorb a lot of torque because the loading is distributed over several trunnions. In French the pode joint is called a "joint coulissant", i.e. sliding joint (Sect. 2.1) [2.7].

1.4 Development of the Pode-Joints

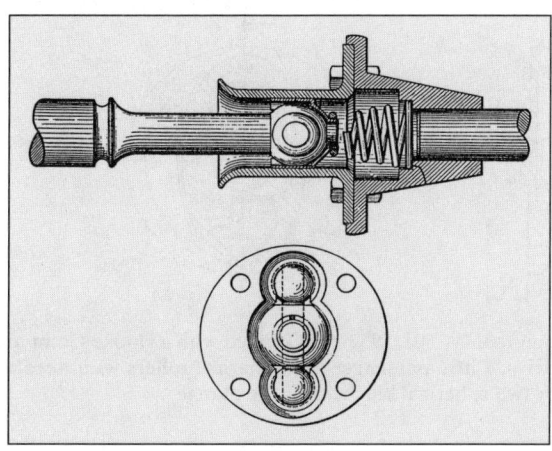

Fig. 1.41. Bipode-joint of John B. Flick 1921, used in Model T four wheel drive Fords (US Patent 1 512 840)

Robert Schwenke took the first step towards today's pode joints in 1902 with his inboard bipode joint (Sect. 1.3.2) [2.7]. In 1935 John W. Kittredge observed that shafts to be coupled were frequently not in true alignment, due to defective workmanship or settling of foundations. This led to undesirable strains and vibrations, especially at high speeds. His solution was a radially constrained tripode joint (US patent 2 125 615/1935).

Bipode Joint. This had already appeared by the middle of the 19th century. It was supposed to split the torque and transmit it in a homokinetic plane. John B. Flick 1921, was the first to give it a substantial plunge capacity and the rollers which became the norm, rather than blocks (Fig. 1.41).

In 1928 Richard Bussien showed in his front wheel drive patent a driveshaft which consisted of a quasi-homokinetic double bipode joint at the wheel and a bipode plunging joint at the differential (Fig. 1.35). This was the first bipode joint to be used in a steer axle [1.19].

Karl Kutzbach showed the basic design of a tripode driveshaft [5.13]. The joints have tracks with parallel flanks (Fig. 1.43). On the left side, in addition to the three balls, a fourth ball touches the centre of the outer body of the joint. The left-hand joint is therefore *fixed*, while the right-hand joint is a *free* tripode joint, which works

Table 1.1. Bipode-Joints 1902–52

1902	Robert Schwenke	DRP 155 834
1907	Sté. des Automobiles Brasier [1.35]	
1910	Mercedes (DMG) + NAG [1.35]	
1921	John B. Flick	USA 1 512 840
1928	Richard Bussien	DRP 456 925
1930	Archibald A. Warner	USA 1 854 873
1931	Fritz Hirschfeld	DRP 568 825
1932	Kugelfischer (FAG)	DRP 565 825
1952	Bayer. Motoren-Werke (BMW)	Bild 1.42

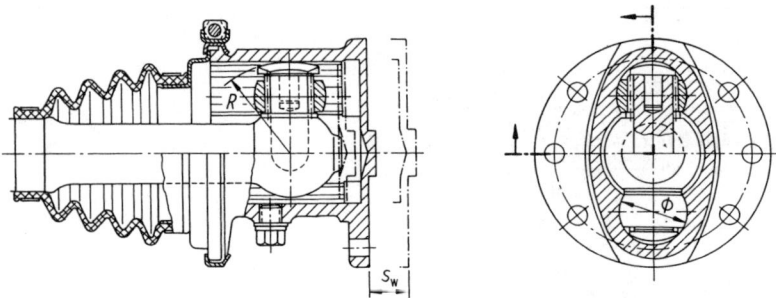

Fig. 1.42. Inboard dipode plunging joint, BMW 501, 1952–58. Coupled with a Hooke's joint to form the halfshaft of a rear wheel drive, 2 litre passenger car. Spherical rollers with needle bearings; connecting shaft centred by two spherical blocks. s_w is the plunge

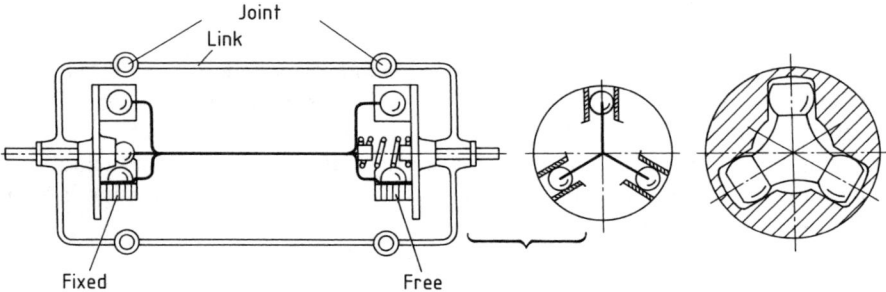

Fig. 1.43. Basic design of a tripode driveshaft by Karl Kutzbach 1937. The right-hand joint is free, the left is fixed

in a statically defined manner; it adjusts itself automatically to distribute the tangential evely to all three transmitting balls. This is an isostatic transmission in accordance with (4.78).

Because of the isostatic load transmission, the tripode spider nutates about the centre of the joint with each revolution. However, notwithstanding the oscillation, the loads on the rollers remain equal (Sect. 4.5.2.4). The movement has the shape of a trefoil with eccentricity

$$\lambda = \frac{x_{11} - R}{R} = \frac{1}{2} \frac{1}{\cos \beta - 1}$$

from Eq. (a) in (4.81), and a displacement

$$s = \frac{l_1 + l_2 + l_3}{R} = \frac{3}{2} \frac{1}{\cos \beta + 1}$$

from Eq. (b) of (4.81), which is three times as great as the eccentricity.

The path to the tripode joint was shown by the Fritz Hirschfeld's 1931 bipode joint, tried out by the firm Voran Automobilbau AG 1932. Bored through balls, which rolled in cylindrical guideways, slid on two trunnions. It had to be mechanically centred (Fig. 1.44).

1.4 Development of the Pode-Joints

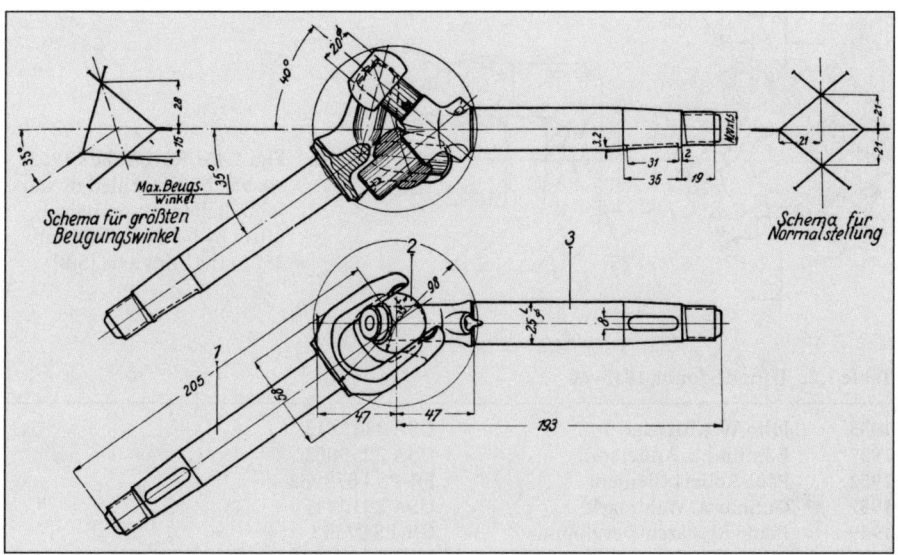

Fig. 1.44. Bipode joint with plunge capability for front wheel drive passenger cars, after Fritz Hirschfeld 1931 (German Patent 568825). From Autom. Techn. Zs. 35 (1932) p. 552

Tripode Joint. In 1935 John W. Kittredge invented a three roller joint with spherical rollers. Three balls fixed onto the three trunnions of the driveshaft, transmitted the torque directly to closed tracks in the outer body of the joint, but without the possibility of plunging.

1937 the Borg Warner engineer Edmund B. Anderson reported on developments for motor vehicles, where high volumes and low cost are the order of the day. His tripode joints (US patent 2 235 002) have the trunnions on pivots to allow oscillation in the plane of rotation of the supporting yokes. In some of his designs the ball races are formed from sheet metal, including fixed joints with arcuate races which restrain radial displacement. Other concepts have elongated ball races which allow the joint to plunge.

The 1963 six-ball star joint of Kurt Enke (French pat. 1353407) consists of an outer three-pointed star and an inner three armed inner race (Fig. 1.45) [5.11]. There are three pairs of parallel tracks in these two parts, which run on the surface of a 50° cone, symmetrically about the axis of rotation of the joint. The six balls, 18 mm in diameter, are steered firmly by the track intersections of both ball yokes: this means that they lie in the plane of symmetry π. The torque is transmitted by only three balls, depending on the direction of rotation, while the remaining three centre and hold it together. The balls are preloaded because this cageless joint only works fully without play. The joint is completely encapsulated and thus protected from falling apart. It is lubricated for life with SAE 90 hypoid oil and is sealed against water by a very strong boot.

Gunnar A. Wahlmark and the Blade Research Development (BRD) company came close to the advanced form developed by Michel Orain, 1960 (Table 1.2). Wahlmark thought he had invented a fully homokinetic pode joint, suitable for speeds up to 30,000 rpm. Its inner and outer parts plunged relative to one another under articu-

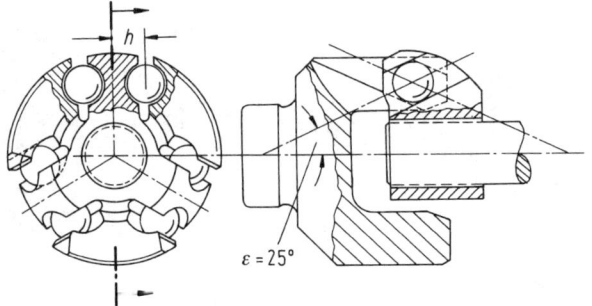

Fig. 1.45. Kurt Enke's 1963 six ball conical ruled or star joint. Ball tracks in pairs offset by h and inclined at 25° to the joint axis [5.9]

Table 1.2. Tripode-Joints 1935–60

1935	John W. Kittredge	USA 2 125 615
1937	Edmund B. Anderson	USA 2 235 002
1952	Paul-Robert Clément	FR-PS 1 078 962
1957	Gunnar A. Wahlmark	USA 2 910 845
1959	Blade Research Developm.	GB-PS 27 513
1960	Michel Orain	FR-PS 1 272 530

lation, with the inner part moving in a circle around the outer part. Michel Orain declared in 1960 that one could not rely on the connecting rollers moving on their own into the correct radial positions on the trunnion shaft. His tripode joint is homokinetic, can work independent of rotational speed, and is practical for articulation angles up to 45° (Sect. 4.5.2).

Mass production of Orain's fixed tripode joint GE[4] started as early as 1958 at Glaenzer-Spicer in Poissy. This permanently lubricated joint for front and rear axle drive passenger cars permitted a maximum articulation angle (β_{max}) of 45°. At $\beta = 4°$ 2000 rpm was possible (Fig. 1.47). It was sealed with a boot to Orain's design (French Patent 1300386/1966). In 1970 the tripode plunging joint GI[4] followed. In 1982 the pre-loaded NG-joint, with needle bearings, increased the articulation angle to 47°, with higher transmission capacity.

Quattropode Joint. This is much older than the tripode. It was invented in 1913 by the American, Victor Lee Emerson. He arranged spherical trunnions which slid along tracks with parallel axes (Fig. 1.48). In 1929 Henri Wouter Jonkhoff took up this idea and designed a quattropode into the worm drive of a heavy vehicle. He fitted rollers that worked against smooth conical wheels on a single ball. Thus, two roller systems acted against each other. With this he hoped to rule out jamming at all articulation angles because new joints of contact were always being made. The joint had no trunnions, the rollers sat directly on the ball (Fig. 1.46). Cylindrical rollers looked ahead to the later trunnions. In 1932, Jonkhoff wanted to use this joint for heavy commercial vehicles. It seemed clear to him that pode joints could be a practical possibility for heavy drives (Fig. 1.46).

---

[4] GE = Glaenzer Exterieur, GI = Glaenzer Interieur, NG = New Generation.

1.4 Development of the Pode-Joints

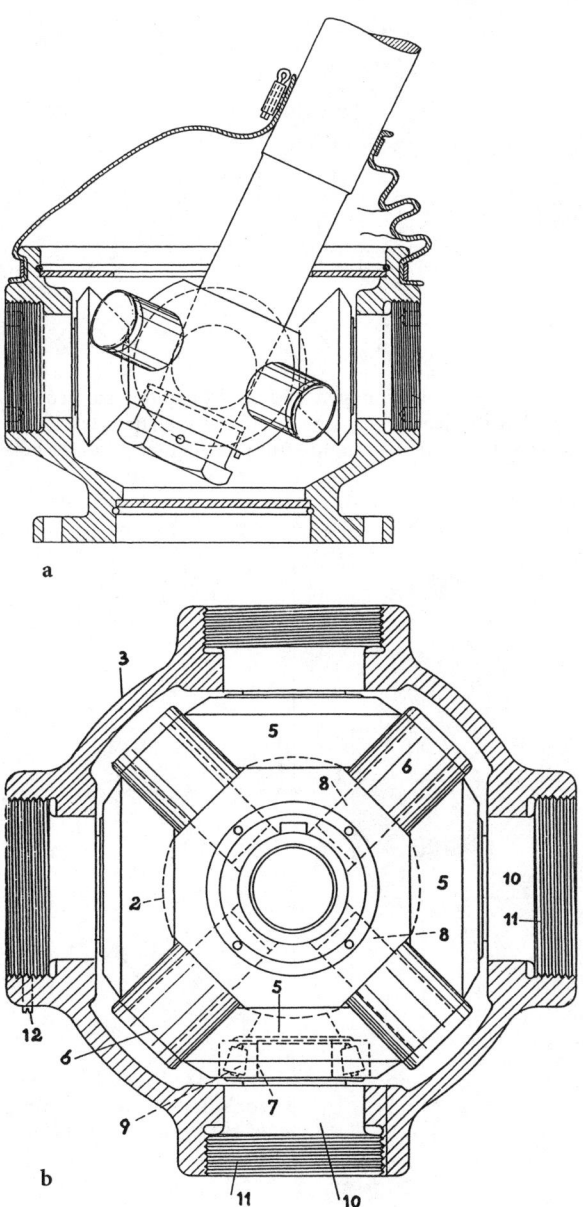

Fig. 1.46a, b. Quattropode joint for commercial vehicles by Henri W. Jonkhoff 1932 (DRP 586776)

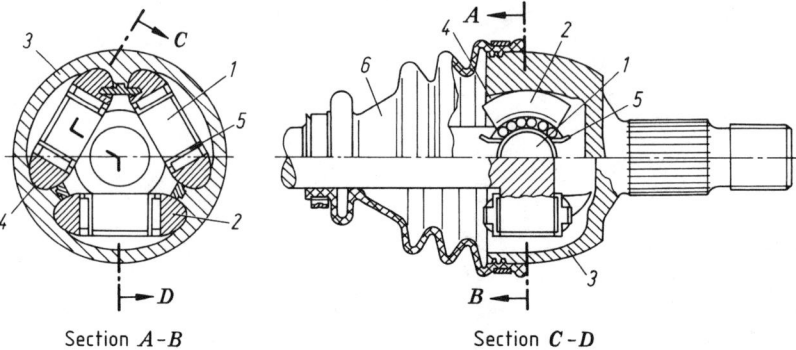

Section A-B Section C-D

Fig. 1.47. Play free NG tripode 47° fixed joint. Bell shaped design with spider connection. Glaenzer Spicer design. Two half-shells with needle bearings, circular arc-shaped track, two and a half times greater torque capacity. *1* Tripode trunnion; *2* annular segment; *3* outer bell; *4* tracks; *5* bearing shell; *6* boot

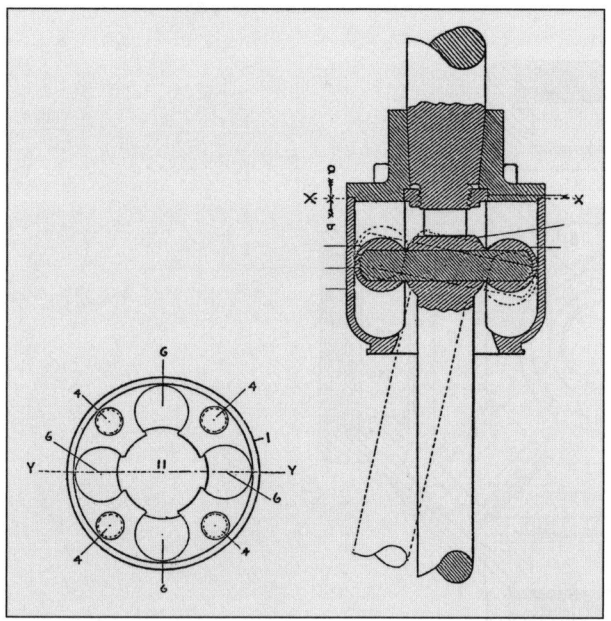

Fig. 1.48. Victor Lee Emerson's first Quattropode joint of 1913, with four spherical trunnions (GB-PS 14129)

S. G. Wingqvist tried out a quattropode joint for motor vehicle drives in 1944–46. Finally, in 1987, Michel Orain at Glaenzer Spicer developed an offset-free quattropode joint capable of high torque transmission t articulation angles up to 10° and speeds of 6000 rpm. It is centred by rocking tracks and distributes the tangential loads onto four roller assemblies at all articulation angles (Fig. 1.49). However as long as the advantages of the quattropode do not surpass those of the classical Hooke's joint, it is not used.

1.2 Theory of the Transmission of Rotational Movements by Hooke's Joints

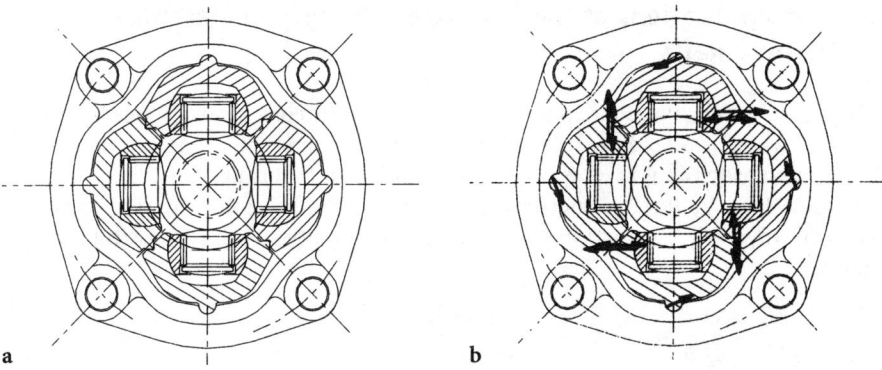

Fig. 1.49. Michel Orain's 1987 quattropode joint for sport and commercial vehicles features large articulation angle and very smooth running, **a** cross section, **b** fourfold torque distribution between the tracks. Designed by GKN Automotive 1991–1997

Table 1.3. Quattropode-Joints 1913–94

1913	Victor Lee Emerson	British Patent 14 129
1929	Henri Wouter Jonkhoff	German Patent 562 108
1932	Henri Wouter Jonkhoff	German Patent 586 776
1944–46	Sven Gustaf Wingqvist	US Patent 2 532 433/434
1951	Ludw. v. Roll'sche Eisenwerke	German Patent 904 256
1984	Michel Alexandre Orain	French Patent 8 410 473
1991	Michel Alexandre Orain	European Patent 0 520 846 A 1
1994	Werner Krude	German Patent 4 430 514 C 2

Table 1.4. Raised demands to the ball joints by the customers

Automotive design feature	Joint requirement
Continually increasing engine performance	Higher torque capacity
Faster vehicles	Higher speed, lower friction losses
All-terrain vehicles and construction machines with all-wheel drive and locking differentials	High torque transmission at large articulation angles, high shock resistance under rotation and articulation
High level of drive comfort	Quiet, vibration-free running, no shudder on start-up
Longitudinal engine with automatic transmission	Able to accommodate axial vibration at standstill
Cost reduction	No finish machining, easier tolerances, simple assembly

1.5 First Applications of the Science of Strength of Materials to Driveshafts

Robert Hooke's second contribution, in 1678, to the theory of joints was his law of elasticity, which is one of the basic principles of the science of strength of materials. However, nobody thought of applying this to joints at that time.

In 1826 Jean-Victor Poncelet found that in the Hooke's joint the moments and forces are inversely proportional to the angular velocities [1.8; 1.9]. Using (1.2) he obtained

$$\frac{M_2}{M_1} = \frac{Q_2 R}{Q_1 R} = \frac{\omega_1}{\omega_2} = \frac{1 - \sin^2\beta \sin^2\varphi_1}{\cos\beta}, \qquad (1.5)$$

where Q_1 denotes the constant driving force and φ_1 the constant driving angle. Poncelet obtained from (1.5) the forces on the cross for

$$\varphi_1 = 0 \quad Q_2 = Q_1 \frac{1 - 0 \cdot \sin^2\beta}{\cos\beta} = Q_1 \frac{1}{\cos\beta} \quad \text{as the maximum} \qquad (1.6)$$

$$\varphi_1 = 90° \quad Q_2 = Q_1 \frac{1 - 1 \cdot \sin^2\beta}{\cos\beta} = Q_1 \frac{\cos^2\beta}{\cos\beta} = Q_1 \cdot \cos\beta \quad \text{as the minimum} \qquad (1.7)$$

With this he created a realistic basis for calculating the strength of Hooke's joints.

1.5.1 Designing Crosses Against Bending

In 1842 the mechanical engineering teacher, Wilhelm Salzenberg, calculated the diameter of the joint trunnion needed to withstand the bending forces, using the encastre beam theories of Galilei, Jakob Bernoulli, Euler and Coulomb (1638–1776).

In Figure 1.46 the bending moment at the imaginary joint of constraint is

$$M_b = Q_1 a = W_b \sigma_b \qquad (1.8)$$

The constant force Q_1 on the drive trunnion is calculated from the equation

$$P_{\text{eff}} = Q_1 v = \frac{Q_1 t}{2R\pi} \qquad (1.9)$$

or from the torque

$$M_d = Q_1 h = Q_1 \cdot 2R \qquad (1.10)$$

and the maximum force Q_2 on the driven trunnion from (1.6) is inserted into (1.8) as

$$\frac{Q_1 a}{\cos\beta} = \frac{\pi d^3}{32} \sigma_b.$$

1.5 First Applications of the Science of Strength of Materials to Driveshafts

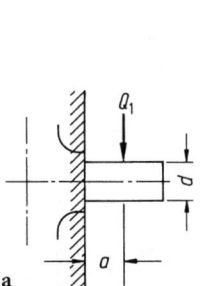
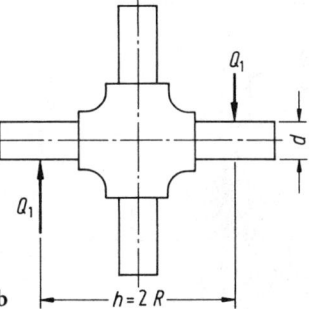

Fig. 1.50 a, b. Forces on the joint cross. **a** Joint trunnion as an encastre beam, **b** pair of forces on the cross

From this the trunnion diameter d is obtained as (Fig. 1.50b)

$$d = \sqrt[3]{\frac{32\,Q_1 a}{\pi \cos\beta\,\sigma_b}}$$

or with $Q_1 = M_d/2R$

$$d = \sqrt[3]{\frac{32\,M_d a}{\pi 2R \cos\beta\,\sigma_b}}. \tag{1.11}$$

This formula is still used today.

Salzenberg's example of a Hooke's joint originated from a capstan which two horses turned every four minutes (Fig. 1.51). He took the continuous output of a horse as 255 foot pounds per second (0.34 kW)[5]. For the lever arm he took the most unfavourable case, that the trunnion only supports at the outer end, giving an effective radius $R = 8$ inches $= 2/3$ foot (203 mm).

In 1842 Salzenberg calculated the trunnion force Q_1 from Eq. (1.9):

$$Q_1 = \frac{Nt}{2R} = \frac{255 \cdot 4 \cdot 60}{2 \cdot \frac{2}{3}\pi} = 14\,618 \text{ Pounds } (73\,900 \text{ N}).$$

He considered that, for a small articulation angle β, the increase in the force on the drive trunnion was compensated by the pessimistic assumption of the effective radius R, so that $\cos\beta = 1$ may be inserted into (1.11)[6].

$$d = \sqrt[3]{\frac{32 \cdot 14\,618 \cdot 3}{\pi \frac{1}{4} \cdot 46\,080}} = 3{,}38 \text{ inches (86 mm)}.$$

---

[5] The figures in brackets are conversions to SI units.

[6] $\sigma_{\text{Fracture}} = 46\,080$ Pounds/inch^2 for cast iron is deduced from the breaking load of 640 pounds given by Peter W. Barlow for an encastre beam with a length of 1 foot and a cross sectional area of 1 square inch.

$640 \cdot 12 = \dfrac{l^3}{6}\sigma_{\text{Fracture}} \Rightarrow \sigma_b = 1/4\sigma_{\text{Fracture}} = 11\,520$ pounds per square inch, an appropriate value compared with that for present day GG15 where $\sigma_b = 150$ N/mm^2.

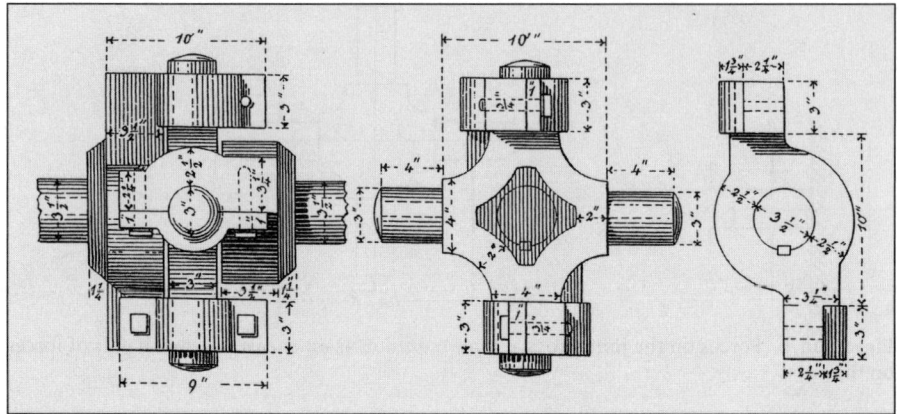

Fig. 1.51. Universal joint and cross of a capstan at the beginning of the 19th century [1.18]. Repro: M. Kunath

Rheinmetall AG of Sömmerda (Thur.) gave their customers standard sheets with maximum values for the circumferential force $Q_1 = M_d/R$ [1.19], which, together with $a = l/2$ in (1.11), gives

$$\frac{1}{2}\frac{Q_1}{\cos\beta}\frac{l}{3} = \frac{\pi}{2}d^3\sigma_b. \tag{1.12}$$

The length of the trunnions is obtained from the equation

$l/d \leqq 1.25$

1.5.2 Designing Crosses Against Surface Stress

The surface *pressure p* is generally understood to be the part of the force dQ on the surface dA, that is $p = dQ/dA$, from which it follows that $Q = \int dQ = \int p dA$.

Let it be assumed that the surface of a body in accordance with Fig. 1.50 has arisen through rotation of the meridian line AB about the z-axis. The arbitrary point P on its surface is fixed by the co-ordinate z and by a meridian plane which forms the angle φ_1 with the xz-plane. Through two further values $z + dz$ and $\varphi_1 + d\varphi_1$ the element of the surface belonging to the point P is defined as

$$dA = \overline{PR}\, d\varphi_1\, ds = x(z)\cos\varphi_1\, d\varphi_1\, ds. \tag{a}$$

In addition

$$\cos\varphi_2 = dz/ds \Rightarrow ds = dz/\cos\varphi_2. \tag{b}$$

Substituting Eq. (b) into (a) gives

$$dA = x(z)\cos\varphi_1\, d\varphi_1\, dz/\cos\varphi_2. \tag{c}$$

1.5 First Applications of the Science of Strength of Materials to Driveshafts

The force on the body acting in the direction of the x-axis

$$Q_x = \int p_x \, dA = \int p \cos \varphi_2 \, dA. \qquad (d)$$

Because of the inclined surface this also results in a component in the direction of the z-axis

$$Q_z = \int p_z \, dA = \int p \sin \varphi_2 \, dA. \qquad (e)$$

Substituting from (c) into (d) it follows that

$$Q_x = \int p x(z) \cos \varphi_1 \, d\varphi_1 \, dz. \qquad (f)$$

If Eq. (f) is expanded with $\cos \varphi_1 \cos \varphi_2 / \cos \varphi_1 \cos \varphi_2$ we get

$$Q_x = \frac{p}{\cos \varphi_1 \cos \varphi_2} \int \cos^2 \varphi_1 \, d\varphi_1 \int x(z) \cos \varphi_2 \, dz$$

or the general equation for the surface pressure given by Carl Bach in 1891 [1.20]

$$p = \frac{Q_x \cos \varphi_1 \cos \varphi_2}{\int \cos^2 \varphi_1 \, d\varphi_1 \int x(z) \cos \varphi_2 \, dz} \qquad (g)$$

The surface pressure for the special cases of cylindrical and ball-shaped trunnions can be derived from this.

If the pin is only half enclosed by the bearing, φ_1 has the limits $-\pi/2$ and $+\pi/2$. This means that

$$\int \cos^2 \varphi_1 \, d\varphi_1 = 1/4 \sin 2\varphi_1 \, \varphi_1/2 + C = \frac{\pi}{2}. \qquad (h)$$

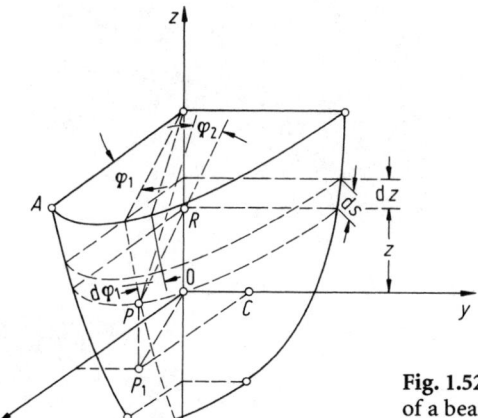

Fig. 1.52. General shape of the surface of a bearing trunnion. According to Carl Bach [1.20]

With $\cos\varphi = \cos\varphi_1 \cos\varphi_2$ then

$$p = \frac{2}{\pi} \frac{Q_x \cos\varphi}{\int x(z) \cos\varphi_2\, dz}. \qquad (1.13)$$

For the cylindrical trunnion (Fig. 1.26c) $x = r$ and $\varphi_2 = 0$ at the extremes $-l/2$ and $+l/2$, that is

$$\int r\, dz = rz + C = r[l/2 - (-l/2)] = rl.$$

The maximum surface pressure is then

$$p = \frac{2}{\pi} \frac{Q_x \cos\varphi_1}{rl}$$

or

$$p_0 = \frac{4}{\pi} \frac{Q_x}{dl}. \qquad (1.14)$$

For the ball-shaped trunnion (Fig. 1.26b)

$$x = r\cos\varphi_2, \quad z = r\sin\varphi_2, \quad \frac{dz}{d\varphi_2} = r\cos\varphi_2,$$

$$\int x(z) \cos\varphi_2\, dz = \int r\cos\varphi_2\, r\cos\varphi_2\, d\varphi_2 = r^2 \int \cos^2\varphi_2\, d\varphi_2$$

$$= \frac{1}{4}\sin 2\varphi_2 + \frac{\varphi_2}{2} + C.$$

At the limits of $-\pi/4$ and $+\pi/4$ we get

$$\varphi_2 = \frac{1}{2}\sin\frac{\pi}{2} + \frac{\pi}{4} = \frac{2+\pi}{4}.$$

The maximum surface pressure is then

$$p = \frac{8 Q_x \cos\varphi}{\pi r^2 (2+\pi)}$$

or when $\varphi = 0°$ and $r = d/2$

$$p_0 = \frac{32}{\pi} \frac{Q_x}{d^2 (2+\pi)}. \qquad (1.15)$$

It is independent of the angle φ_2.

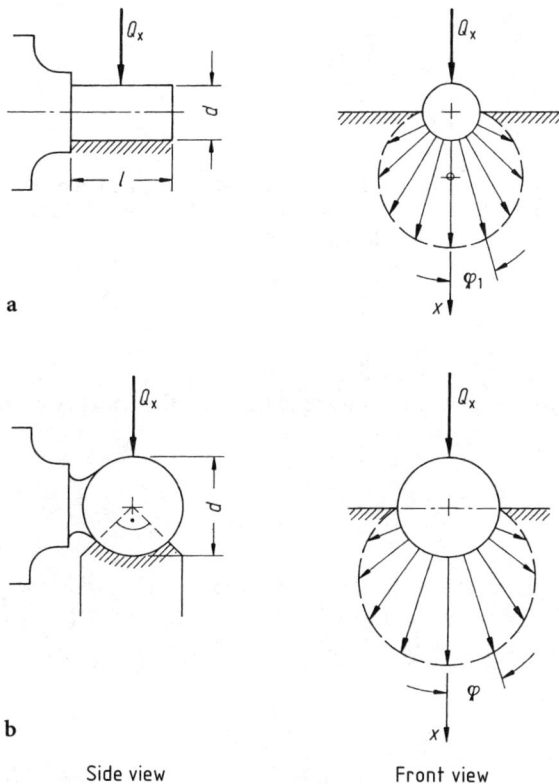

Side view   Front view

Fig. 1.53 a, b. Surface pressure on cylindrical and ball-shaped bearing trunnions according to Carl Bach [1.20]. **a** Cylindrical trunnion (1.14), **b** ball trunnion (1.15). The left hand diagrams give the side view with loading by a single force Q_x. On the right are the front view and the distribution of the surface pressure p_0 from the single force Q_x

This theory was based on a geometrically complete matching of the bearing diameter d_L with the trunnion diameter d_Z. In practice this is not realisable because the diameter of the bearing must always be greater than that of the trunnion if this is to rotate. The ratio of the two diameters or radii is the conformity (Fig. 1.54).[7] The radius of curvature of a curved surface of a body is positive if the centre of curvature lies inside the body (a convex surface).

Then for the concave case it follows that $\dfrac{+r_Z}{-r_L} = -\varkappa,$ (1.16)

for the convex case $+\dfrac{r_Z}{r_L} = +\varkappa.$ (1.17)

---

[7] Richard STRIBECK introduced this idea 1907 [1.44].

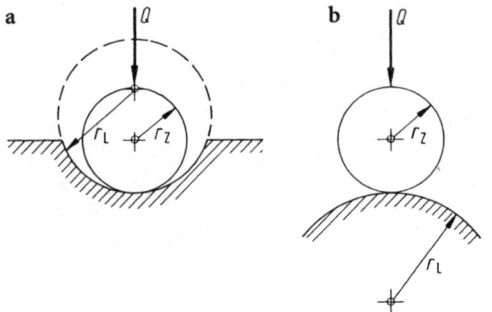

Fig. 1.54a, b. Mutual osculation of curved surfaces. a Concave case (1.16), b convex case (1.17)

The flat surface when $r_L = \infty$ is then clearly the borderline between the concave and convex cases

$$\frac{+r_Z}{\infty} = \varkappa = 0.$$

Equation (1.14) becomes more inaccurate as the diameter of the bearing increases relative to that of the trunnion. This case was convered by Heinrich Hertz 1881 in his study of the surface *stress* between bodies with curved surfaces [1.21]. This involved the "line contact" of two bodies

$$p_0 = 270 \sqrt{\frac{Q \Sigma \varrho}{l_w \cdot 2}},$$

two bodies which Hertz handled in addition to the equally important case of "point contact"

$$p_0 = \frac{858}{\mu \nu} \sqrt[3]{Q (\Sigma \varrho)^2}.$$

The two cases are dealt with in Chap. 3, see (3.54) to (3.57).

When Wilfred Spicer started to fit the trunnions of his universal joints with needle[8] bearings in 1928-30 he was able to refer to empirical and analytical studies of needle bearings. These had been carried out by Richard Stribeck in 1898-99 [1.22] and around 1918 by Waloddi Weibull, Arvid Palmgren, Knut Sundberg and Axel Danielsson, who built on Hertz's work and developed it mathematically [1.23-1.26]. Those studies and new work in 1943 supplied information which also enabled the capacity of ball joints to be calculated [1.27].

It was necessary however to present to engineers in a comprehensible way the rather inaccessible theory of Hertz. It did not therefore become a general engineering tool until after the first quarter of this century. In 1948 Constantin Weber made

---

[8] Needles are in use since 1888 because their need of space is approx. the same as a sliding bearing.

the Hertzian theory easier to understand through an analogy [1.28] which no longer had as a pre-requisite the potential theory of the triaxial ellipsoid.

1.5.3 Designing Driveshafts for Durability

Around the turn of the century designers sought to prevent fracture of the driveshaft or the joint. Since the beginning of the 1930s steps had to be taken to prevent pitting on the trunnions, because this often caused damage to the joints. The engineers recognised that the rolling pairs of transmission elements did not behave according to the Woehler function but that the endurance strength was the determining factor. The joint contact fatigue tests carried out by SKF AB Gøteborg in 1913–1935 yielded the inverse cubic law between the durability L and the load P which is outlined below [1.27, 1.29, 1.30].

It should be noted that in bearings the compressive force Q is not necessarily normal to the surface and it may vary with time. In these cases an equivalent dynamic force for the radial and axial loading directions must be determined [3.3, p. 158]. In a universal joint however the force on the balls or rollers is constant and normal to the mutual contact area (Fig. 4.3). The complication here is that the bearing elements do not rotate at a fixed speed but oscillate or even stand still (when they reverse). Therefore we must introduce an "equivalent force P" in Palmgren's general formula

$$L = \left(\frac{\text{load capacity}}{\text{load}}\right)^3 = \left(\frac{C}{P}\right)^3 \text{ in millions of revolutions} \tag{1.18}$$

Arvid Palmgren recognised as early as 1924 [1.30] that for variable amplitude loading the life of rolling elements cannot be determined simply as an arithmetical mean, because of the non-linear damage relationship between the load and the number of load cycles. According to his hypothesis, after N million load cycles the fraction N/L of the life is used up if the rolling elements have alife of L million load cycles for the same load P.

If the rolling elements are subjected to N_1 load cycles corresponding to a life L_1, N_2 load cycles corresponding to a life L_2 etc, the fatigue damage accumulates to

$$\frac{N_1}{L_1} + \frac{N_2}{L_2} + \ldots + \frac{N_n}{L_n} = 1. \tag{1.19}$$

The relationship between the life L and the mean, equivalent loading P_m is, from Palmgren's Eq. (1.18), $L\,(C/P_m)^3$.

If the $N_1, N_2, \ldots, N_n$ load cycles for the loads $P_1, P_2, \ldots, P_n$ are split up into the fractions $a_1 L, a_2 L, \ldots, a_n L$ of the total service life is follows from (1.19) that

$$\frac{a_1 L}{L_1} + \frac{a_2 L}{L_2} + \ldots + \frac{a_n L}{L_n} = 1, \tag{a}$$

$$\frac{a_1}{L_1} + \frac{a_2}{L_2} + \ldots + \frac{a_n}{L_n} = \frac{1}{L}. \tag{1.20}$$

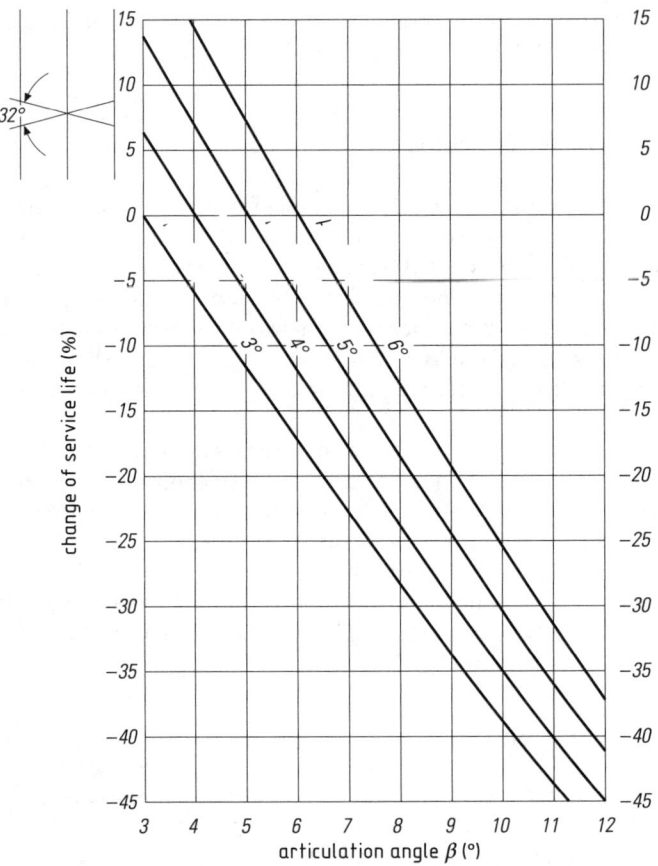

Fig. 1.55. Influence of the joint's articulation angle β on its service life. Exemple: A change of β from 3° to 4,5° reduces the service life of the joint to 9,2%

Equation (1.18) substituted in (1.20) gives:

$$\frac{a_1 P_1^3}{C^3} + \frac{a_2 P_2^3}{C^3} + \ldots + \frac{a_n P_n^3}{C^3} = \frac{1 \cdot P_m^3}{C^3}, \tag{b}$$

$$P_m^3 = a_1 P_1^3 + a_2 P_2^3 + \ldots + a_n P_n^3, \tag{1.21}$$

or

$$P_m = \sqrt[3]{a_1 P_1^3 + a_2 P_2^3 + \ldots + a_n P_n^3}. \tag{1.22}$$

Using the relationship between P_m and the total service life L (1.18) yields:

$$L = \frac{C^3}{a_1 P_1^3 + a_2 P_2^3 + \ldots + a_n P_n^3}. \tag{1.23}$$

1.5 First Applications of the Science of Strength of Materials to Driveshafts

Fig. 1.56. Alfred Hans Rzeppa (1885–1965), son of the well-known Austrian railway construction engineer, Emil Rzeppa. His cousin, Fritz Hückel (1884–1973), hat maker and automobile pioneer in Bohemia, got him interested in automotive engineering. After studying at the technical college in Vienna, he went to the USA in 1913 and worked at Ford in Detroit. In 1923 he opened his own engineering firm. In 1927 he invented the centric ball fixed joint with concentric meridian grooves; 1933/34 with offset. These inventions helped the breakthrough of front wheel drive in cars. He manufactured constant velocity joints in his own company which was bought by Dana Corporation of Toledo, Ohio in 1959

If the speed also changes during the loading a mean speed n_m is applied [1.31]:

$$n_m = q_1 n_1 + q_2 n_2 + \ldots + q_n n_n. \quad (1.24)$$

The above form the basis of the theories of damage accumulation which emerged from the Palmgren/Miner rule of 1945 [1.29, 1.32].

1.6 Literature to Chaper 1

1.1 Feldhaus, F. M.: Die Technik der Vorzeit, der Geschichtlichen Zeit und der Naturvölker (Engineering in prehistoric times, in historical times and by primitive man). Leipzig: Engelmann 1914, p. 678, 869–870
1.2 Cardano, H.: Opera Omnia, tome X opuscala miscellanea ex fragmentis et paralipomenis, p. 488–489. De Armillarum instrumento. Lugduni (Lyon) 1663
1.3 Schott, C.: Technica curiosa sive mirabilia artis, Pars II, Liber Nonus, Mirabilia Chronometrica, propositio XIX. Nuremberg 1664, p. 618, 664–665, 727 and Table VII, Fig. 32

1.4 Hooke, R.: Animadversions on the first part of the Machina Coelestis of the Hon., Learned and deservedly Famous Astronomer Johannes Hevelius Consul of Dantzick, Together with an Explication of some Instruments. London: John Martyn Printer 1674, Tract II
1.5 Hooke, R.: A Description of Helioscopes, and some other Instruments. London: John Martyn Printer 1676, Tract III
1.6 Hooke, R.: Lectiones Cutlerianae or a Collection of Lectures: Physical, Mechanical, Geographical and Astronomical. Made before the Royal Society on several Occasions at Gresham College. London: John Martyn Printer 1679. 6 Tracts
1.7 Gunther, R. Th.: Early Science in Oxford. Vol. VIII. The Cutler Lectures of Robert Hooke. Oxford 1931, Tab. II, Fig. 10
1.8 Poncelet, J. V.: Course of Mécanique appliquée aux Machines, deuxième section: Du Joint brisé ou universel (Mechanics course applied to machines, second section: Universal or Cardan Joint), Paris: Gauthier-Villars 1836 and 1874
1.9 Poncelet, J. V.: Lehrbuch der Anwendung der Mechanik auf Maschinen (Textbook on the application of mechanics to machines). German version by C. H. Schnuse. Darmstadt: Leske 1845 and 1848
1.10 Szabó, I.; Wellnitz, K.; Zander, W.: Mathematik-Hütte, 2nd edition, Berlin: Springer 1974, p. 86, 100, 177
1.11 Willis, R.: Principles of Mechanism. London: Longmans, Green and Co. 1841 and 1870
1.12 d'Ocagne, M.: Cours de géométrie pure et appliquée de l'Ecole Polytechnique (Course of pure and applied geometry of the Ecole Polytechnique), vol. 2, Ch. VI Cinematique Appliquée (Applied Kinematics), No. 170–172 Joint universel de Cardan (Universal Cardan Joint), p. 52–57. Paris: Gauthier-Villars 1918
1.13 Duditza, F.: Transmissions par Cardan (Cardan transmissions). Paris: Editions Eyrolles 1971. German version: Kardangelenkgetriebe und ihre Anwendungen (Cardan joint drives and their applications). Düsseldorf: VDI-Verlag 1973
1.14 Grégoire, J. A.: L'automobile et la traction avant (The automobile and front wheel drive). La Technique Automobile 21 (1930) 46–53. German version by G. Prachtl, Automobiltech. Z. 35 (1932) p. 390–393, 422–424
1.15 Treyer, A.: Le joint Tracta (The Tracta joint). SIA J. 38 (1964) 221–225
1.16 Bouchard, R.: Transmission Economique pour véhicule à traction avant (Economic transmission for front wheel drive vehicles). 6th Fisita-Congress May 1956
1.17 Bouchard, R.: Joints mécaniques de transmission (Mechanical transmission joints). SIA J. 38 (1964) p. 77–86
1.18 Salzenberg, W.: Vorträge über Maschinenbau (Lectures on Mechanical Engineering). Berlin: Realschulbuchhandlung (Duncker and Humblot) 1842
1.19 Bussien, R. (Publ.): Automobiltechnisches Handbuch (Automotive Engineering Handbook), 11th edition Berlin: Krayn 1925, p. 233–235
1.20 Bach, C.: Die Machinen-Elemente (Machine Elements). 1st Vol., 2nd edition, Stuttgart: Cotta 1891, p. 293
1.21 Hertz, H.: Über die Berührung fester elastischer Körper (Contact of solid elastic bodies). J. reine u. angew. Math. 92 (1881) p. 156–171
1.22 Stribeck, R.: Kugellager für beliebige Belastungen (Ball bearings for any load). VDI Z. 45 (1901) 73–79, p. 118–125
1.23 Weibull, W.: Experimentell prövning av Hertz' kontaktformler för elastika kroppar (Experimental testing for Hertzian formulae for the contact of elastic bodies). Tek. Tidskr. Mekanik 49 (1919) p. 30–34
1.24 Weibull, W.: Ytspänningarna vid elastika kroppars beröring (Surface stresses for contact of elastic bodies). Tek. Tidskr. Mekanik 49 (1919) p. 160–163
1.25 Palmgren, A.; Sundberg, K.: Spörsmal rörande kullagrens belastningsförmaga (Questions relating to the loading capacity of ball bearings). Tek. Tidskr. Mekanik 49 (1919) p. 57–67
1.26 Danielsson, A.: Elastika kroppars sammantryckning i beröringsytorna (Compression of elastic bodies at the contact surfaces). Tek. Tidskr. Mekanik 49 (1919) p. 101–103

1.6 Literature to Chaper 1

1.27 Palmgren, A.; Lundberg, O.; Bratt, E.: Statische Tragfähigkeit von Wälzlagern (Static load bearing capacity of roller bearings). Kugellager 18 (1943) p. 33–41
1.28 Weber, C.: Die Hertzsche Gleichung für elliptische Druckflächen (The Hertzian equation for elliptical bearing surface areas). Z. angew. Math. Mech. 28 (1948) p. 94–95
1.29 Palmgren, A.: Tragfähigkeit und Lebensdauer der Wälzlager (Load carrying capacity and service life of roller bearings). Tek. Tidskr. Mekanik 66 (1936) No. 2, Kugellager 12 (1937) 50–60, and Jürgensmeyer, W.: Die Wälzlager (Roller bearings), Berlin: Springer 1937, p. 159
1.30 Palmgren, A.: Die Lebensdauer von Kugellagern (The durability of ball bearings). VDI Z. 68 (1924) p. 339–341
1.31 Dubbel: Taschenbuch für den Maschinenbau (Mechanical engineering handbook) 15th edition, Sect. 1.6.3. p. 267–268, and Sect. 5.3.3, p. 420. Berlin: Springer 1983
1.32 Miner, M. A.: Cumulative Damage in Fatigue, J. Appl. Mech. 12 (1945) A 159–164
1.33 Automobil-Revue Bern 47 (1951) No. 24, p. 23
1.34 Vanderbeek, H.: The Limitations of the Universal Joint. Trans. SAE 3 (1908) p. 23–29
1.35 Heller, A.: Motorwagen und Fahrzeugmaschinen für flüssigen Brennstoff (Motor cars and vehicles engines for liquid fuel). Berlin: Springer 1912 and 1922, p. 348. Reprint Moers: Steiger 1985
1.36 Roloff, H.; Matek, W.: Maschinenelemente (Machine Elements), 8th edition, part I. Brunswick Vieweg 1983, p. 399
1.37 Schmidt, Fr.: Berechnung und Gestaltung von Wellen (Analysing and designing shafts). Design Handbooks, Vol. 10, 2nd edition, Berlin: Springer 1967, p. 19, Fig. 23 and p. 68, Fig. 78
1.38 Shapiro, J.: Universal Joints, in: The Engineers' Digest, Survey No. 6, 1958
1.39 Langer, F. (1937) Fatigue Failure of Varying Amplitude. Trans ASME 59: 160–162
1.40 Bencini, M. (1967) La Trazione Anteriore. Rom: L'Editrice dell'Automobile
1.41 Meisl, Ch. (1954) L'Aventure Automobile. Paris
1.42 Weiss, C. W.: Universal Joint Redesigned to Facilitate Production. Automotive Industries, vol 58, 1928, No 19, p. 738/739
1.43 Rzeppa (1930) Universal Joint has 40 Deg. Maximum Angularity. Automotive Industries, vol 63, No. 3, p. 83/84
1.44 Stribeck, R. (1907) Prüfverfahren für gehärteten Stahl u. Berücks. der Kugelform. VDI Z. 51, 2. Teil, S 1500f
1.45 Norbye, J. P. (1979) The Complete Handbook of Front Wheel Drive Cars. TAB Modern Automotive Series No. 2052. Blue Ridge Summit/Pa: TAB Books, p. 18
1.46 Grégoire, J. A. (1954) L'Aventure Automobile. (Übersetzt von Chas. Meisl). Best Wheel Forward London: Lowe & Brydone (Printers) 1966
1.47 Sturges, E. (1924) A mechanical continuous-torque variable-speed transmission. Autom. Industries 15, July, p. 86 (Weiss joint)

Summarized descriptions: 1.35, 1.38, 1.40, 1.41, 1.45, 1.46.

2 Theory or Constant Velocity Joints

According to the principles of kinematics the universal joints of a driveshaft belong to the family of *spherical crank mechanisms*. They arise from the planar four-bar linkage (Fig. 2.1a) if the axes of rotation 1 to 4 are placed such that they meet at the point 0 (Fig. 2.1b). The four-bar linkage remains movable and positively actuated as before. It is called a

Conical or Spherical Four-bar Linkage

Its links *a* to *c* lie on a sphere about 0 and are parts of great circles. If three of these links are selected to be at right angles then the

Right-angled, Spherical Four-bar Linkage

occurs, or the right-angled, spherical universal joint drive (Fig. 2.2). This is an example of a *kinematic chain* thus named for the first time in 1875 by Franz Reuleaux [2.1; 2.2].

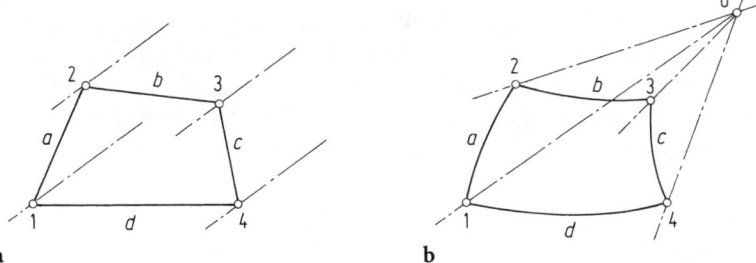

Fig. 2.1 a, b. The four-bar linkage. **a** Planar four-bar linkage, **b** spherical four-bar linkage

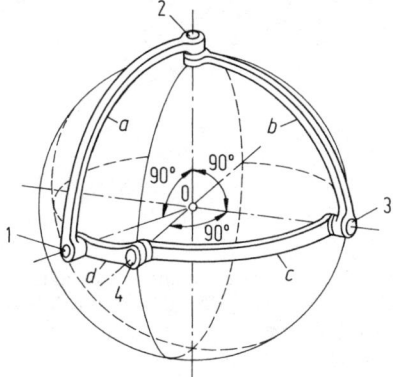

Fig. 2.2. Right-angled spherical Hooke's joint chain [2.13]

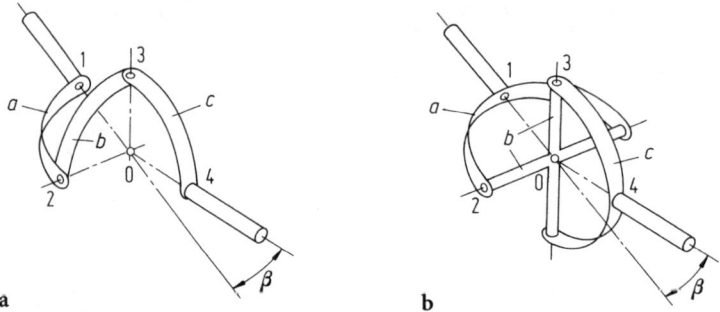

Fig. 2.3 a, b. Development of the rotating Hooke's joint. **a** right-angled spherical Hooke's joint chain, **b** by transforming link *b* into a cross and fixing link *d* (drawn without support *d*)

It is particularly significant for driveshafts because, by fixing link *d* at an angle not equal to 90°, the

Rotating Hooke's Joint

Arises (Fig. 2.3), see Sect. 1.2.1.

2.1 The Origin of Constant Velocity Joints

By arranging two Hooke's joints back to back it is possible to do away with the non-uniformity of the single Hooke's joint described in Sect. 1.2.2, that is, to make a constant velocity joint (Fig. 2.4a). If one lets the length of the intermediate shaft become zero (Fig. 2.4b) one obtains a single compact joint (Fig. 2.4c) which has furthermore the angle bisecting plane π.

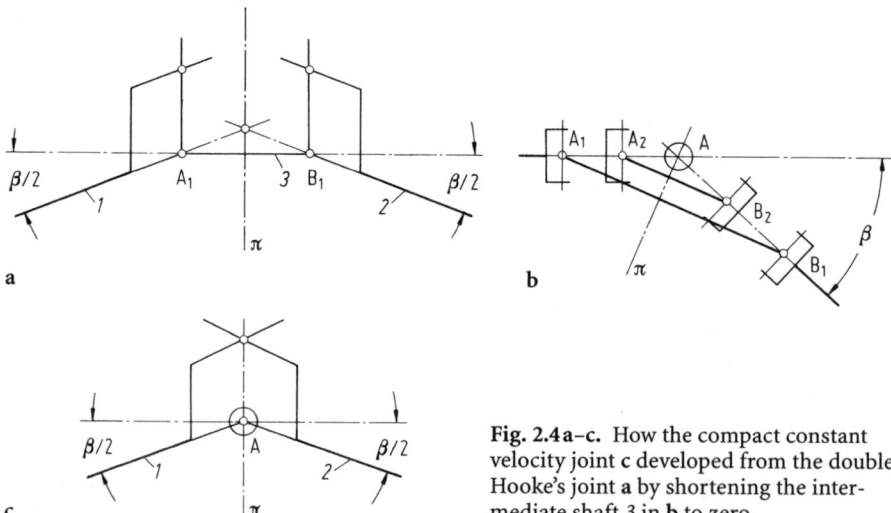

Fig. 2.4 a–c. How the compact constant velocity joint **c** developed from the double Hooke's joint **a** by shortening the intermediate shaft *3* in **b** to zero

2.1 The Origin of Constant Velocity Joints

The task is now to find suitable elements which are able to transmit the torque from one shaft to the other. For this one turns for help to the kinematic chain. Martin Gruebler [2.3] and Maurice d'Ocagne [1.12] wrote in 1917 and 1918 respectively an equation for kinematic chain with f degrees of freedom[1]

$$f = 6(n-1) - \sum_{1}^{i}(6-f_i). \tag{a}$$

Using this it is possible to calculate the *sum of the degrees of freedom of the links of the chain* $\sum f_i$ by turning round equation (a):

$$\sum_{1}^{i} f_i = 6(i-n) + (f+6). \tag{b}$$

In the above the number of the links n of a chain is equal to the number i of its joints. One arrives at the sum of the degrees of freedom $\sum f_i$ of the joints which a chain with $f = 3$ must have in order to permit the transmission of rotational movement at an angle, as

$$\sum_{1}^{i} f_i = 6\underbrace{(i-n)}_{0} + (3+6) = 9. \tag{c}$$

There are many solutions for the chain with $\sum f_i = 9$, among which the multi-bar linkage is the simplest (Fig. 2.5a, as shown by William Steeds in 1937 [2.4] and Jochen Balken in 1981 [2.5]. For the purpose of simplification however Balken did not look at the whole joint, only the angled part which was important for transmitting torque.

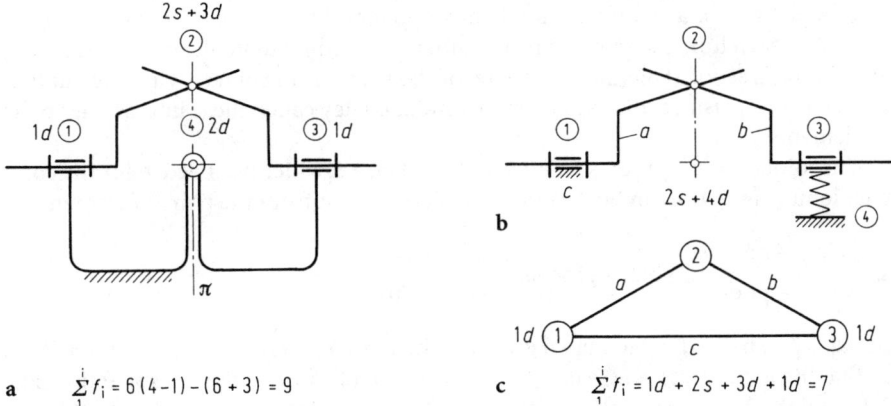

Fig. 2.5 a–c. Examples of bar linkages with four degrees of freedom between the input and output. **a** William Steed's 1937 four-bar linkage [2.4]; **b** simplified three-bar linkage according to Jochen Balken 1981 [2.5]; **c** kinematic chain with three links and three joints

---

[1] The following definitions apply: f degree of freedom (degree of movement) of the kinematic chain, n number of links, i number of joints, f_i degree of freedom of the ith joint.

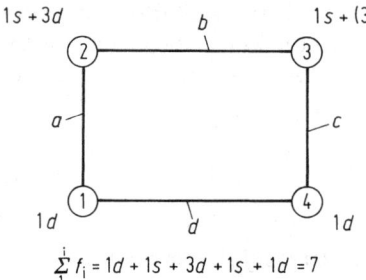

Fig. 2.6. Diagram of a kinematic chain with four links and four joints. Identical degrees of freedom ( ) are only counted once. Letters a to d correspond to the links, numbers 1 and 4 to the joints of the kinematic chain

Because of this the chain is reduced by one link and only has 1 degree of freedom for the rotation about the joint axes. From Fig. 2.5b and c therefore

$$\sum_{1}^{i} f_i = 6\,(\underbrace{i-n}_{0}) + (1+6) = 7.$$

Linkages cannot of course transmit large loads but they are useful for theoretical studies. If we let the diameters of the bars go towards zero, they can be perceived as the centre lines of the paths on which the torque is transmitted. In order to cover all useful joints, in the following these lines can be any desired curves in space. If one removes both degrees of freedom of joints a and b about their axes (Fig. 2.5b), then $\sum f_i = 5$ still remain for transmitting the torque from the input to the output. This is for example the case for point contact between curved surfaces.

It is clear that two round bars only touch at one point. There are however possibilities got point load transmissions which can carry large loads. For this the chain has to be extended by one element – a rolling body between the input and output members – giving a chain with four links a to d and four joints 1 to 4 (Fig. 2.6). In this chain each link is supported by the others. According to Reuleaux 1875 [2.1; 2.2] and Franz Grashof 1883 [2.6] the shape of the surfaces in contact with one another is important (prism, cylinder, sphere or similar). They called these interacting "pairs of elements".

The joints 2 and 3 of links a and b and c in Fig. 2.6, which are needed for the compact joint (Fig. 2.4c), can be designed as joints chosen from the pairs of elements:

$$\left.\begin{array}{l}\text{– groove}\\ \text{– pode}\end{array}\right\} > \text{ball – groove.} \quad \begin{array}{l}\text{(Fig. 2.7a)}\\ \text{(Fig. 2.7b)}\end{array}$$

In Fig. 2.7a a ball is guided in two grooves with a roughly semicurcular cross-section so that the locus of its centre is defined by the intersection of the lines s_1 and s_2 (Fig. 2.8a). Joints 2 and 3 are rolling pairs with $\sum f_i = 5$ (joint contact). Joints with these pairs of elements belong to the family of

ball joints.

In Fig. 2.7b a pode (a through-bored ball on a cylindrical pin) is guided in a slotted hollow cylinder: the locus of the centre of the ball is thus defined by the inter-

2.1 The Origin of Constant Velocity Joints

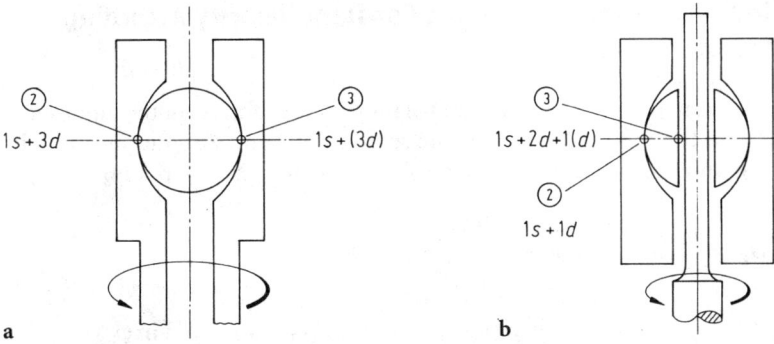

Fig. 2.7 a, b. Degrees of freedom of rolling bodies between input and output members. **a** Groove-ball-groove pairing of elements; **b** pode-ball-groove pairing of elements

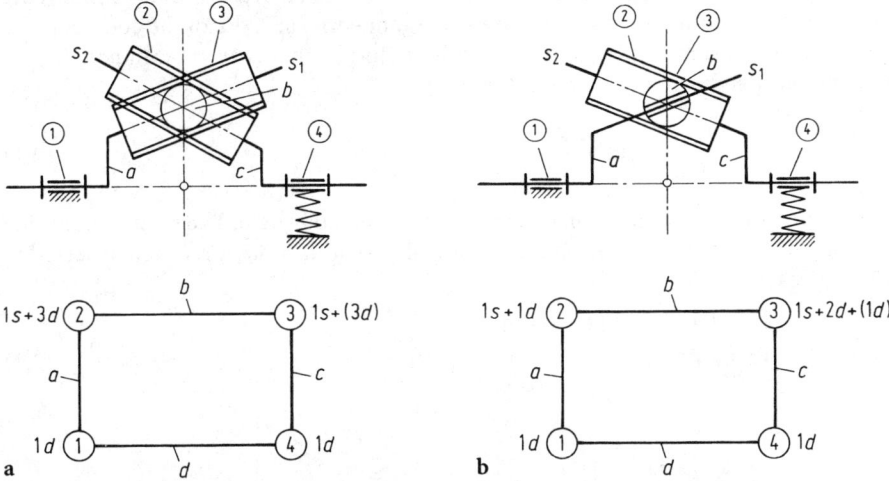

Fig. 2.8 a, b. Basic diagrams of the ball and pode joint families with straight tracks, together with their four-bar linkage kinematic chains. $\sum f_i = 2s + 4d = 6$. **a** Ball joint, **b** pode joint

section of the lines s_1 and s_2 (Fig. 2.8b). 2 is a rolling pair with $\sum f_i = 2$, whereas 3 is a rotating pair with $\sum f_i = 3$ (line contact). Joints from these pairs of elements belong to the family of

pode joints.

Many more combinations of the degrees of freedom f, the displacement s and the rotation d giving $\sum f_i = 5$ are possible at 2 and 3. Hence many more compact joints could therefore be developed [2.5]. The *torque capacity* (Chap. 4) and the *configuration* (Chap. 5) of the various constant velocity joints are determined by the types of pairs of elements used and the geometry of the paths on which they are guided (Fig. 2.8).

2.2 First Indirect Method of Proving Constant Velocity According to Metzner[2]

In Fig. 2.9a, let $x_1 y_1 z_1$ and $x_2 y_2 z_2$ be the (right-handed) co-ordinates of the input 1 and the output 2 with the common origin 0. The rotation of 1 and 2 about the z-axis by $\varphi_1 = \varphi_2$ corresponds analytically to the transformation matrix [2.8, p. 159].

$$D_{\varphi 1} = \begin{pmatrix} \cos\varphi_1 & -\sin\varphi_1 & 0 \\ \sin\varphi_1 & \cos\varphi_1 & 0 \\ 0 & 0 & 1 \end{pmatrix},$$

$$\begin{pmatrix} x'_1 \\ y'_1 \\ z'_1 \end{pmatrix} = \begin{pmatrix} x_1 \cos\varphi_1 - y_1 \sin\varphi_1 \\ x_1 \sin\varphi_1 + y_2 \cos\varphi_1 \\ z_1 \end{pmatrix} ; \quad \begin{pmatrix} x'_2 \\ y'_2 \\ z'_2 \end{pmatrix} = \begin{pmatrix} x_2 \cos\varphi_1 - y_2 \sin\varphi_1 \\ x_2 \sin\varphi_1 + y_1 \cos\varphi_1 \\ z_2 \end{pmatrix}.$$

The coordinates relative to the bodies are converted into the coordinates 1' and 2'.

Since the transmission elements (Fig. 2.9b) do not penetrate one another and are constrained peripherally not to separate, the tangential speed v of the contact point $P(x, y, z)$ is the same on both curved paths s_1 and s_2, that is $v_2 = v_1$. In general the peripheral speed is $v = r\omega = r\, d\omega/dt$, that is

$$r_2(z') \frac{d\varphi_2}{dt} = r_1(z') \frac{d\varphi_1}{dt}, \tag{2.1}$$

where the radius r is a function of z which is expressed by $r(z)$. The requirement that $\varphi_1 = \varphi_2$ throughout the rotation also means that $d\varphi_2/dt = d\varphi_1/dt$, which, inserted in (2.1), gives

$$r_2(z')\, d\varphi_2/dt = r_1(z')\, d\varphi_1/dt,$$
$$r_2(z') = r_1(z'). \tag{2.2}$$

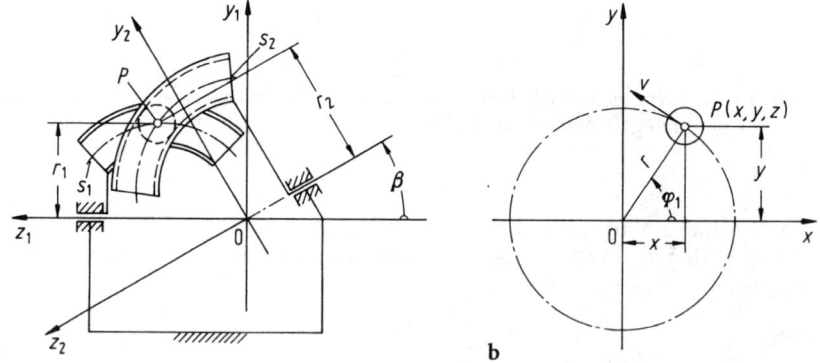

Fig. 2.9a, b. Coordinate systems of the first and second methods for studying the constant velocity conditions of connected shafts. **a** Side view, s_1 and s_2 effective geometry of bodies 1 and 2; **b** section through the contact point P

---

[2] Eberhard Metzner, Design Department, Auto Union GmbH, then Audi AG, [2.7, p. 202], found the important relationship in 1967 (2.8).

2.2 First Indirect Method of Proving Constant Velocity According to Metzner

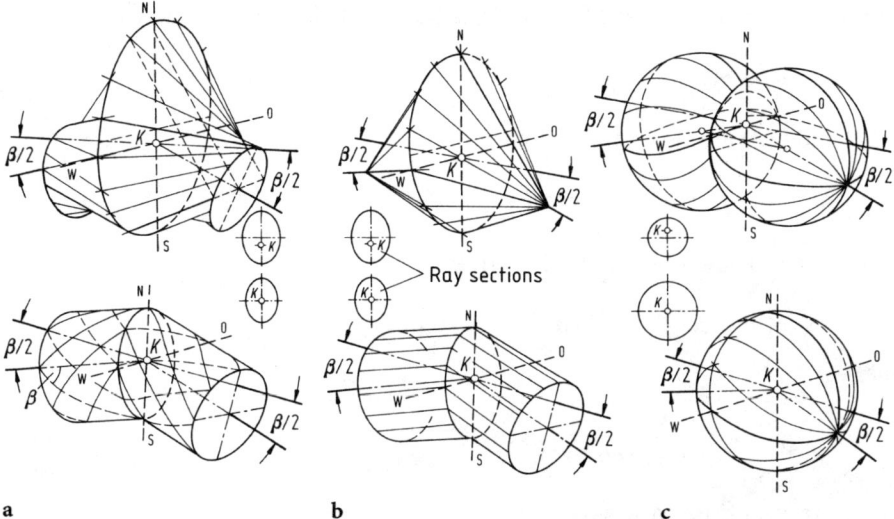

Fig. 2.10a–c. Ruled surfaces generated from the effective geometry of the constant velocity joint, according to Hans Molly 1969. N–S and W–E = elliptical plane of reflection; K pivot point and centre of the joint. **a** Top: hyperboloid joint, bottom: tangent joint; **b** top: conical joint, bottom: cylindrical joint; **c** spherical meridian joint: top: with offset, bottom: without offset [2.9]

This means that when the joint is articulated the distances r_2 and r_1 from their axes z_2 and z_1 must be equal.

The articulation of the body 2 through the angle β about the x-axis corresponds to the transformation of the system $2'$ with the matrix D_β into the $1'$ system:

$$\begin{pmatrix} x'_1 \\ y'_1 \\ z'_1 \end{pmatrix} = \begin{pmatrix} 1 & 0 & 0 \\ 0 & \cos\beta & -\sin\beta \\ 0 & \sin\beta & \cos\beta \end{pmatrix} \begin{pmatrix} x'_2 \\ y'_2 \\ z'_2 \end{pmatrix} = \begin{pmatrix} x'_2 \\ y'_2 \cos\beta - z'_2 \sin\beta \\ y'_2 \sin\beta + z'_2 \cos\beta \end{pmatrix}$$

or, written in coordinates [2.8, p. 62]

$$x'_1 = x'_2, \tag{2.3}$$

$$y'_1 = y'_2 \cos\beta - z'_2 \sin\beta, \tag{2.4}$$

$$z'_1 = y'_2 \sin\beta + z'_2 \cos\beta. \tag{2.5}$$

Rotation of the guide paths s_1 and s_2 about their axes z'_1 and z'_2 generates rotationally symmetrical envelopes (Fig. 2.16) for which, using Fig. 2.9b and Pythagoras's theorem:

$$s_1: \; x'^2_1 + y'^2_1 = r^2_1(z'), \tag{2.6}$$

$$s_2: \; x'^2_2 + y'^2_2 = r^2_2(z'). \tag{2.7}$$

If (2.6) and (2.7) are inserted into the quadratic equation (2.2) one gets

$$x_2'^2 + y_2'^2 = x_1'^2 + y_1'^2.$$

(2.3) and (2.4) are also inserted giving

$$x_2'^2 + y_2'^2 = x_2'^2 + (y_2' \cos \beta - z_2' \sin \beta)^2.$$

Taking the square root of both sides gives

$$y_2' = y_2' \cos \beta - z_2' \sin \beta$$

or

$$y_2' = -z_2' \frac{\sin \beta}{1 - \cos \beta} = -z_2' \cos \beta/2 \tag{2.8}$$

see [1.10, Sect. 3.1.4.10, p. 86]

If (2.8) is inserted in the remaining Eq. (2.5) if follows that

$$z_1' = \left\{ -z_2' \frac{\sin \beta}{1 - \cos \beta} \right\} \sin \beta + z_2' \cos \beta = -z_2' \left\{ \frac{1 - \cos \beta}{1 - \cos \beta} \right\}, \Rightarrow z_1' = -z_2'. \tag{2.9}$$

The requirement $\varphi_2 = \varphi_1$ of the indirect method therefore gives the two conditions in (2.8) and (2.9). They demand that the effective geometries of the input and output members are *exact spatial mirror images* in the plane of symmetry π.

This can be explained simply using a hypothetical experiment.

In Fig. 2.11 the input member is placed (with its geometry) in front of half-silvered mirror and its image, lying symmetrical to it, appears in the mirror. If the real geometry of the output member is positioned behind the mirror and rotated in the same direction, the real and the imaginary geometry can be made to coincide and be followed through the mirror. The rotation of the two geometries here is homokinetic, as one sees in real life. When positioning the output member it can be seen that the mirror stands on the angle bisecting line between the two axes A and A_1 as well as perpendicular to the common plane through these axes. The significance of Eq. (2.9) is also made clear by the experiment:

Since the z' axes are equidirectional in both right-handed systems $1'$ and $2'$, the z_2'-coordinates have negative values because of the mirror symmetry.

With articulation of the output member 2 through the angle β the movement of the contact point P is on a spherical surface (Fig. 2.9a and b) given by the equation

$$x_2'^2 + y_2'^2 + z_2'^2 = r_2^2.$$

Its intersection with the plane of symmetry π

$$y_2' = -z_2' \cot \beta/2 \Rightarrow z_2' = -y_2' \tan \beta/2$$

2.2 First Indirect Method of Proving Constant Velocity According to Metzner

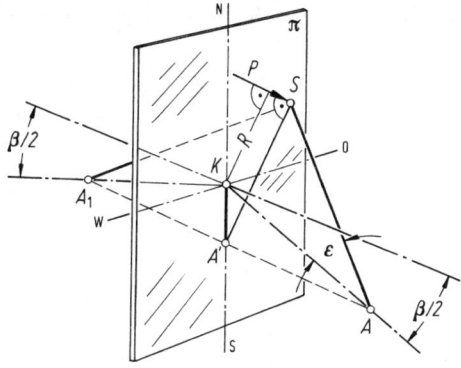

Fig. 2.11. Half-silvered mirror in the conical ruled surface joint to prove the reflective symmetry of the effective geometry. P Equivalent compressive force in the mirror plane perpendicular to the surface $A - A_1 - S$; R effective radius of the joint; K pivot point; $\beta/2$ semi-articulation angle [2.9]

gives the equation of the curve of intersection

$$x_2'^2 + y_2'^2 + (-y_2' \tan \beta/2)^2 = r_2^2$$

or

$$\frac{x_2'^2}{r_2^2} + \frac{y_2'^2}{r_2^2 \cos^2 \beta/2} = 1. \tag{2.10}$$

That is an ellipse in the $x'_2 y'_2$ plane, in which the projection of the point of contact P revolves. The tests using the indirect method showed therefore that the generators s_1 and s_2 of a constant velocity joint must be *mirror images in space*.

It now has to be shown for straight and curved tracks that the transmitting elements rotate in the angle bisecting plane π of the axes 1 and 2 if these are mirror images.

2.2.1 Effective Geometry with Straight Tracks

r is a straight generator which is skew to its axis of rotation z' (Fig. 2.12). If it rotates through the angle φ_1 about the axis z, one gets a hyperboloid as the envelope. The projection of the straight line r onto the $y'z'$-plane is inclined by the angle ε, and onto the $x'z'$-plane by the angle γ with respect to the z'-axis.

In addition, it intersects the y-axis at a distance b from the origin O. Hans Molly and Oezdemir Bengisu [2.9] describe the joints arising from the generating straight line r, as a function of the angles ε, and γ, as ruled surface joints in the forms (Fig. 2.10a–c):

Ruled Surface	ε	γ
Hyperboloid	> 0	> 0
Conical	> 0	$= 0$
Tangent	$= 0$	> 0
Cylindrical	$= 0$	$= 0$

For the general case of the hyperboloid ruled surface joint the function $r^2(z')$ is now derived from Fig. 2.12. It is:

$$r^2(z') = l^2 \tan^2 \gamma + (z' \tan \varepsilon + b)^2. \tag{a}$$

In addition

$$\tan \varepsilon = \frac{y' - b}{z'} \Rightarrow (y' - b) = z' \tan \varepsilon \tag{b}$$

and

$$l^2 = z'^2 + (y' - b)^2. \tag{c}$$

Inserting (a) into (b) gives

$$l^2 = z'^2 + z'^2 \tan^2 \varepsilon = z'^2 (1 + \tan^2 \varepsilon) = \frac{z'^2}{\cos^2 \varepsilon}. \tag{d}$$

From (2.6) one gets

$$x'^2 + y'^2 = r^2(z') = l^2 \tan^2 \gamma + (z' \tan \varepsilon + b)^2. \tag{e}$$

Putting (d) into (e) gives

$$x'^2 + y'^2 = z'^2 \left(\frac{\tan^2 \gamma}{\cos^2 \varepsilon} + \tan^2 \varepsilon \right) + 2bz' \tan \varepsilon + b^2. \tag{f}$$

Proceeding as in Sect. 2.3 from (2.6) and (2.7), gives for input member 1:

$$x_1'^2 = z_1'^2 \left(\frac{\tan^2 \gamma_1}{\cos^2 \varepsilon_1} + \tan^2 \varepsilon_1 \right) + 2z_1' b_1 \tan \varepsilon_1 + b_1^2 - y_1'^2,$$

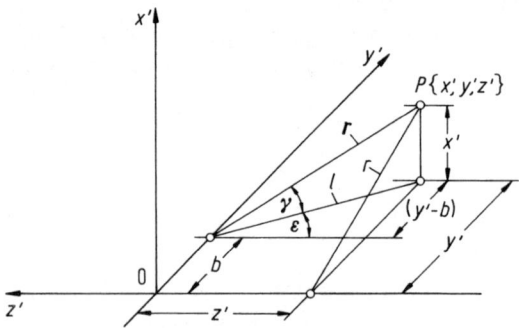

Fig. 2.12. General straight line which gives a hyperboloid as an envelope

2.2 First Indirect Method of Proving Constant Velocity According to Metzner

and for output member 2:

$$x_2'^2 = z_2'^2 \left(\frac{\tan^2 \gamma_2}{\cos^2 \varepsilon_2} + \tan^2 \varepsilon_2\right) + 2z_2' b_2 \tan \varepsilon_2 + b_2^2 - y_2'^2,$$

$$x_2'^2 = x_1'^2,$$

$$(y_1'^2 - y_2'^2) + \left\{ z_2'^2 \left(\frac{\tan^2 \gamma_2}{\cos^2 \varepsilon_2} + \tan^2 \varepsilon_2\right) - z_1'^2 \left(\frac{\tan^2 \gamma_1}{\cos^2 \varepsilon_1} + \tan^2 \varepsilon_1\right) \right\}$$
$$+ 2 (x_2' b_2 \tan \varepsilon_2 - z_1' b_1 \tan \varepsilon_1) = (b_2^2 - b_1^2).$$

The mirror image requirement means that

$$\varepsilon_1 = \varepsilon, \quad \gamma_1 = \gamma,$$
$$\varepsilon_2 = -\varepsilon, \quad \gamma_2 = -\gamma,$$
$$b_1 = b_2 = b.$$

It follows that

$$(y_1'^2 - y_2'^2) + \left\{ z_2'^2 \left(\frac{\tan^2 \gamma}{\cos^2 \varepsilon} + \tan^2 \varepsilon\right) - z_1'^2 \left(\frac{\tan^2 \gamma}{\cos^2 \varepsilon} + \tan^2 \varepsilon\right) \right\}$$
$$+ 2 [z_2' b \tan (-\varepsilon) - z_2' b \tan \varepsilon] = 0.$$

By inserting $\tan \varepsilon = m$ and $\dfrac{\tan^2 \gamma}{\cos^2 \varepsilon} + \tan^2 \varepsilon = n$, as an abbreviation, one gets

$$(y_1'^2 - y_2'^2) + n (z_2'^2 - z_1'^2) - 2mb (z_2' + z_1') = 0.$$

After inserting (2.4) and (2.5) one gets

$$(y_2' \cos \beta - z_2' \sin \beta)^2 - y_2'^2\} + n \{z_2' - (y_2' \sin \beta + z_2' \cos \beta)^2 -$$
$$- 2mb \{z_2' + (y_2' \sin \beta + z_2' \cos \beta)\} = 0. \tag{g}$$

A secondary calculation to prove that

$$(y_2' \cos \beta - z_2' \sin \beta)^2 - y_2'^2 = z_2'^2 - (y_2' \sin \beta + z_2' \cos \beta)^2$$

is given by the following:

$$\text{RHS} = z_2'^2 - y_2'^2 (1 - \cos^2 \beta) - z_2'^2 \cos^2 \beta - 2y_2' z_2' \sin \beta \cos \beta$$
$$= z_2'^2 \sin^2 \beta - y_2'^2 + y_2'^2 \cos^2 \beta - 2y_2' z_2' \sin \beta \cos \beta,$$
$$\text{HS} = \text{RHS} (y_2' \cos \beta - z_2' \sin \beta)^2 - y_2'^2 = (y_2' \cos \beta - z_2' \sin \beta)^2 - y_2'^2.$$

With this one obtains

$$\{(y_2' \cos \beta - z_2' \sin \beta)^2 - y_2'^2\} (1 + n) - 2mb \{z_2' + (y_2' \sin \beta + z_2' \cos \beta)\} = 0. \tag{h}$$

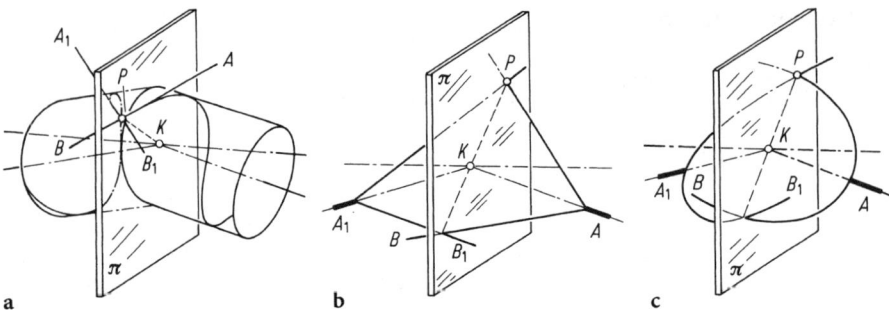

Fig. 2.13a–c. Effective geometry of a constant velocity joint. **a** Helical, tangent and hyperboloid joint; **b** conical joint; **c** spherical meridian joint. B opposing generator to the one starting from point A, interacting at B_1 with that from point A_1; K pivot point for the conical axes [2.9]

If constant velocity is to be obtained with the straight generators r_1 and r_2 the condition given by (2.8) is

$$y'_2 = -z'_2 \frac{1+\cos\beta}{\sin\beta}$$

Equation (g) [1.10, p. 86] must be fulfilled:

$$\left\{ z''^2_2 \left(\frac{\cos\beta + \overbrace{\cos^2\beta + \sin^2\beta}^{1}}{\sin\beta} \right)^2 - z'^2_2 \left(\frac{1+\cos\beta}{\sin\beta} \right)^2 \right\} \\ - zmb \{ z'_2 (1+\cos\beta) - z'_2 (1+\cos\beta) \} = 0 \tag{2.11}$$

It can be seen that both terms in parentheses become equal to nought. The fact that the straight generators r_1 and r_2 are mirror images results in constant velocity because the contact point P rotates in the angle bisecting plane π (Fig. 2.13a and b).

2.2.2 Effective Geometry with Circular Tracks

Figure 2.14a shows the general case of circular generators k which, when rotated through φ_1 about the z-axis, give spherical envelopes. The relationship between the radius r and the z co-ordinate is given by the equation for the circle:

$$y' = r(z') \Rightarrow y'^2 = r^2(z'). \tag{a}$$

It can be seen from the sectional view in Fig. 2.14b that in the top position $P(0, r)$, y' and $r(z')$ are equal:

$$y' = r(z') \Rightarrow y'^2 = r^2(z).$$

2.2 First Indirect Method of Proving Constant Velocity According to Metzner

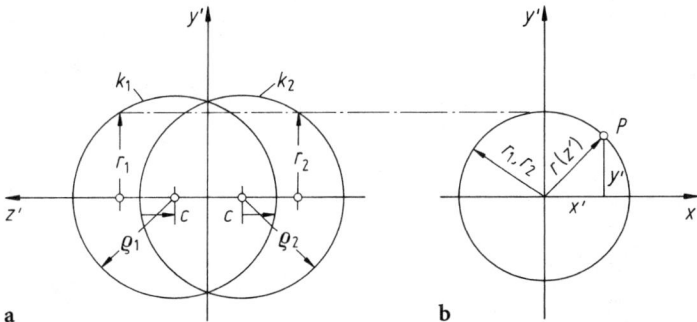

Fig. 2.14a, b. Spherical meridian ruled joint with circular tracks. a Side view, b section through the generating points of the radii r_1 and r_2

In the general position $P(x, y)$

$$x'^2 + y'^2 = r^2(z'). \tag{2.12a}$$

If (a) is inserted in (2.11b) we get

$$x'^2 + y'^2 = \varrho^2 - (z'^2 + c^2 - 2cz'). \tag{2.12b}$$

Proceeding as in Sect. 2.2 from (2.6) and (2.7)

$$x_1'^2 = \varrho_1^2 - z_1'^2 - c_1^2 + 2c_1 z_1' - y_1'^2,$$
$$x_2'^2 = \varrho_2^2 - z_2'^2 - c_2^2 + 2c_2 z_2' - y_2'^2,$$
$$x_2'^2 = x_1'^2,$$
$$(y_1'^2 - y_2'^2) - (z_2'^2 - z_1'^2) + 2(c_2 z_2' - c_1 z_1') = (c_2^2 - c_1^2) + (\varrho_1^2 - \varrho_2^2).$$

In Figure 2.14 the mirror image condition of the effective geometry requires

$$\varrho_1 = \varrho_2 = \varrho \tag{a}$$

and for the

1st case: $c_1 = c, \quad c_2 = -c.$ (b)
2nd case: $c_1 = -c, \quad c_2 = c.$ (c)

From this it follows for the *1st case* from (b) that

$$(y_1'^2 - y_2'^2) - (z_2'^2 - z_1'^2) - 2c(z_2' + z_1') = 0,$$

and for the *2nd case* from (c) that

$$(y_1'^2 - y_2'^2) - (z_2'^2 - z_1'^2) + 2c(z_1'^2 + z_1') = 0.$$

With (2.11) one obtains

$$\pm 2c\,(z_2' + z_1') = 0$$

and, with (2.5),

$$\pm 2c\,(z_2' + y_2' \sin\beta + z_2' \cos\beta) = 0. \tag{d}$$

Condition (2.8) inserted in (d) gives

$$\pm 2c \left\{ z_2' + \left(-z_2' \frac{1 + \cos\beta}{\sin\beta} \sin\beta + z_2' \cos\beta \right) \right\} = 0$$

$$\pm 2c \left\{ z_2'\,(1 + \cos\beta) - z_2'\,(1 + \cos\beta) \right\} = 0.$$

The term in brackets is equal to nought.

Both cases (b) and (c) of the mirror image generators k_1 and k_2 result in constant velocity because the contact point P once again revolves in the angle bisecting plane π. In [2.9] the joint is described as an offset spherical meridian ruled line joint (Fig. 2.13c).

2.3 Second, Direct Method of Proving Constant Velocity by Orain[3]

If the rules of transmission are to be determined for a given joint, the effective geometry of the tracks of the transmission members must be represented analytically.

For straight tracks in space, the following vectors apply (Fig. 2.15)

$$\boldsymbol{r} = \boldsymbol{a} + t\boldsymbol{u}$$

or written in the form of their components [2.8, p. 62]

$$\begin{pmatrix} x \\ y \\ z \end{pmatrix} = \begin{pmatrix} k \\ l \\ m \end{pmatrix} + t \begin{pmatrix} n \\ p \\ q \end{pmatrix}.$$

The starting position of the straight line $\boldsymbol{r}$ is such that it intersects the y-axis at a distance r and is parallel to the z-axis which is expressed according to Fig. 2.16 by the position vector

$$\boldsymbol{a} = \begin{pmatrix} k \\ l \\ m \end{pmatrix} = \begin{pmatrix} 0 \\ r \\ 0 \end{pmatrix}$$

---

[3] Dr. Michel Orain, former Head of Development at Glaenzer-Spicer SA, Paris, his thèses 1976.

2.3 Second, Direct Method of Proving Constant Velocity by Orain

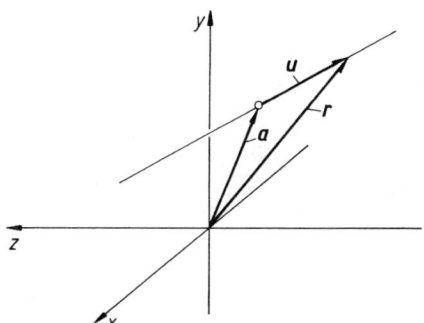

Fig. 2.15. Straight track in space in vector form

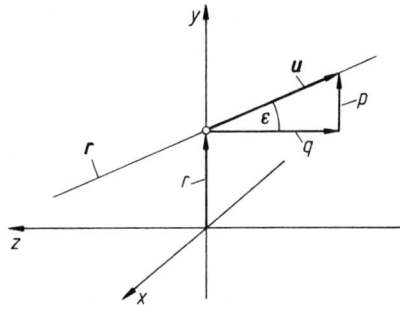

Fig. 2.16. Position of the straight lines for the cone joint

and by the direction vector

$$u = \begin{pmatrix} n \\ p \\ q \end{pmatrix} = \begin{pmatrix} 0 \\ \sin\varepsilon \\ \cos\varepsilon \end{pmatrix}.$$

The case of the hyperboloid ruled line joint is not included since this does not provide any new information.

The same coordinate systems are used as with the first method in Sect. 2.3 and the straight lines r_1 and r_2 are transformed with the matrix D from the body to the spatial systems:

$$r'_1 = D_{\varphi 1} r_1 \quad {}^4$$

$$\begin{pmatrix} x'_1 \\ y'_1 \\ z'_1 \end{pmatrix} = \begin{pmatrix} \cos\varphi_1 & -\sin\varphi_1 & 0 \\ \sin\varphi_1 & \cos\varphi_1 & 0 \\ 0 & 0 & 1 \end{pmatrix} \left\{ \begin{pmatrix} 0 \\ r_1 \\ 0 \end{pmatrix} + t_1 \begin{pmatrix} 0 \\ \sin\varepsilon_1 \\ \cos\varepsilon_1 \end{pmatrix} \right\}$$

$$= \begin{pmatrix} -r_1 \sin\varphi_1 \\ r_1 \cos\varphi_1 \\ 0 \end{pmatrix} + t_1 \begin{pmatrix} -\sin\varphi_1 \sin\varepsilon_1 \\ \cos\varphi_1 \sin\varepsilon_1 \\ \cos\varepsilon_1 \end{pmatrix}$$

$$= \begin{pmatrix} k'_1 \\ l'_1 \\ m'_1 \end{pmatrix} + t_1 \begin{pmatrix} n'_1 \\ p'_1 \\ q'_1 \end{pmatrix}, \qquad (2.13)$$

---

[4] $D_{\varphi 1}$ rotation by an angle φ_1 about the z-axis.

$r'_2 = D_{\varphi_2} r_2$

$$\begin{pmatrix} x'_2 \\ y'_2 \\ z'_2 \end{pmatrix} = \begin{pmatrix} \cos\varphi_2 & -\sin\varphi_2 & 0 \\ \sin\varphi_2 & \cos\varphi_2 & 0 \\ 0 & 0 & 1 \end{pmatrix} \left\{ \begin{pmatrix} 0 \\ r_2 \\ 0 \end{pmatrix} + t_2 \begin{pmatrix} 0 \\ \sin\varepsilon_2 \\ \cos\varepsilon_2 \end{pmatrix} \right\}$$

$$= \begin{pmatrix} -r_2 \sin\varphi_2 \\ r_2 \cos\varphi_2 \\ 0 \end{pmatrix} + t_2 \begin{pmatrix} -\sin\varphi_2 \sin\varepsilon_2 \\ \cos\varphi_2 \sin\varepsilon_2 \\ \cos\varepsilon_2 \end{pmatrix}$$

$$= \begin{pmatrix} k'_2 \\ l'_2 \\ m'_2 \end{pmatrix} + t_2 \begin{pmatrix} n'_2 \\ p'_2 \\ q'_2 \end{pmatrix}. \tag{2.14}$$

The output shaft 2 is now articulated through the angle β:

$r''_2 = D_\beta r'_2$

$$\begin{pmatrix} x''_2 \\ y''_2 \\ z''_2 \end{pmatrix} = \begin{pmatrix} 1 & 0 & 0 \\ 0 & \cos\beta & -\sin\beta \\ 0 & \sin\beta & \cos\beta \end{pmatrix} \left\{ \begin{pmatrix} -r_2 \sin\varphi_2 \\ r_2 \cos\varphi_2 \\ 0 \end{pmatrix} + t_2 \begin{pmatrix} -\sin\varphi_2 \sin\varepsilon_2 \\ \cos\varphi_2 \sin\varepsilon_2 \\ \cos\varepsilon_2 \end{pmatrix} \right\},$$

$$\begin{pmatrix} x''_2 \\ y''_2 \\ z''_2 \end{pmatrix} = \begin{pmatrix} -r_2 \sin\varphi_2 \\ r_2 \cos\varphi_2 \cos\beta \\ r_2 \cos\varphi_2 \sin\beta \end{pmatrix} + t_2 \begin{pmatrix} \sin\varphi_2 \sin\varepsilon_2 \\ \cos\varphi_2 \sin\varepsilon_2 \cos\beta - \cos\varepsilon_2 \sin\beta \\ \cos\varphi_2 \sin\varepsilon_2 \sin\beta + \cos\varepsilon_2 \cos\beta \end{pmatrix}$$

$$= \begin{pmatrix} k''_2 \\ l''_2 \\ m''_2 \end{pmatrix} + t_2 \begin{pmatrix} n''_2 \\ p''_2 \\ k''_2 \end{pmatrix} ^5 \tag{2.15}$$

$k_1 = -r_1 \sin\varphi_1 \Rightarrow \sin\varphi_1 = -k_1/r_1,$

$l_1 = r_1 \cos\varphi_1 \Rightarrow \cos\varphi_1 = l_1/r_1,$

$m_1 = 0;$

$n_1 = -\sin\varphi_1 \sin\varepsilon_1 = \dfrac{k_1}{r_1} \sin\varepsilon_1,$

$p_1 = \cos\varphi_1 \sin\varepsilon_1 = \dfrac{l_1}{r_1} \sin\varepsilon_1,$

$q_1 = \cos\varepsilon_1;$

$k_2 = -r_2 \sin\varphi_2 \Rightarrow \sin\varphi_2 = -k_2/r_2,$

$l_2 = r_2 \cos\varphi_2 \cos\beta \Rightarrow \cos\varphi_2 = \dfrac{l_2}{r_2 \cos\beta},$

$m_2 = r_2 \cos\varphi_2 \sin\beta \Rightarrow \cos\varphi_2 = \dfrac{m_2}{r_2 \sin\beta};$

---

[5] For simplification the prime symbols will be left out after (2.14). Components with the index 1 should be thought of as having *one* prime suffix and components with the index 2 as having *two* prime suffices.

2.3 Second, Direct Method of Proving Constant Velocity by Orain

$$n_2 = -\sin\varphi_2 \sin\varepsilon_2 = \frac{k_2}{r_2}\sin\varepsilon_2,$$

$$p_2 = \frac{l_2}{r_2}\sin\varepsilon_2 - \cos\varepsilon_2 \sin\beta,$$

$$q_2 = \frac{m_2}{r_2}\sin\varepsilon_2 + \cos\varepsilon_2 \cos\beta. \tag{2.16}$$

The transmission of rotational movement demands that the transmitting member is in continuous contact with the contact surfaces. This means in mathematical terms that the straight generators r_1 and r_2 intersect one another and the scalar triple product is equal to zero [1.10, p. 177]

$$[(a_2 - a_1), u_1, u_2] = 0.$$

This equation with the components of the position vectors a and the direction vectors u leads to the determinant

$$\begin{vmatrix} (k_2 - k_1) & n_1 & n_2 \\ (l_2 - l_1) & p_1 & p_2 \\ (m_2 - m_1) & q_1 & q_2 \end{vmatrix} = 0 \tag{2.17}$$

and is transformed according to Pierre-Frederic Sarrus' rule and developed with $m_1 = 0$

$$(k_2 - k_1) p_1 q_2 - (k_2 - k_1) q_1 p_2 + (l_2 - l_1) q_1 n_2 - (l_2 - l_1) n_1 q_2 + m_2 n_1 p_2 - m_2 p_1 n_2 = 0. \tag{2.18}$$

If the simplifications in (2.15) are employed in (2.17) and suitably transformed one obtains the fundamental equation for the rules of transmission of joints with straight ball tracks parallel to the axis of rotation:

$$\left(\frac{m_2}{r_2}\sin\varepsilon_2 + \cos\varepsilon_2 \cos\beta\right)\left\{\left(\frac{k_2 l_1}{r_1} - \frac{k_1 l_2}{r_1}\right)\sin\varepsilon_1\right\}$$

$$\cdot \left(\frac{l_2}{r_2}\sin\varepsilon_2 - \cos\varepsilon_2 \sin\beta\right)\left\{(k_2 - k_1)\cos\varepsilon_1 - \frac{k_1}{r_1}m_2 \sin\varepsilon_1\right\}$$

$$+ (l_2 - l_1)\frac{k_2}{r_2}\cos\varepsilon_1 \sin\varepsilon_2 - \frac{k_2 l_1}{r_1 r_2}m_2 \sin\varepsilon_1 \sin\varepsilon_2 = 0. \tag{2.19}$$

If the mirror image conditions are inserted in (2.19)

$$r_1 = r_2 = r,$$
$$\varepsilon_1 = \varepsilon \quad \text{and} \quad \varepsilon_2 = -\varepsilon$$

or

$$\varepsilon_1 = -\varepsilon \quad \text{and} \quad \varepsilon_2 = \varepsilon,$$

then it can be proved that the constant velocity condition $\varphi_2 = \varphi_1$ is fulfilled.

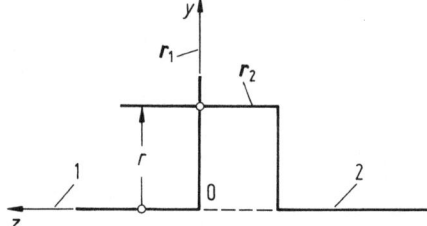

Fig. 2.17. Effective geometry of the monopode joint for the case $\varepsilon_1 = 90°$, $\varepsilon_2 = 0°$

This is however only the inversion of the result of the first, indirect method by Metzner (Sect. 2.2). It is more important to determine the rules of transmission for the case of the *monopode joint* (Fig. 2.17) where

$$\varepsilon_1 = 90°, \qquad \varepsilon_2 = 0°,$$
$$\sin \varepsilon_1 = 1, \qquad \sin \varepsilon_2 = 0,$$
$$\cos \varepsilon_1 = 0, \qquad \cos \varepsilon_2 = 1.$$

If these values are inserted in the fundamental equation (2.19) we get

$$-\frac{l_2}{r}(k_2 - k_1) + (l_2 - l_1)\frac{k_2}{r} = 0$$

$$k_2 l_1 - k_1 l_2 = 0.$$

If the simplifications from equation (2.16) are also employed then:

$$[-r \sin \varphi_2 r \cos \varphi_1 - (-r \sin \varphi_1 r \cos \varphi_2 \cos \beta)] = 0,$$
$$\tan \varphi_2 \cot \varphi_1 = \cos \beta,$$
$$\tan \varphi_2 = \cos \beta \tan \varphi_1. \tag{2.20}$$

The monopode joint shown in Fig. 2.18 does not therefore have constant velocity properties but follows the same rules of transmission as the simple Hooke's joint described in (1.1).

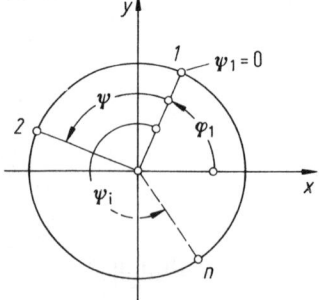

Fig. 2.18. Position of the straight generators in the section through the centre of the polypode joint

2.3 Second, Direct Method of Proving Constant Velocity by Orain

2.3.1 Polypode Joints

The rules deduced for the monopode joint are now extended to n transmitting members. Their position is given here by the fixed angles (Fig. 2.18) obtained from

$$\psi_i = (i-1)\, 2\pi/n . \tag{2.21}$$

The n straight generators r, given in the body system of coordinates in (2.21), depend in the spatial system on the angles of rotation

$$\varphi_{1i} = \varphi_1 + \psi_i,$$
$$\varphi_{2i} = \varphi_2 + \psi_i.$$

The equations in the body system are as follows for the driving member 1:

$$r_{1i} = \begin{pmatrix} x_{1i} \\ y_{1i} \\ z_{1i} \end{pmatrix} = \begin{pmatrix} 0 \\ 0 \\ 0 \end{pmatrix} + t_i \begin{pmatrix} \cos\varphi_{1i} \\ \sin\varphi_{1i} \\ 0 \end{pmatrix}$$

and for the driven member 2:

$$r_{2i} = \begin{pmatrix} x_{2i} \\ y_{2i} \\ z_{2i} \end{pmatrix} = \begin{pmatrix} r_{2i}\cos\varphi_{2i} \\ r_{2i}\sin\varphi_{2i} \\ 0 \end{pmatrix} + t_i \begin{pmatrix} 0 \\ 0 \\ 1 \end{pmatrix}.$$

They are transformed back into the spatial systems, marked with one prime, using the matrix D

$$r'_{1i} = D_{\varphi 1} r_{1i},$$

$$r'_{1i} = \begin{pmatrix} x'_{1i} \\ y'_{1i} \\ z'_{1i} \end{pmatrix} = \begin{pmatrix} \cos\varphi_1 & -\sin\varphi_1 & 0 \\ \sin\varphi_1 & \cos\varphi_1 & 0 \\ 0 & 0 & 1 \end{pmatrix} \left\{ \begin{pmatrix} 0 \\ 0 \\ 0 \end{pmatrix} + t_i \begin{pmatrix} \cos\psi_i \\ \sin\psi_i \\ 0 \end{pmatrix} \right\}$$

$$= \begin{pmatrix} 0 \\ 0 \\ 0 \end{pmatrix} + t_i \begin{pmatrix} \cos(\varphi_1 + \psi_i) \\ \sin(\varphi_1 + \psi_i) \\ 0 \end{pmatrix} = \begin{pmatrix} k'_{1i} \\ l'_{1i} \\ m'_{1i} \end{pmatrix} + t_i \begin{pmatrix} n'_{1i} \\ p'_{1i} \\ q'_{1i} \end{pmatrix}, \tag{2.22}$$

$$r'_{2i} = D_{\varphi 2} r_{2i},$$

$$r'_{2i} = \begin{pmatrix} \cos\varphi_2 & -\sin\varphi_2 & 0 \\ \sin\varphi_2 & \cos\varphi_2 & 0 \\ 0 & 0 & 1 \end{pmatrix} \left\{ \begin{pmatrix} r_{2i}\cos\psi_i \\ r_{2i}\sin\psi_i \\ 0 \end{pmatrix} + t_i \begin{pmatrix} 0 \\ 0 \\ 1 \end{pmatrix} \right\}$$

$$= \begin{pmatrix} r_{2i}\cos(\varphi_2 + \psi_i) \\ r_{2i}\sin(\varphi_2 + \psi_i) \\ 0 \end{pmatrix} + t_i \begin{pmatrix} 0 \\ 0 \\ 1 \end{pmatrix}.$$

The driven member 2 is now articulated further through the angle β:

$$r''_{2i} = D_\beta r'_{2i},$$

$$r''_{2i} = \begin{pmatrix} 1 & 0 & 0 \\ 0 & \cos\beta & -\sin\beta \\ 0 & \sin\beta & \cos\beta \end{pmatrix} \left\{ \begin{pmatrix} r_{2i} \cos(\varphi_2 + \psi_i) \\ r_{2i} \sin(\varphi_2 + \psi_i) \\ 0 \end{pmatrix} + t_i \begin{pmatrix} 0 \\ 0 \\ 1 \end{pmatrix} \right\}$$

$$= \begin{pmatrix} r_{2i} \cos(\varphi_2 + \psi_i) \\ r_{2i} \sin(\varphi_2 + \psi_i) \cos\beta \\ r_{2i} \sin(\varphi_2 + \psi_i) \sin\beta \end{pmatrix} + t_i \begin{pmatrix} 0 \\ -\sin\beta \\ \cos\beta \end{pmatrix}$$

$$= \begin{pmatrix} k''_{2i} \\ l''_{2i} \\ m''_{2i} \end{pmatrix} + t_i \begin{pmatrix} n''_{2i} \\ p''_{2i} \\ q''_{2i} \end{pmatrix}. \tag{2.23}$$

The components k to q from (2.22) and (2.23) are inserted in the main equation (2.18). The components for index 1 should be thought of as having one prime suffix and those for index 2 as having two prime suffices. Since

$$k_{1i} = l_{1i} = m_{1i} = q_{1i} = n_{2i} = 0$$

it follows that

$$k_{2i} p_{1i} q_{2i} - l_{2i} n_{1i} q_{2i} + m_{2i} n_{1i} p_{2i} = 0$$

or:

$$-\sin(\varphi_2 + \psi_i) \cos(\varphi_1 + \psi_i) \overbrace{(\sin^2\beta + \cos^2\beta)}^{1} +$$
$$+ \cos(\varphi_2 + \psi_i) \sin(\varphi_1 + \psi_i) \cos\beta = 0.$$

Dividing by $\cos(\varphi_2 + \psi_i) \sin(\varphi_1 + \psi_i)$ gives

$$\tan(\varphi_2 + \psi_i) = \cos\beta \tan(\varphi_1 + \psi_i). \tag{2.24}$$

Equation (2.24) is the basis for understanding bipode and tripode joints. As anticipated one gets (2.19) once again for $\psi_i = 0$.

The bipode joint is the oldest joint after the classical Hooke's joint; it appeared in the 18th century. With $n = 2$ and $i = 1$ and 2 one obtais for the bipode joint from (2.21)

$$\psi_1 = (1 - 1) \frac{2\pi}{2} = 0 \text{ (i.e. } 0°\text{)},$$

$$\psi_2 = (2 - 1) \frac{2\pi}{2} = \pi \text{ (i.e. } 180°\text{)}.$$

These two angles, inserted in (2.24) give

$$\tan\varphi_2 = \cos\beta \tan\varphi_1, \tag{a}$$
$$\tan(\varphi_2 + 180°) = \cos\beta \tan(\varphi_1 + 180°),$$
$$\tan\varphi_2 = \cos\beta \tan\varphi_1. \tag{b}$$

2.3 Second, Direct Method of Proving Constant Velocity by Orain

The two identical equations (a) and (b) demonstrate that the velocity is not constant; instead there is the same non-uniformity as with the *single Hooke's joint* described by Poncelet's equation (1.1a).

The ball and trunnion and ring universal joints (Fig. 2.19a and b) have the same effective geometry as the bipode joint applies (2.24), with $\psi_i = 0$ leading again to Poncelet's single Hooke's joint (1.1a–c, 1.2).

For the *tripode joint*, fixed at the coordinate origin, $n = 3$ and $i = 1, 2, 3$ apply. With (2.21) one gets

$$\psi_1 = \underbrace{(1-1)}_{0} \frac{2\pi}{3} = 0 \text{ (i.e. } 0°),$$

$$\psi_2 = \underbrace{(2-1)}_{1} \frac{2\pi}{3} = \frac{2}{3}\pi \text{ (i.e. } 120°),$$

$$\psi_3 = \underbrace{(3-1)}_{2} \frac{2\pi}{3} = \frac{4}{3}\pi \text{ (i.e. } 240°).$$

If the input and output members of the tripode joint remain fixed, as is the case with the bipode joint, constant velocity does not occur. Since, however, a joint with $n = 3$ podes can centre itself[6] the fixing of the shafts *1* and *2* at the origin is relaxed so that shaft *1* is able to move freely in the *xy*-plane. Displacement along the *z*-axis does not after the geometrical relationships.

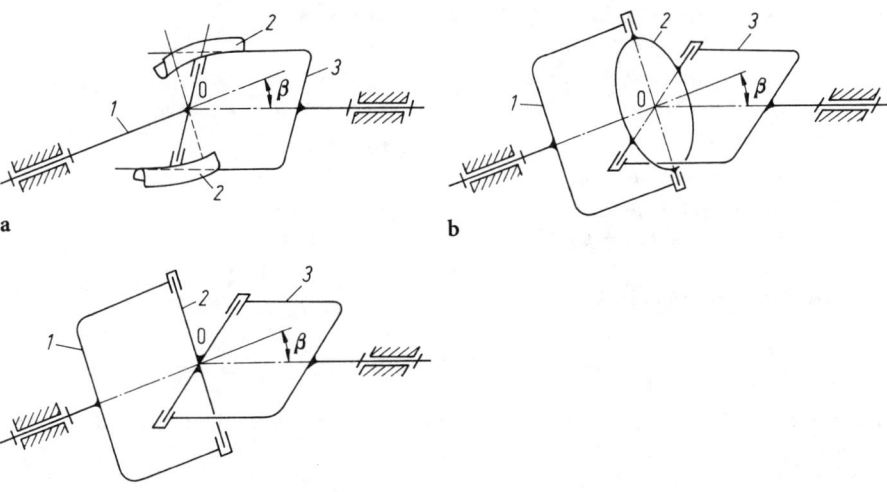

Fig. 2.19a–c. The most important structures of Hooke's joints shown diagrammatically [2.12, 2.15]. **a** ball track; **b** ring cross; **c** peg cross. All three shapes are based on the same "effective geometry" which leads to nonuniformity in transmission as given in (1.2). *1* input member; *2* coupling member: *3* output member; *0* support

---

[6] Isostatic connection, see Sect. 4.5.5, Eq. (4.81).

From this Michel Orain recognised in 1960 the possibility for a free constant velocity joint which he named the tripode joint (French patent 1 272 530, German patent 1 194 205). In 1964 he proved its constant velocity characteristics [2.11]. This proof is repeated here. The authors have however followed a new approach by making use of vector geometry.

The new equation, analogous to (2.22), reads

$$r'_{1i} = D_{\varphi 1}(r_{1i} - v_{1i}),$$

or written in component form

$$r'_{1i} = \begin{pmatrix} \cos\varphi_1 & -\sin\varphi_1 & 0 \\ \sin\varphi_1 & \cos\varphi_1 & 0 \\ 0 & 0 & 1 \end{pmatrix} \left\{ \begin{pmatrix} 0 \\ 0 \\ 0 \end{pmatrix} + t_{1i}\begin{pmatrix} \cos\psi_i \\ \sin\psi_i \\ 0 \end{pmatrix} - \begin{pmatrix} a \\ b \\ 0 \end{pmatrix} \right\}$$

$$= \begin{pmatrix} -a\cos\varphi_1 + b\sin\varphi_1 \\ -a\sin\varphi_1 - b\cos\varphi_1 \\ 0 \end{pmatrix} + t_{1i}\begin{pmatrix} \cos(\varphi_1 + \psi_i) \\ \sin(\varphi_1 + \psi_i) \\ 0 \end{pmatrix} = \begin{pmatrix} k'_{1i} \\ l'_{1i} \\ m'_{1i} \end{pmatrix} + t_{1i}\begin{pmatrix} n'_{1i} \\ p'_{1i} \\ q'_{1i} \end{pmatrix}. \quad (2.25)$$

(2.22) becomes (2.25),

$$r'_{2i} = \begin{pmatrix} r_{2i}\cos(\varphi_2 + \psi_i) \\ r_{2i}\sin(\varphi_2 + \psi_i)\cos\beta \\ r_{2i}\sin(\varphi_2 + \psi_i)\sin\beta \end{pmatrix} + t_{2i}\begin{pmatrix} 0 \\ -\sin\beta \\ \cos\beta \end{pmatrix} = \begin{pmatrix} k''_{2i} \\ l''_{2i} \\ m''_{2i} \end{pmatrix} + t_{2i}\begin{pmatrix} n''_{2i} \\ p''_{2i} \\ q''_{2i} \end{pmatrix}. \quad (2.26)$$

The components from (2.25 and 2.26) are as follows[7]

$$k_{1i} = -a\cos\varphi_1 + b\sin\varphi_1, \qquad n_{1i} = \cos(\varphi_1 + \psi_i),$$
$$l_{1i} = -a\sin\varphi_1 - b\cos\varphi_1, \qquad p_{1i} = \sin(\varphi_1 + \psi_i),$$
$$m_{1i} = 0, \qquad q_{1i} = 0,$$
$$k_{2i} = r_{2i}\cos(\varphi_2 + \psi_i), \qquad n_{2i} = 0,$$
$$l_{2i} = r_{2i}\sin(\varphi_2 + \psi_i)\cos\beta, \qquad p_{2i} = -\sin\beta,$$
$$m_{2i} = r_{2i}\sin(\varphi_2 + \psi_i)\sin\beta, \qquad q_{2i} = \cos\beta.$$

Inserted in (2.18) this gives

$$(k_{2i} - k_{1i})p_{1i}q_{2i} - (l_{2i} - l_{1i})n_{1i}q_{2i} + m_{2i}n_{1i}p_2 = 0.$$

The fundamental equation for the free pode joint, after some development, is then

$$r_{2i}\cos(\varphi_2 + \psi_i)\sin(\varphi_1 + \psi_i) - r_{2i}\frac{\sin(\varphi_2 + \psi_i)}{\cos\beta}\cos(\varphi_1 + \psi_i) +$$
$$+ (a\cos\varphi_1 - b\sin\varphi_1)\sin(\varphi_1 + \psi_i) - (a\sin + b\cos\varphi_1)\cos(\varphi_1 + \psi_i) = 0. \quad (2.27)$$

---

[7] The components with the index 1 should be considered as having one prime suffix, and those with the index 2 as having two prime suffices.

2.3.2 The Free Tripode Joint

For angular positions of the bearing surfaces $\psi_1 = 0°$, $\psi_2 = 120°$, $\psi_3 = 240°$ relative to the body and the joint radius $r_{2i} = r_2$, it follows from the fundamental equation for the free pode joint (2.27):

For the 1st transmitting element at $\psi_1 = 0°$:

$$r_2 \cos\varphi_2 \sin\varphi_1 - r_2 \frac{\sin\varphi_2}{\cos\beta} \cos\varphi_1 + (a\cos\varphi_1 - b\sin\varphi_1)\sin\varphi_1 -$$
$$- (a\sin\varphi_1 + b\cos\varphi_1)\cos\varphi_1 = 0,$$

$$r_2 \cos\varphi_2 \sin\varphi_1 - r_2 \frac{\sin\varphi_2}{\cos\beta} \cos\varphi_1 - b = 0. \qquad (2.28)$$

For the 2nd transmitting element at $\psi_2 = 120°$.

$$r_2 (\cos\varphi_2 \cos 120° - \sin\varphi_2 \sin 120°)(\sin\varphi_2 \cos 120° + \cos\varphi_2 \sin 120°) -$$
$$- r_2 \frac{(\sin\varphi_2 \cos 120° + \cos\varphi_2 \sin 120°)}{\cos\beta} (\cos\varphi_1 \cos 120° - \sin\varphi_1 \sin 120°) +$$
$$+ (a\cos\varphi_1 - b\sin\varphi_1)(\sin\varphi_1 \cos 120° + \cos\varphi_1 \sin 120°) -$$
$$- (a\sin\varphi_1 + b\cos\varphi_1)(\cos\varphi_1 \cos 120° - \sin\varphi_1 \sin 120°) = 0.$$

Since
$$\sin 120° = \sin(90° + 30°) = \cos 30° = \frac{1}{2}\sqrt{3}$$

and
$$\cos 120° = \cos(90° + 30°) = -\sin 30° = -\frac{1}{2}$$

one gets

$$r_2 \left(-\frac{1}{2}\cos\varphi_2 - \frac{1}{2}\sqrt{3}\sin\varphi_2\right)\left(-\frac{1}{2}\sin\varphi_1 + \frac{1}{2}\sqrt{3}\cos\varphi_1\right) -$$
$$- r_2 \frac{-\frac{1}{2}\sin\varphi_2 + \frac{1}{2}\sqrt{3}\cos\varphi_2}{\cos\beta}\left(-\frac{1}{2}\cos\varphi_1 - \frac{1}{2}\sqrt{3}\sin\varphi_1\right) +$$
$$+ (a\cos\varphi_1 - b\sin\varphi_1)\left(-\frac{1}{2}\sin\varphi_1 + \frac{1}{2}\sqrt{3}\cos\varphi_1\right) -$$
$$- (a\sin\varphi_1 + b\cos\varphi_1)\left(-\frac{1}{2}\cos\varphi_1 - \frac{1}{2}\sqrt{3}\sin\varphi_1\right) = 0.$$

Auxiliary calculation to the last equation:

$$\frac{1}{2}a\sqrt{3}\underbrace{(\sin^2\varphi_1+\cos^2\varphi_1)}_{1}+\frac{1}{2}b\underbrace{(\sin^2\varphi_1+\cos^2\varphi_1)}_{1}=\frac{1}{2}a\sqrt{3}+\frac{1}{2}b.$$

Inserted gives:

$$r_2\left(\sin\varphi_2+\frac{\cos\varphi_2}{\sqrt{3}}\right)\left(\frac{\sin\varphi_1}{\sqrt{3}}-\cos\varphi_1\right)-$$

$$-r_2\frac{\frac{\sin\varphi_2}{\sqrt{3}}-\cos\varphi_2}{\cos\beta}\left(\sin\varphi_1+\frac{\cos\varphi_1}{\sqrt{3}}\right)+\frac{2a}{\sqrt{3}}+\frac{2b}{3}=0. \quad (2.29)$$

For the 3rd transmitting element at $\psi_3 = 240°$

$$r_2(\cos\varphi_2\cos 240°-\sin\varphi_2\sin 240°)(\sin\varphi_1\cos 240°+\cos\varphi_1\sin 240°)-$$

$$-r_2\frac{\sin\varphi_2\cos 240°+\cos\varphi_2\sin 240°}{\cos\beta}(\cos\varphi_1\cos 240°-\sin\varphi_1\sin 240°)+$$

$$+(a\cos\varphi_1-b\sin\varphi_1)(\sin\varphi_1\cos 240°+\cos\varphi_1\sin 240°)-$$

$$-(a\sin\varphi_1+b\cos\varphi_1)(\cos\varphi_1\cos 240°-\sin\varphi_1\sin 240°)=0.$$

Since

$$\sin 240°=\sin(180°+60°)=-\sin 60°=-\frac{1}{2}\sqrt{3}$$

and

$$\cos 240°=\cos(180°+60°)=-\cos 60°=-\frac{1}{2}$$

one gets:

$$r_2\left(-\frac{1}{2}\cos\varphi_2+\frac{1}{2}\sqrt{3}\sin\varphi_2\right)\left(-\frac{1}{2}\sin\varphi_1-\frac{1}{2}\sqrt{3}\cos\varphi_1\right)-$$

$$-r_2\frac{-\frac{1}{2}\sin\varphi_2-\frac{1}{2}\sqrt{3}\cos\varphi_2}{\cos\beta}\left(-\frac{1}{2}\cos\varphi_1+\frac{1}{2}\sqrt{3}\sin\varphi_1\right)+$$

$$+(a\cos\varphi_1-b\sin\varphi_1)\left(-\frac{1}{2}\sin\varphi_1-\frac{1}{2}\sqrt{3}\cos\varphi_1\right)-$$

$$-(a\sin\varphi_1+b\cos\varphi_1)\left(-\frac{1}{2}\cos\varphi_1+\frac{1}{2}\sqrt{3}\sin\varphi_1\right)=0.$$

2.3 Second, Direct Method of Proving Constant Velocity by Orain

Auxiliary calculation to the last equation:

$$-\frac{1}{2}a\sqrt{3}\underbrace{(\sin^2\varphi_1 + \cos^2\varphi_1)}_{1} + \frac{1}{2}b\underbrace{(\sin^2\varphi_1 + \cos^2\varphi_1)}_{1} = -\frac{1}{2}a\sqrt{3} = -\frac{1}{2}b.$$

Inserted gives

$$-r_2\left(\sin\varphi_2 + \frac{\cos\varphi_2}{\sqrt{3}}\right)\left(\frac{\sin\varphi_1}{\sqrt{3}} - \cos\varphi_1\right) +$$

$$+ r_2 \frac{\frac{\sin\varphi_2}{\sqrt{3}} - \cos\varphi_2}{\cos\beta}\left(\sin\varphi_1 + \frac{\cos\varphi_1}{\sqrt{3}}\right) + \frac{2a}{\sqrt{3}} + \frac{2b}{3} = 0. \quad (2.30)$$

Adding (2.29) and (2.30)

$$r_2\left(\sin\varphi_2 + \frac{\cos\varphi_2}{\sqrt{3}}\right)\left(\frac{\sin\varphi_1}{\sqrt{3}} - \cos\varphi_1\right) -$$

$$- r_2 \frac{\frac{\sin\varphi_2}{\sqrt{3}} - \cos\varphi_2}{\cos\beta}\left(\sin\varphi_1 + \frac{\cos\varphi_1}{\sqrt{3}}\right) + \frac{2a}{\sqrt{3}} + \frac{2b}{\sqrt{3}} -$$

$$- r_2\left(\sin\varphi_2 + \frac{\cos\varphi_2}{\sqrt{3}}\right)\left(\frac{\sin\varphi_1}{\sqrt{3}} + \cos\varphi_1\right) +$$

$$+ r_2 \frac{\frac{\sin\varphi_2}{\sqrt{3}} + \cos\varphi_2}{\cos\beta}\left(\sin\varphi_1 - \frac{\cos\varphi_1}{\sqrt{3}}\right) + \frac{2a}{\sqrt{3}} + \frac{2b}{\sqrt{3}} = 0.$$

Multiplying out the brackets and inserting b from (2.27) gives

$$-\sin\varphi_2 \cos\varphi_1 \underbrace{\left(2 + \frac{6}{3\cos\beta}\right)}_{2 + \frac{2}{\cos\beta}} + \cos\varphi_2 \sin\varphi_1 \underbrace{\left(\frac{6}{3} + \frac{2}{\cos\beta}\right)}_{2 + \frac{2}{\cos\beta}} = 0,$$

$$-\sin\varphi_2 \cos\varphi_1 + \cos\varphi_2 \sin\varphi_1 = 0,$$

$$\sin\varphi_2 \cos\varphi_1 = \cos\varphi_2 \sin\varphi_1,$$

dividing through by $\cos\varphi_2 \sin\varphi_1$

$$\tan\varphi_2 \cot\varphi_1 = 1,$$

$$\tan\varphi_2 = \tan\varphi_1 \Rightarrow \varphi_2 = \varphi_1. \quad (2.31)$$

It was thus proved by Michel Orain that a joint whose transmitting elements are not guided on mirror image tracks is also able to fulfil the conditions of constant velocity.

In 1975 Duditza and Diaconescu showed with their equation

$$\varphi_2 = \varphi_1 + \frac{R}{2l} \tan\beta \tan^2 \frac{\beta}{2} \cos 3\varphi_1,$$

that the result of (2.31) only applies to *infinitely long* driveshafts [2.12 Eq. (11)]. However the length of a driveshaft is usually more than 10 times the effective joint radius and hence the angular difference $\Delta\varphi = \varphi_2 - \varphi_1$ is negligibly small.

For a motor vehicle being driven fast on *good roads*, the articulation angles are less than 10°. The error from the compensating movements of the tripode joint is

in radians: $\quad \Delta\varphi = \varphi_2 - \varphi_1 = 1/2 \cdot 1/10 \cdot 0.1763 \cdot 0.0875^2 \cdot 1 = 6.75 \cdot 10^{-5}$,
in degrees: $\quad \Delta\varphi = 180/\pi \cdot 60 \cdot 60 \cdot 6.75 \cdot 10^{-5} = 14''$.

These 14 seconds are within the manufacturing tolerances of constant velocity joints and can be disregarded.

Since motor vehicles cannot be driven fast on *bad roads*, the speed of the joints is 100 to 200 rpm. The resulting larger angular errors $\Delta\varphi$ will therefore have little effect on the torque transmission.

2.4 Literature to Chapter 2

2.1 Reuleaux, F.: Theoretische Kinematik (Theoretical Kinematics). Vol. I, Brunswick: Vieweg 1875, p. 179 onwards
2.2 Reuleaux, F.: Der Construkteur (The Designer). 3rd edition Brunswick: Vieweg 1869, p. 260–267
2.3 Gruebler, M. F.: Getriebelehre (Theory of Kinematics). Berlin: Springer 1917, p. 6–39
2.4 Steeds, W.: Universal Joints. Automob. Eng. 27 (1937) No. 354, p. 10–12
2.5 Balken, J.: Systematische Entwicklung von Gleichlaufgelenken (Systematic development of constant velocity joints). Diss. TU Munich 1981
2.6 Grashof, F.: Theoretische Maschinenlehre (Theoretical Engineering). Vol. 2: Theorie der Getriebe etc. (Theory of power transmission etc.) Leipzig: Voss 1883
2.7 Schmelz, F.: Count Seherr-Thoß, H. Ch.: Die Entwicklung der Gleichlaufgelenke für den Frontantrieb (Development of constant velocity joints for front wheel drive). VDI-Report 418 (1981) 197–207
2.8 Jeger, M.: Eckmann, B.: Einführung in die vektorielle Geometrie und lineare Algebra (Introduction to vector geometry and linear algebra). Basle: Birkhäuser 1967
2.9 Molly, H.: Bengisu, O.O Das Gleichgang-Gelenk im Symmetriespiegel (The constant velocity joint in the mirror of symmetry). Automob. Ind. 14 (1969) No. 2, p. 45–54
2.10 Guimbretière, P.: La Traction Avant (Front wheel drive): Orain, M.: Démonstration de l'homocinétie du joint GI (Demonstration of the homokinetics of the GI joint). Both in: Joints mécaniques de transmission. SIA J. 38 (1964) p. 213–227
2.11 Orain, M.: Die Gleichlaufgelenke, allgemeine Theorie und experimenteller Forschung (Constant velocity joints, general theory and experimental research). German translation by H. W. Günther. Paris: Glaenzer-Spicer 1976
2.12 Duditza, Fl.: Diaconescu, D.: Zur Kinematik und Dynamik von Tripode-Gelenkgetrieben (The kinematics and dynamics of tripode jointed drivelines) Konstruction 27 (1975) p. 335–341

2.4 Literatture to Chapter 2

2.13 Beyer, R.: Technische Kinematic (Engineering Kinematics). Leipzig: Barth 1931, p. 88, Fig. 176
2.14 Norbye, J. P.: The Complete Handbook of Front Wheel Drive Cars. TAB Modern Automotive Series No. 2052. Blue Ridge Summit/Pa.: TAB Books 1979, p. 18
2.15 Richtlinien VDI 2722 (VDI Directives): Homokinetische Kreuzgelenk-Getriebe einschließlich Gelenkwellen (Homokinetic universal jointed transmissions including driveshafts). Düsseldorf: VDI-Verlag 1978

3 Hertzian Theory and the Limits of Its Application

In Chap. 2 the simplified, three-bar kinematic chain was extended to a four-bar linkage by inserting a roller body between the input and output members. This gave rise to joints *2* and *3* in Fig. 2.7a and b. A *point or line* contact occurs at these points from the reciprocal pressure on their curved surfaces due to the torque being transmitted. Heinrich Hertz's theory of 1881 [1.21] allowed the deformations, surface stresses and compressive forces occurring here to be calculated for the first time. This theory also enables the surface stresses in Hooke's joints with trunnions in roller bearings and in ball joints to be analysed. As early as 1878 Hertz found that the determination of the surface stress of rolling bodies was only approximate and was hampered by uncertain empirical values. Hence in 1881 he worked out a strict solution [1.21].

However as stated in Chap. 1 Hertz's theory is not easily accessible to practising engineers because he used the potential function of a layer with distributed density and imagined an infinitely flattened ellipsoid with uniform mass distribution between the two bodies pressed together. Mechanical engineers could not immediately understand the potential theory of the triaxial ellipsoid.

In 1948 Constantin Weber simplified the understanding of the Hertzian theory with an analogy which used instead the sectional area of an arbitrary stress distribution over the surface of the half-space inside the stress ellipse and compared the loaded elliptical face with a circle [1.29].

Weber's analogy is still the generally accepted method of analysis.

Hertz's solutions [1.21] are valid under the following conditions:

- the bodies are isotropic and homogeneous,
- the dimensions of the contact area are very small compared with the radii of curvature,
- the contact area is completely smooth,
- the contacting surfaces behave linearly elastically.

The *first condition* is not really applicable for metallic materials. The structure of steel consists of many irregularly arranged crystals which behave very differently along their crystalline axes. However on a macroscopic scale the microstructural irregularities generally even out, and common engineering practice is to work within the regime of linear elasticity.

The *second condition* is not fulfilled in the case of high conformity. The contact face is then clearly curved and its dimensions are no longer small compared with the radii of curvature. Hertz gives the limit for the radius a of the circular contact face [1.21, p. 167] as

$$a \leqq 0.1 r.$$

The *third condition* is equivalent to normal forces only being allowed to act to the contact face. Hertz therefore rules out frictional forces between the faces and assumes completely smooth surfaces. In practice this condition will not be completely fulfilled.

The *fourth condition* is not broken in the case of rolling pairs because one then gets into the area of substantial, permanent deformation in which rolling bodies would quickly become unusable.

3.1 Systems of Coordinates

Mathematically both bodies are in contact at the point O, the co-ordinate origin. The common tangential plane is the xy-plane, the positive z-axes of the *body* systems *1* and *2* are normal lines parallel to the load Q and directed into each of the bodies (Fig. 3.1).

The minimum curvatures $\varrho_{11} = 1/r_{11}$ and $\varrho_{21} = 1/r_{21}$ of the xz-plane lie in the principal plane of curvature (Fig. 3.1b). The maximum curvatures $\varrho_{12} = 1/r_{12}$ and $\varrho_{22} = 1/r_{22}$ of the yz-plane lie in the principal plane of curvature 2 (Fig. 3.1c). The first index denotes the body, the second index the principal plane of curvature.

There is another orthogonal system without index where the x-axis forms the angle ω_1 with the x_1-axis, the angle ω_2 with the x_2-axis and where the positive z-axis is directed into body *1*.

By assigning the minimum principal curvature *1* to the xz-plane, the larger a-axis lies on the x-axis of the contact ellipse (Fig. 3.2).

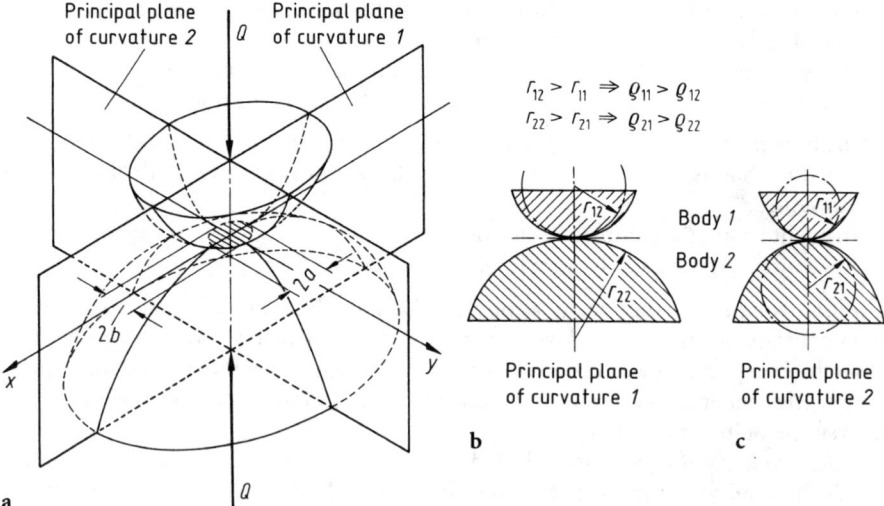

Fig. 3.1a–c. Principal planes of curvature and radii of curvature of two curved bodies [3.3, p. 101]. **a** Contact between curved bodies, **b** principal plane of curvature *1*, **c** principal plane of curvature *2*

3.2 Equations of Body Surfaces

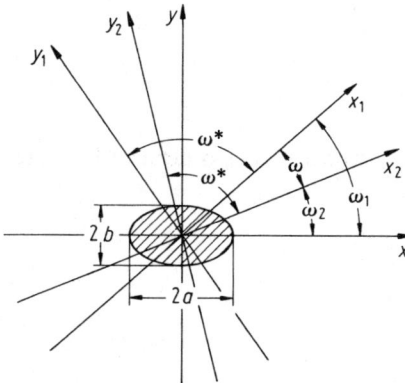

Fig. 3.2. Orthogonal co-ordinate systems, the positive z-axis of which is directed into body 1. The principal curvature 1 is allocated to the xz-plane

Since we are dealing with rotational bodies (spheres, cylinders, cones, toruses etc.) the principal curvatures of which are perpendicular to one another ($\omega^* = 90°$), the two systems 1 and 2 are also orthogonal systems.

3.2 Equations of Body Surfaces

Near the contact area the real surfaces are represented by homogeneous functions[1] of the second degree in x and y. Points which are at the same distance from the xy-plane then lie on an ellipse. The following thus applies for bodies 1 and 2

$$z_1 = A_1 x_1^2 + B_1 y_1^2, \tag{3.1}$$

$$z_2 = A_2 x_2^2 + B_2 y_2^2. \tag{3.2}$$

The constants A and B can be represented by the principal curvatures; they are considered positive if the corresponding centre of curvature lies inside the body. The following generally applies [1.10, p. 190]

$$\varrho = \frac{f''(x)}{[1 + f'(x)^2]^{3/2}}.$$

In the $x_1 z_1$-plane therefore we get

$$\varrho_{11} = \frac{2A_1}{[1 + (2A_1 x_1)^2]^{3/2}}$$

---

[1] A function is homogeneous if its terms with respect to the unknown parameter are of the same degree.

and in the y_1z_1-plane

$$\varrho_{12} = \frac{2B_1}{[1+(2B_1y_1)^2]^{3/2}}.$$

Since A and B are small values [1.21], the terms of a higher order can be disregarded[2]. We therefore get

$$\varrho_{11} = 2A_1 \quad \text{and} \quad \varrho_{12} = 2B_1.$$

The same applies for body 2

$$\varrho_{21} = 2A_2 \quad \text{and} \quad \varrho_{22} = 2B_2.$$

The equations transforming into the xy-system are

$$\begin{aligned} x' &= x\cos\omega' + y\sin\omega' \\ y' &= -x\sin\omega' + y\cos\omega', \end{aligned} \tag{3.3}$$

where the indices 1 and 2 are used for the values having prime suffices.

Robert Mundt's approximations for the body surfaces [3.1] give

$$z_1 = \frac{1}{2}\varrho_{11}x_1^2 + \frac{1}{2}\varrho_{12}y_1^2.$$

Equation (3.3) then gives

$$z_1 = \frac{1}{2}\varrho_{11}(x^2\cos^2\omega_1 + y^2\sin^2\omega_1 + 2xy\sin\omega_1\cos\omega_1) +$$
$$+ \frac{1}{2}\varrho_{12}(x^2\sin^2\omega_1 + y^2\cos^2\omega_1 - 2xy\sin\omega_1\cos\omega_1)$$

and

$$z_2 = \frac{1}{2}\varrho_{21}(x^2\cos^2\omega_2 + y^2\sin^2\omega_2 + 2xy\sin\omega_2\cos\omega_2) +$$
$$+ \frac{1}{2}\varrho_{22}(x^2\sin^2\omega_2 + y^2\cos^2\omega_2 - 2xy\sin\omega_2\cos\omega_2).$$

Arranging in terms of x^2, y^2 and xy gives

$$z_1 = x^2\left(\frac{1}{2}\varrho_{11}\cos^2\omega_1 + \frac{1}{2}\varrho_{12}\sin^2\omega_1\right) + y^2\left(\frac{1}{2}\varrho_{11}\sin^2\omega_1 + \frac{1}{2}\varrho_{12}\cos^2\omega_1\right) +$$
$$+ xy\left(\frac{1}{2}\varrho_{11}\sin\omega_1\cos\omega_1 - \frac{1}{2}\varrho_{12}\sin\omega_1\cos\omega_1\right)$$

---

[2] Here the Hertzian limit is a $\leq 0.1\,r$.

and

$$z_2 = x^2 \left(\frac{1}{2} \varrho_{21} \cos^2 \omega_2 + \frac{1}{2} \varrho_{22} \sin^2 \omega_2 \right) + y^2 \left(\frac{1}{2} \varrho_{21} \sin^2 \omega_2 + \frac{1}{2} \varrho_{22} \cos^2 \omega_2 \right) +$$
$$+ xy \left(\frac{1}{2} \varrho_{21} \sin \omega_2 \cos \omega_2 - \frac{1}{2} \varrho_{22} \sin \omega_2 \cos \omega_2 \right).$$

By comparing coefficients with (3.1) one obtains

$$A_1 = \frac{1}{2} \varrho_{11} \cos^2 \omega_1 + \frac{1}{2} \varrho_{12} \sin^2 \omega_1,$$

$$B_1 = \frac{1}{2} \varrho_{11} \sin^2 \omega_1 + \frac{1}{2} \varrho_{12} \cos^2 \omega_1,$$

$$C_1 = \frac{1}{2} (\varrho_{11} - \varrho_{12}) \sin \omega_1 \cos \omega_1$$

and similarly with (3.2)

$$A_2 = \frac{1}{2} \varrho_{21} \cos^2 \omega_2 + \frac{1}{2} \varrho_{22} \sin^2 \omega_2,$$

$$B_2 = \frac{1}{2} \varrho_{21} \sin^2 \omega_2 + \frac{1}{2} \varrho_{22} \cos^2 \omega_2,$$

$$C_2 = \frac{1}{2} (\varrho_{21} - \varrho_{22}) \sin \omega_2 \cos \omega_2.$$

3.3 Calculating the Coefficient cos τ

The sum $(z_1 + z_2)$ gives the distance between corresponding points on the surfaces of the bodies

$$(z_1 + z_2) = (A_1 + A_2) x^2 + (B_1 + B_2) y^2 + (C_1 + C_2) xy = Ax^2 + By^2 + Cxy. \tag{3.4}$$

To simplify this equation the following are substituted

$$A_1 + A_2 = A,$$
$$B_1 + B_2 = B,$$
$$C_1 + C_2 = C.$$

The sum $(A + B)$ then becomes

$$(A + B) = \frac{1}{2} \varrho_{11} \overbrace{(\sin^2 \omega_1 + \cos^2 \omega_1)}^{1} + \frac{1}{2} \varrho_{12} \overbrace{(\sin^2 \omega_1 + \cos^2 \omega_1)}^{1} +$$

$$+ \frac{1}{2} \varrho_{21} \underbrace{(\sin^2 \omega_2 + \cos^2 \omega_2)}_{1} + \frac{1}{2} \varrho_{22} \underbrace{(\sin^2 \omega_2 + \cos^2 \omega_2)}_{1}$$

$$(A + B) = \frac{1}{2} (\varrho_{11} + \varrho_{12} + \varrho_{21} + \varrho_{22}) . \tag{3.5}$$

And the difference $(A - B)$ becomes:

$$(A - B) = \frac{1}{2} \varrho_{11} (\cos^2 \omega_1 - \sin^2 \omega_1) - \frac{1}{2} \varrho_{12} (\cos^2 \omega_1 - \sin^2 \omega_1) +$$

$$+ \frac{1}{2} \varrho_{21} (\cos^2 \omega_2 - \sin^2 \omega_2) - \frac{1}{2} \varrho_{22} (\cos^2 \omega_2 - \sin^2 \omega_2)$$

$$(A - B) = \frac{1}{2} \{(\varrho_{11} - \varrho_{12}) \cos 2\omega_1 + (\varrho_{21} - \varrho_{22}) \cos 2\omega_2\} . \tag{3.6}$$

The position of the xy-system with respect to body *1* has not yet been fixed. It is rotated so that the xy terms fall away. On then gets

$$C = C_1 + C_2 = \frac{1}{2} \{(\varrho_{11} - \varrho_{12}) \sin 2\omega_1 + (\varrho_{21} - \varrho_{22}) \sin 2\omega_2\} = 0 . \tag{3.7}$$

With $\omega_2 + \omega = \omega_1$ (Fig. 3.2) it follows from (3.7) that

$$C = C_1 + C_2 = (\varrho_{11} - \varrho_{12}) \sin 2\omega_1 + (\varrho_{21} - \varrho_{22}) \sin (2\omega_1 - 2\omega) = 0 ,$$

$$\tan 2\omega_1 \{(\varrho_{11} - \varrho_{12}) + (\varrho_{21} - \varrho_{22}) \cos 2\omega\} = (\varrho_{21} - \varrho_{22}) \sin 2\omega ,$$

$$\tan 2\omega_1 = \frac{(\varrho_{21} - \varrho_{22}) \sin 2\omega}{(\varrho_{11} - \varrho_{12}) + (\varrho_{21} - \varrho_{22}) \cos 2\omega} = \frac{Z}{N} . \tag{3.8}$$

In (3.8) Z and N are abbreviations for numerator and denominator, it being assumed that ω is known. From (3.8) it follows that [1.10, p. 85]

$$\tan 2\omega_1 = \frac{Z}{N} = \frac{\sqrt{1 - \cos^2 2\omega_1}}{\cos 2\omega_1} ,$$

$$\frac{Z^2}{N^2} \cos^2 2\omega_1 = 1 - \cos^2 2\omega_1 ,$$

$$\cos^2 2\omega_1 = \frac{N^2}{Z^2 + N^2} \Rightarrow \cos 2\omega_1 = \frac{N}{\sqrt{Z^2 + N^2}} . \tag{3.9}$$

3.3 Calculating the Coefficient cos τ

ω_2 from (3.6) must now be obtained

$$\cos 2\omega_2 = \cos(2\omega_1 - 2\omega) = \frac{N}{\sqrt{Z^2 + N^2}} \cos 2\omega + \sqrt{1 + \frac{N^2}{Z^2 + N^2}} \sin 2\omega$$

$$\cos 2\omega_2 = \frac{N \cos 2\omega + Z \sin 2\omega}{\sqrt{Z^2 + N^2}}. \tag{3.10}$$

And thus (3.6) becomes

$$(A - B) = \frac{1}{2}\left\{(\varrho_{11} - \varrho_{12})\frac{N}{\sqrt{Z^2 + N^2}} + (\varrho_{21} - \varrho_{22})\frac{N \cos 2\omega + 2 \sin 2\omega}{\sqrt{Z^2 + N^2}}\right\}.$$

From (3.8) the following can be deduced

$$(\varrho_{11} - \varrho_{12}) + (\varrho_{21} - \varrho_{22}) \cos 2\omega = \frac{N}{Z}(\varrho_{21} - \varrho_{22}) \sin 2\omega,$$

$$(\varrho_{11} - \varrho_{12}) = \frac{N \sin 2\omega - Z \cos 2\omega}{\sin 2\omega} \tag{3.11}$$

and also

$$(\varrho_{21} - \varrho_{22}) \sin 2\omega = \frac{Z}{N}\left\{\underbrace{(\varrho_{11} - \varrho_{12}) + (\varrho_{21} - \varrho_{22}) \cos 2\omega}_{N}\right\},$$

$$(\varrho_{21} - \varrho_{22}) = \frac{Z}{\sin 2\omega}. \tag{3.12}$$

The values from (3.9) to (3.12) inserted in (3.6) give

$$(A - B) = \frac{1}{2}\left\{\frac{N \sin 2\omega - Z \cos 2\omega}{\sin 3\omega}\frac{N}{\sqrt{Z^2 + N^2}} + \frac{Z}{\sin 2\omega}\left(\frac{N \cos 2\omega + Z \sin 2\omega}{\sqrt{Z^2 + N^2}}\right)\right\},$$

$$(A - B) = \frac{1}{2}\left\{\frac{Z^2 + N^2}{\sqrt{Z^2 + N^2}}\right\} = \frac{1}{2}\sqrt{Z^2 + N^2}.$$

Inserting the values Z and N from (3.8) gives

$$(A - B) = \frac{1}{2}\sqrt{[(\varrho_{21} - \varrho_{22}) \sin^2 2\omega]^2 + [(\varrho_{11} - \varrho_{12}) + (\varrho_{21} - \varrho_{22}) \cos 2\omega]^2}$$

$$(A - B) = \frac{1}{2}\sqrt{(\varrho_{11} - \varrho_{12})^2 + (\varrho_{21} - \varrho_{22})^2 + 2(\varrho_{11} - \varrho_{12})(\varrho_{21} - \varrho_{22}) \cos 2\omega}. \tag{3.13}$$

To simplify this further Hertz introduced the coefficient $\cos \tau$:

$$\cos \tau = \frac{A-B}{A+B} = \frac{\sqrt{(\varrho_{11} - \varrho_{12})^2 + (\varrho_{21} - \varrho_{22})^2 + 2(\varrho_{11} - \varrho_{12})(\varrho_{21} - \varrho_{22})\cos 2\omega}}{\varrho_{11} + \varrho_{12} + \varrho_{21} + \varrho_{22}} \tag{3.14}$$

If the minimum (1) and the maximum (2) curvatures of both bodies lie in the same plane (e.g. Fig. 3.1), $\omega = 0$ and $\cos 2 \cdot 0 = 1$. This almost always applies for the rolling pairs used in engineering (3.14) thus becomes [1.21]

$$\cos \tau = \frac{(\varrho_{11} - \varrho_{12}) + (\varrho_{21} - \varrho_{22})}{\Sigma \varrho}. \tag{3.15}$$

The sign of Eq. (3.15) can be + or –. This is of no importance because it is determined by the principal planes of curvature, see Sect. 3.1.

By using the relationship between functions for multiple angles [1.10, p. 86] one obtains the following equations needed in Sect. 3.7:

$(A + B) \cos \tau = A - B$

$$(A + B)\left(2\cos^2 \frac{\tau}{2} - 1\right) = A - B \implies A = (A + B)\cos^2 \frac{\tau}{2}, \tag{3.16}$$

$$(A + B)\left(1 - 2\sin^2 \frac{\tau}{2}\right) = A - B \implies B = (A + B)\sin^2 \frac{\tau}{2}. \tag{3.17}$$

3.4 Calculating the Deformation δ at the Contact Face

If the two bodies are pressed together, the point contact becomes an area contact and a common contact surface arises. Through this pressing together the corresponding points *1* and *2* become displaced by δ_1 and δ_2 relative to the original surfaces

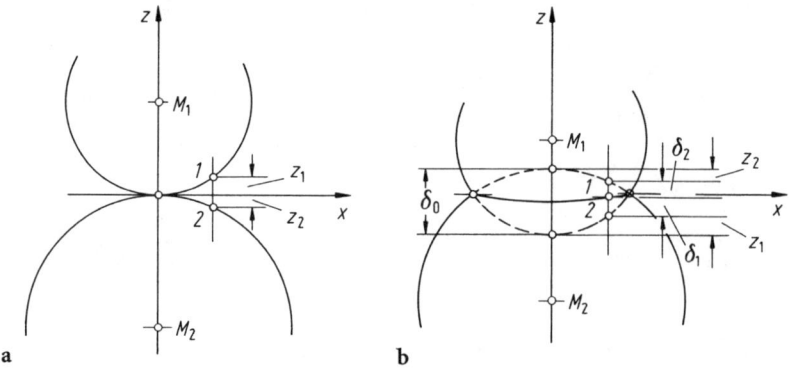

Fig. 3.3a, b. Contact between curved bodies. **a** Before deformation, **b** after deformation

3.4 Calculating the Deformation δ at the Contact Face

(Fig. 3.3), while the original crowns of the bodies move further apart by δ_0. The centres M_1 and M_2 have moved closer to one another by the amount of the flattening

$$\delta_0 = (z_1 + \delta_1) + (z_2 + \delta_2) \tag{3.18}$$

From this one gets

$$\delta_1 + \delta_2 = \delta_0 - (z_1 + z_2)$$

and with (3.4), noting that $C = 0$, one gets the following

$$\delta_1 + \delta_2 = \delta_0 - (Ax^2 + By^2). \tag{3.19}$$

In 1885 Joseph-Valentin Boussinesq found [3.1; 3.2] the deformation δ of a semi-infinite body which is loaded by a point load Q (Fig. 3.4)

$$\delta_{z=0} = \frac{1-m^2}{\pi E} \frac{Q}{r}. \tag{3.20}$$

From this one obtains for an arbitrary load distribution in terms of polar coordinates $p(r, \varphi)$

$$\delta = \frac{1-m^2}{\pi E} \int_{(A)} \frac{p\,dA}{r}$$

where m is Poisson's ratio = 0.3 and $E = 2.08 \times 10^5$ N/mm².

With Hertz's abbreviation

$$\vartheta = \frac{4}{E}(1-m^2)$$

the equation becomes

$$\delta = \frac{\vartheta}{4\pi} \int_{(A)} \frac{p\,dA}{r}. \tag{3.21}$$

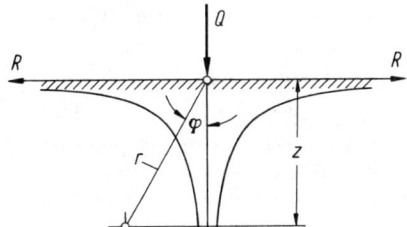

Fig. 3.4. Displacement δ in a semi-infinite body. According to J. V. Boussinesq 1885 [3.1, 3.2]

For an arbitrary load distribution (Fig. 3.5c) the following applies

$$Q = p\,dA = p\,\underbrace{dr}_{\text{length}}\,\underbrace{r\,d\varphi}_{\text{width}}.$$

Inserted in (3.21) this gives

$$\delta = \frac{\vartheta}{4\pi}\int_{(A)}\frac{p\,dr\,r\,d\varphi}{r} = \frac{\vartheta}{4\pi}\int_0^\pi\left(\int_0^r p\,dr\right)d\varphi \qquad (3.22)$$

The integral $\int_0^r p\,dr$ is the area of the arrowed surface in Fig. 3.5a, through point A at the angle φ (Fig. 3.5b). In order to work out the integral, the loaded elliptical surface is

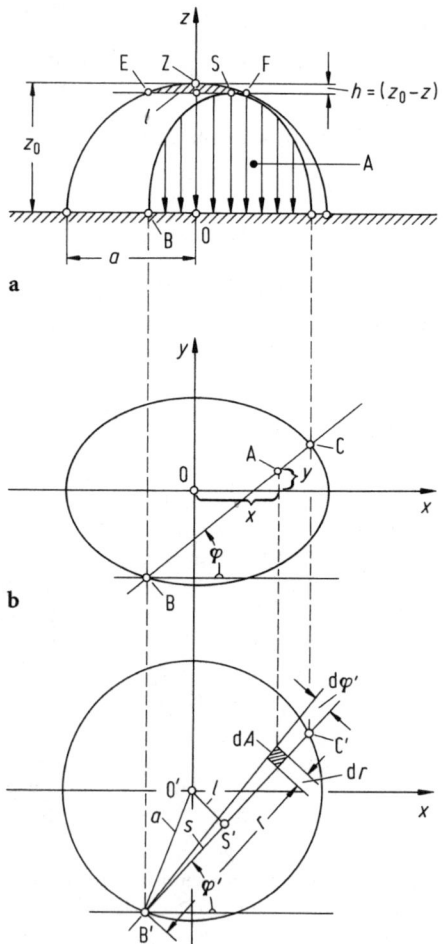

Fig. 3.5 a–c. Loaded elliptical and circular surfaces, according to Constantin Weber 1948 [1.28].

a Stress distribution from $\int_0^r p\,dr$ for the arrowed surface A; b elliptical section of stress distribution through point A at an angle φ; c circle with radius a by stretching out the ellipse in the y-direction

3.4 Calculating the Deformation δ at the Contact Face

compared with a circular surface of radius a, according to Constantin Weber [1.28]. The circular surface results from stretching the elliptical surface in the ratio of its semiaxes $a : b$ in the y-direction. There is thus a point of the circle corresponding to each point of the elliptical surface. All elements of the circular surface are given the same designations as the elliptical surface, excepting that the letters marked with a prime, that is

$$x' = x \quad \text{und} \quad y' = \frac{a}{b} y.$$

$A'(x', y')$ in the circle corresponds to the point $A(x, y)$. The corresponding sectioning straight line has the angle φ'. Here

$$\tan \varphi = \frac{b}{a} \tan \varphi',$$

$$\sqrt{1 - \cos^2 \varphi} = \frac{b}{a} \frac{\sin \varphi'}{\cos \varphi'} \cos \varphi,$$

$$1 = \cos^2 \varphi \left(1 + \frac{b^2}{a^2} \frac{\sin^2 \varphi'}{\cos^2 \varphi'}\right),$$

$$\cos \varphi = \frac{\cos \varphi'}{\sqrt{1 - \sin^2 \varphi' \left(1 - \frac{b^2}{a^2}\right)}} \qquad (3.23)$$

The term $\left(1 - \frac{b^2}{a^2}\right)$ is equal to k'^2. k is the modulus of the elliptical integral as in [1.10, p. 54]. One then gets

$$b^2/a^2 = 1 - k'^2 = k^2. \qquad (3.24)$$

Equation (3.24) substituted in (3.23) gives

$$\cos \varphi = \frac{\cos \varphi'}{\sqrt{1 - k'^2 \sin^2 \varphi'}}. \qquad (3.25)$$

The distance l from the middle point $0'$ of the circle to the straight line through A' is represented in Hesse's standard form [1.10, p. 174]

$$x' \cos(-\alpha) + y' \sin(-\alpha) - l = 0, \quad \alpha = (90° - \varphi')$$

$$x' \cos \alpha - y' \sin \alpha - l = 0,$$

$$x' \sin \varphi' - y' \cos \varphi' = l. \qquad (3.26)$$

In the triangle O′S′B′ (Fig. 3.5c) the following applies

$$a^2 = s^2 + l^2.$$

$$s = \sqrt{a^2 - l^2} = a\sqrt{1 - \left(\frac{x'}{a}\sin\varphi' - \frac{y'}{a}\cos\varphi'\right)^2}. \tag{3.27}$$

The stress p in Section A is proportional to the z-ordinate

$$p^* = z = p/c_p, \tag{3.28}$$

where c_p has the dimension N/mm³.

With $h = (z_0 - z)$ it follows from the relationship to the hatched spherical segment EZF [1.10, p. 312] (Fig. 3.5a) that

$$l^2 = h(2z_0 - h) = (z_0 - z)[2z_0 - (z_0 - z)] = z_0^2 - z^2 \Rightarrow z^2 = z_0^2 - l^2. \tag{3.29}$$

Since $z_0 = a$ is apparent in Fig. 3.5a,

$$z^2 = z_0^2 - l^2 = a^2 - l^2$$

and using (3.26)

$$z = a\sqrt{1 - \left(\frac{x'}{a}\sin\varphi' - \frac{y'}{a}\cos\varphi'\right)^2}.$$

With (3.29) one then gets

$$z = z_0\sqrt{1 - \left(\frac{x'}{a}\sin\varphi' - \frac{y'}{a}\cos\varphi'\right)^2}.$$

and using (3.28) for p

$$p = p_0\sqrt{1 - \left(\frac{x'}{a}\sin\varphi' - \frac{y'}{a}\cos\varphi'\right)^2}. \tag{3.30}$$

The section of the stress distribution is a semicircle (Fig. 3.5a and c) with radius $s = p$, its area A is therefore $1/2\pi s^2$, $1/2\pi p^2$ or

$$A = 1/2\,\pi\,s\,p. \tag{3.31}$$

With (3.27) and (3.31) one obtains for the area of the semicircle

$$A = \frac{\pi}{2} a\sqrt{1 - \left(\frac{x'}{a}\sin\varphi' - \frac{y'}{a}\cos\varphi'\right)^2}\, p_0\sqrt{1 - \left(\frac{x'}{a}\sin\varphi' - \frac{y'}{a}\cos\varphi'\right)^2}$$

$$= \frac{\pi}{2} a\, p_0\left[1 - \left(\frac{x'}{a}\sin\varphi' - \frac{y'}{a}\cos\varphi'\right)^2\right].$$

3.4 Calculating the Deformation δ at the Contact Face

The equivalent section for the ellipse is then

$$A = \int_0^r p\,dr = \frac{\pi}{2} a\, p_0 \left[1 - \left(\frac{x'}{a} \sin\varphi' - \frac{y'}{a} \cos\varphi'\right)^2\right] \frac{\cos\varphi'}{\cos\varphi}. \tag{3.32}$$

If (3.32) is substituted into (3.22) one gets

$$\delta = \frac{\vartheta}{4\pi} \int_0^\pi \left(\frac{\pi}{2} a\, p_0 \left[1 - \left(\frac{x'}{a} \sin\varphi' - \frac{y'}{a} \cos\varphi'\right)^2\right] \frac{\cos\varphi'}{\cos\varphi}\right) d\varphi.$$

The solve this integral the following are used

$$x' = x \quad \text{und} \quad y' = \frac{a}{b} y$$

and instead of φ the variable φ' is introduced. From (3.23) it follows through differentiation that

$$d\varphi = -\frac{b}{a} \frac{\cos^2\varphi}{\cos^2\varphi'} d\varphi',$$

$\cos\varphi$ is inserted from (3.25) so that we get

$$\delta = \frac{\vartheta}{8} b\, p_0 \int_0^\pi \left[1 - \left(\frac{x}{a} \sin\varphi' - \frac{y}{b} \cos\varphi'\right)^2\right] \frac{d\varphi'}{\sqrt{1 - k'^2 \sin^2\varphi}}. \tag{3.33}$$

By squaring the term in the square brackets in (3.33), we get

$$\left[1 - \left(\frac{x^2}{a^2} \sin^2\varphi' + \frac{y^2}{b^2} \cos^2\varphi' - 2\frac{xy}{ab} \sin\varphi' \cos\varphi'\right)\right]$$

and the integral of (3.33) can be split up into a sum of single integrals as

$$\delta = \frac{\vartheta}{8} b\, p_0 \left\{ \underbrace{\int_0^\pi \frac{d\varphi'}{\sqrt{1 - k'^2 \sin^2\varphi'}}}_{J_1} - \frac{x^2}{a^2} \underbrace{\int_0^\pi \frac{\sin^2\varphi'\, d\varphi'}{\sqrt{1 - k^2 \sin^2\varphi'}}}_{J_2} - \frac{y^2}{b^2} \underbrace{\int_0^\pi \frac{\cos^2\varphi'\, d\varphi'}{\sqrt{1 - k^2 \sin^2\varphi'}}}_{J_3} + 2\frac{xy}{ab} \underbrace{\int_0^\pi \frac{\sin\varphi' \cos\varphi'\, d\varphi'}{\sqrt{1 - k^2 \sin^2\varphi'}}}_{J_4} \right\}. \tag{3.34}$$

3.5 Solution of the Elliptical Single Integrals J_1 to J_4

The integrals J_1 to J_4 are elliptical integrals in the normal form of Adrien Marie Legendre 1786 [1.10, p. 138]. They are solved as follows

$$J_1 = \int_0^{\pi} \frac{d\varphi'}{\sqrt{1-k'^2\sin^2\varphi'}} = 2\int_0^{\pi/2} \frac{d\varphi'}{\sqrt{1-k'^2\sin^2\varphi'}} = 2K'. \qquad (3.35)$$

In (3.35) K' is the complete elliptical integral of the 1st kind, see Table 3.1.
The elliptical integral

$$J_2 = \int_0^{\pi} \frac{\sin^2\varphi'\,d\varphi'}{\sqrt{1-k^2\sin^2\varphi'}} = 2\int_0^{\pi/2} \frac{\sin^2\varphi'\,d\varphi'}{\sqrt{1-k^2\sin^2\varphi'}} \qquad (a)$$

can be solved by an algebraic transformation

$$\frac{\sqrt{1-(k't)^2}}{\sqrt{1-(k't)^2}} = \sqrt{1-(k't)^2}. \qquad (b)$$

The first part is split into the two parts

$$\frac{1}{\sqrt{1-(kt)^2}} - \frac{(k't)^2}{\sqrt{1-(kt)^2}} = \sqrt{1-(k't)^2}. \qquad (c)$$

The second part is taken out of Eq. (c)

$$\frac{t^2}{\sqrt{1-(k't)^2}} = \frac{1}{k'^2}\left\{\frac{1}{\sqrt{1-(k't)^2}} - \sqrt{1-(k't)^2}\right\}. \qquad (d)$$

With Eq. (d) and $t = \sin\varphi'$, Eq. (a) for J_2 can be rearranged to give

$$J_2 = \frac{1}{k'^2}\left\{\int_0^{\pi} \frac{d\varphi'}{\sqrt{1-k'^2\sin^2\varphi'}} - 2\int_0^{\pi}\sqrt{1-k'^2\sin^2\varphi'}\,d\varphi'\right\}$$

$$= \frac{1}{k'^2}\left\{2\int_0^{\pi/2} \frac{d\varphi'}{\sqrt{1-k'^2\sin^2\varphi'}} - 2\int_0^{\pi/2}\sqrt{1-k'^2\sin^2\varphi'}\,d\varphi'\right\}$$

$$J_2 = \frac{1}{k'^2}[2K' - 2E'] = \frac{2}{k'^2}[K' - E']. \qquad (3.36)$$

3.5 Solution of the Elliptical Single Integrals J_1 to J_4

Here J_2 is the complete elliptical integral of the 2nd kind.

$$J_3 = \int_0^\pi \frac{\cos^2\varphi' d\varphi'}{\sqrt{1 - k'^2\sin^2\varphi'}} = \int_0^\pi \frac{(1 - \sin^2\varphi') d\varphi'}{\sqrt{1 - k'^2\sin^2\varphi'}}$$

$$= \int_0^\pi \frac{d\varphi'}{\sqrt{1 - k'^2\sin^2\varphi'}} = \int_0^\pi \frac{\sin^2\varphi' d\varphi'}{\sqrt{1 - k'^2\sin^2\varphi'}}.$$

The solutions of this analysis are known from (3.35) and (3.36). The integral J_3 is solved as follows

$$J_3 = J_1 - J_2 = 2K' - \frac{2}{k'^2}[K' - E'] = \frac{2}{k'^2}[K'(k'^2 - 1) + E']. \tag{3.37}$$

$$J_4 = \int_0^\pi \frac{\sin\varphi' \cos\varphi' d\varphi'}{\sqrt{1 - k'^2 \sin^2\varphi'}} = 0. \tag{3.38}$$

Because it is for

$$\varphi' = 0 \Rightarrow \sin 0 = 0,$$
$$\varphi' = \pi \Rightarrow \sin \pi = 0.$$

Unconcerned about computing

$$K = \int_0^{\pi/2} (1 - k^2 \sin^2\varphi) \quad \text{with} \quad k^2 = \frac{a^2}{b^2} \quad \text{or}$$

$$K' = \int_0^{\pi/2} (1 - k'^2 \sin^2\varphi' \quad \text{with} \quad k'^2 = 1 - k^2.$$

After this we get, f.i.

for $\quad b^2/a^2 = k^2 = 0.01 \Rightarrow K = 1.5747$
or $\quad\quad\quad k'^2 = 0.99 \Rightarrow K = 1.5747$.

With equal values we can conclude after Constantin Weber's Definition [1.28]

to $\quad k^2\quad$ belong to the integrals K and E,
to $\quad k'^2\quad$ belong to the integrals K' and E'.

Table 3.1 is computed with $k^2 = b^2/a^2$.

Table 3.1. Complete elliptical integrals by A. M. Legendre 1786. [1.10, p. 54, 3.10]

$$K(k) = \int_0^{\pi/2} d\psi/\sqrt{1 - k^2 \sin^2\psi}, \quad E(k) = \int_0^{\pi/2} \sqrt{1 - k^2 \sin^2\psi} \, d\psi,$$

$$k' = \sqrt{1 - k^2}, k^2 + k'^2 = 1, K' = K(k'), E' = E(k')$$

k^2	K	E	E'	K'	k'^2
0.00	1.5708	1.5708	1.0000	∞	1.00
0.01	1.5747	1.5669	1.0160	3.6956	0.99
0.02	1.5787	1.5629	1.0286	3.3541	0.98
0.03	1.5828	1.5589	1.0399	3.1559	0.97
0.04	1.5869	1.5550	1.0505	3.0161	0.96
0.05	1.5910	1.5510	1.0605	2.9083	0.95
0.06	1.5952	1.5470	1.0700	2.8208	0.94
0.07	1.5994	1.5429	1.0791	2.7471	0.93
0.08	1.6037	1.5389	1.0879	2.6836	0.92
0.09	1.6080	1.5348	1.0965	2.6278	0.91
0.10	1.6124	1.5308	1.1048	2.5781	9.90
0.11	1.6169	1.5267	1.1129	2.5333	0.89
0.12	1.6214	1.5226	1.1207	2.4926	0.88
0.13	1.6260	1.5184	1.1285	2.4553	0.87
0.14	1.6306	1.5143	1.1360	2.4209	0.86
0.15	1.6353	1.5101	1.1434	2.3890	0.85
0.16	1.6400	1.5059	1.1507	2.3593	0.84
0.17	1.6448	1.5017	1.1578	2.3314	0.83
0.18	1.6497	1.4975	1.1648	2.3052	0.82
0.19	1.6546	1.4933	1.1717	2.2805	0.81
0.20	1.6596	1.4890	1.1785	2.2572	0.80
0.21	1.6647	1.4848	1.1852	2.2351	0.79
0.22	1.6699	1.4805	1.1918	2.2140	0.78
0.23	1.6751	1.4762	1.1983	2.1940	0.77
0.24	1.6804	1.4718	1.2047	2.1748	0.76
0.25	1.6858	1.4675	1.2111	2.1565	0.75
0.26	1.6912	1.4631	1.2173	2.1390	0.74
0.27	1.6967	1.4587	1.2235	2.1221	0.73
0.28	1.7024	1.4543	1.2296	2.1059	0.72
0.29	1.7081	1.4498	1.2357	2.0904	0.71
0.30	1.7139	1.4454	1.2417	2.0754	0.70
0.31	1.7198	1.4409	1.2476	2.0609	0.69
0.32	1.7258	1.4364	1.2535	2.0469	0.68
0.33	1.7319	1.4318	1.2593	2.0334	0.67
0.34	1.7381	1.4273	1.2650	2.0203	0.66
0.35	1.7444	1.4227	1.2707	2.0076	0.65
0.36	1.7508	1.4181	1.2763	1.9953	0.64
0.37	1.7573	1.4135	1.2819	1.9834	0.63
0.38	1.7639	1.4088	1.2875	1.9718	0.62
0.39	1.7706	1.4041	1.2930	1.9605	0.61
0.40	1.7775	1.3994	1.2984	1.9496	0.60
0.41	1.7845	1.3947	1.3038	1.9389	0.59
0.42	1.7917	1.3899	1.3092	1.9285	0.58
0.43	1.7989	1.3851	1.3145	1.9184	0.57
0.44	1.8063	1.3803	1.3198	1.9085	0.56
0.45	1.8139	1.3754	1.3250	1.8989	0.55
0.46	1.8216	1.3705	1.3302	1.8895	0.54
0.47	1.8295	1.3656	1.3354	1.8804	0.53
0.48	1.8375	1.3606	1.3405	1.8714	0.52
0.49	1.8457	1.3557	1.3456	1.8626	0.51
0.50	1.8541	1.3506	1.3506	1.8541	0.50

3.6 Calculating the Elliptical Integrals K and E

$$K = \int_0^{\pi/2} (1 - k^2 \sin^2 \varphi)^{-1/2} d\varphi, \tag{a}$$

$$E = \int_0^{\pi/2} (1 - k^2 \sin^2 \varphi)^{1/2} d\varphi. \tag{b}$$

Both bracketed terms in the integrals for K and E can be expanded using the binomial theorem which applies for positive, negative and fractional exponents [1.10, Sect. 2.1.2]

$$(1 - k^2 \sin^2 \varphi)^{-1/2} = 1 + 1\frac{1}{2} k^2 \sin^2 \varphi + \frac{1}{8} k^4 \sin^4 \varphi + \frac{15}{48} k^6 \sin^6 \varphi + \ldots,$$

$$(1 - k^2 \sin^2 \varphi)^{1/2} = 1 - \frac{1}{2} k^2 \sin^2 \varphi - \frac{1}{8} k^4 \sin^4 \varphi - \frac{15}{48} k^6 \sin^6 \varphi - \ldots.$$

By substituting into (a) and (b), K and E can be split up into the sums of integrals

$$K = \int_0^{\pi/2} d\varphi + \frac{1}{2} k^2 \int_0^{\pi/2} \sin^2\varphi \, d\varphi + \frac{1}{8} k^4 \int_0^{\pi/2} \sin^4\varphi \, d\varphi + \ldots,$$

$$= \varphi + \frac{1}{2} k^2 \left(\frac{1}{2} \varphi - \frac{1}{4} \sin 2\varphi \right) + \frac{3}{8} k^4 \left(\frac{3}{8} \varphi - \frac{1}{4} \sin 2\varphi + \frac{1}{32} \sin^4\varphi \right) +$$

$$+ \ldots \Big|_0^{\pi/2} = \frac{\pi}{2} \left[1 + \frac{1}{4} k^2 + \frac{9}{64} k^4 + \ldots \right].$$

$$E = \frac{\pi}{2} \left[1 - \frac{1}{4} k^2 - \frac{9}{64} k^4 + \ldots \right].$$

See [1.10, Sect. 4.4.7.10, p. 138]. In order to see how the values for the elliptical integrals K and E are calculated for different values of k^2 in Table 3.1, when $k = 0.1$ or $k^2 = 0.01$.

$$K = \frac{\pi}{2} \left[1 + \frac{1}{4} 0.01 + \frac{9}{64} 0.01^2 + \ldots \right],$$

$$= \frac{\pi}{2} [1 + 0.0025 + 0.000014 + \ldots],$$

$$K = \frac{\pi}{2} \cdot 1.002514 = 1.57474, \quad \text{which is}$$

rounded off to $K_{\text{Calc}} = 1.5747$ and $K_{\text{Tab}} = 1.5747$ [1.10, p. 54, Tab. 1–12].

$$E_{\text{Calc}} = \frac{\pi}{2}[1 - 0.0025 - 0.000014 - \ldots] = \frac{\pi}{2} \cdot 0.997486 = 1.566847, \text{ which is}$$

rounded off to $E_{\text{Calc}} = 1.5668$ and $E_{\text{Tab}} = 1.5669$.

The elliptical integrals K' and E' are used in exactly the same way. The equation $k^2 = 1 - k'^2$ is used for k or $k' > 0.5$ because there is better convergence.

3.7 Semiaxes of the Elliptical Contact Face for Point Contact

It can be seen from (3.19) with (3.34)

$$\frac{\vartheta_1 + \vartheta_2}{8} b p_0 \left\{ J_1 - \frac{J_2}{a^2} x^2 - \frac{J_3}{b^2} y^2 \right\} = \delta_0 - Ax^2 - By^2. \tag{3.39a}$$

Instead of the stress p_0, the compressive force Q must be used in (3.39a). If one imagines the compressive force Q to be uniformly distributed over the whole contact surface A_E, the mean stress is

$$p_m = Q/A_E. \tag{a}$$

From (3.33) the semi-ellipsoidal distribution of the stress is like that for the deformation δ (Fig. 3.6a). In order to obtain the value of the maximum stress p_0, the following approach is adopted: the volume of the semiellipsoid V_H over the contact face must be equal to the volume of an elliptical prism V_E with height p_m.

$$V_H = V_E,$$

$$\frac{1}{2}\left(\frac{4}{3} a p_0^2\right) = \pi a\, p_0\, p_m, \tag{b}$$

$$p_0 = \frac{3}{2} p_m. \tag{c}$$

Or with Eq. (a) and (c)

$$p_0 = \frac{3}{2} \frac{Q}{A_E} = \frac{3}{2} \cdot \frac{Q}{2\pi ab}. \tag{3.39b}$$

Putting this into (3.39a) gives

$$\frac{\vartheta_1 + \vartheta_2}{8} b \frac{3}{2} \frac{Q}{\pi ab} \left\{ J_1 - \frac{J_2}{a^2} x^2 - \frac{J_3}{b^2} y^2 \right\} = \delta_0 - Ax^2 - By^2. \tag{3.39c}$$

3.7 Semiaxes of the Elliptical Contact Face for Point Contact

Comparing terms with the same coefficients on both sides of Eq. (3.39c) gives

$$\delta_0 = \frac{3Q(\vartheta_1 + \vartheta_2)}{\pi ab \cdot 2 \cdot 8} bJ_1 = C\frac{J_1}{a}, \tag{3.40}$$

$$A = \frac{3Q(\vartheta_1 + \vartheta_2)}{\pi ab \cdot 2 \cdot 8} b\frac{J_2}{a^2} = C\frac{J_2}{a^3}, \tag{3.41}$$

$$B = \frac{3Q(\vartheta_1 + \vartheta_2)}{\pi ab \cdot 2 \cdot 8} b\frac{J_3}{b^2} = C\frac{J_3}{ab^2}. \tag{3.42}$$

Approximations for δ_0 and the semiaxes a and b of the contact ellipse can be calculated from these equations. From (3.16), one obtains by inserting (3.41)

$$\cos^2\frac{\tau}{2} = \frac{A}{A+B} = \frac{J_2/a^2}{J_2/a^2 + J_3/b^2 \cdot a} \cdot \frac{b^3}{b^3}. \tag{a}$$

The ratio of the semiaxes of the contact ellipses is fixed in (3.24) with $k = b/a$. It is now put into (a) to give

$$\cos^2\frac{\tau}{2} = \frac{k^2 J_2}{k^2(J_2 + J_3)}. \tag{b}$$

By inserting (3.5) and (3.16) into (3.41), one gets

$$A = C\frac{J_2}{a^3} = \frac{1}{2}\Sigma\varrho \cos^2\frac{\tau}{2}.$$

From this we get

$$a^3 = \frac{2C}{\Sigma\varrho} \frac{J_2}{\cos^2\tau/2} = \frac{2C}{\Sigma\varrho} J_2 \frac{k'^2(J_2 + J_3)}{k'^2 J_2}. \tag{c}$$

The solutions for J_2 and J_3 from (3.36) and (3.37) inserted in (c) give:

$$a^3 = \frac{2C}{k^2 \Sigma\varrho}\left\{k^2\left[\frac{2}{k'^2}(K' - E')\right] + \frac{2}{k'^2}[K'(\overbrace{k'^2 - 1}^{-k^2}) + E']\right\} =$$

$$= \frac{3Q(\vartheta_1 + \vartheta_2)}{8 \cdot \Sigma\varrho} \underbrace{\frac{2E'}{\pi k^2}}_{\mu^3} = \mu^3 \frac{3Q(\vartheta_1 + \vartheta_2)}{8 \cdot \Sigma\varrho}.$$

$$a = \mu \sqrt[3]{\frac{3Q(\vartheta_1 + \vartheta_2)}{8 \cdot \Sigma\varrho}} = \sqrt[3]{\underbrace{\frac{3(\vartheta_1 + \vartheta_2)}{8}}_{C_e}} \mu \sqrt[3]{\frac{Q}{\Sigma\varrho}}$$

$$a = C_e \mu \sqrt[3]{\frac{Q}{\Sigma\varrho}}. \tag{3.43}$$

Using (3.17) in the same way:

$$\sin^2 \tau/2 = \frac{B}{A+B} = \frac{J_3}{k^2(J_2+J_3)}.\tag{a}$$

If (3.4) and (3.17) are inserted into (3.42) one obtains

$$B = C\frac{kJ_3}{a \cdot b^2} = \frac{1}{2}\Sigma\varrho \sin^2 \tau/2,\tag{b}$$

$$b^3 = \frac{2C}{\Sigma\varrho}\frac{kJ_3}{\sin^2 \tau/2} = \frac{2C}{\Sigma\varrho}kJ_3\frac{k^2J_2+J_3}{J_3}.\tag{c}$$

The solutions for J_2 and J_3 from (3.36) and (3.37) inserted in (c) give

$$b^3 = \frac{2Ck}{\Sigma\varrho}\left\{k^2\left[\frac{2}{k'^2}(K'-E')\right] + \frac{2}{k'^2}[K'(\overbrace{k'^2-1}^{-k^2})+E']\right\}$$

$$b^3 = \frac{3Q(\vartheta_1+\vartheta_2)}{8\cdot\Sigma\varrho}\underbrace{\frac{2\,kE'}{\pi}}_{v^3} = v^3\frac{3Q(\vartheta_1+\vartheta_2)}{8\cdot\Sigma\varrho}.$$

$$b = v\sqrt[3]{\frac{3Q(\vartheta_1+\vartheta_2)}{8\cdot\Sigma\varrho}} = \underbrace{\sqrt[3]{\frac{3(\vartheta_1+\vartheta_2)}{8}}}_{C_e}v\sqrt[3]{\frac{Q}{\Sigma\varrho}}$$

$$b = C_e\,v\sqrt[3]{\frac{Q}{\Sigma\varrho}}.\tag{3.44}$$

If both bodies are made of steel, with a modulus of elasticity $E = 2.08 \cdot 10^5$, Poisson's ratio $m = 0.3$ and using the Hertzian abbreviation $\vartheta = \frac{4}{E}(1-m^2)$, it follows for C_e from (3.43) and (3.44)

$$C_e = \sqrt[3]{\frac{3[2\cdot 4/E(1-0{,}3^2)]}{8}} = \frac{2{,}3588}{10^2},$$

$$2a = 2C_e\,\mu\sqrt[3]{\frac{Q}{\Sigma\varrho}} = \frac{4{,}72}{10^2}\mu\sqrt[3]{\frac{Q}{\Sigma\varrho}},\tag{3.45}$$

$$2b = 2C_2\,v\sqrt[3]{\frac{Q}{\Sigma\varrho}} = \frac{4{,}72}{10^2}v\sqrt[3]{\frac{Q}{\Sigma\varrho}}.\tag{3.46}$$

3.8 The Elliptical Coefficients μ and ν

The elliptical integrals K, K', E and E' can be calculated from the corresponding values of k or k'. The following derivation shows their relationship with the coefficient $\cos \tau$.

If (3.41) and (3.42) are substituted into (3.14), then by considering (3.24) one gets

$$\cos \tau = \frac{A-B}{A+B} = \frac{k^2(J_2 - J_3)}{k^2(J_2 + J_3)} = \frac{k^2 \left[\frac{2}{k'^2}(K'-E')\right] - \frac{2}{k'^2}\overbrace{[K'(k'^2-1)+E']}^{-k'^2}}{k^2 \left[\frac{2}{k'^2}(K'-E')\right] + \frac{2}{k'^2}[K'(k'^2-1)+E']}$$

$$\cos \tau = \frac{2k^2 K' - (1+k^2)E'}{(1-k^2)E'} . \tag{3.47}$$

A relationship has been found between the ratio of the semiaxes and the coefficient $\cos \tau$. Hence using (3.43) and (3.44) gives

$$k = \frac{b}{a} = \frac{C_e \nu \sqrt[3]{\frac{Q}{\Sigma \varrho}}}{C_e \mu \sqrt[3]{\frac{Q}{\Sigma \varrho}}} = \frac{\nu}{\mu} \tag{3.48}$$

Table 3.2 can be compiled for values of k' and k, giving values of the elliptical coefficients $\cos \tau, \mu, \nu, \mu \cdot \nu$ and $2K/\pi\mu$ [3.3]. μ and ν were introduced as abbreviations in the derivations of (3.43) and (3.44)

$$\mu = \sqrt[3]{\frac{2E'}{\pi k^2}}, \tag{3.49}$$

$$\nu = \sqrt[3]{\frac{2kE'}{\pi}}, \tag{3.50}$$

The elliptical coefficients for $k' = 0.1$ and $k'^2 = 0.01$ were calculated individually in Table 3.3 for $\cos \tau = 0.9480$ or 0.9460. For K look at Table 3.1.

3.9 Width of the Rectangular Contact Surface for Line Contact

If the two contacting bodies are cylinders then $\varrho_{12} = \varrho_{22} = 0$, $\varrho_{11} = \varrho_1$ and $\varrho_{21} = \varrho_2$. Thus it follows from (3.5) that

$$A + B = \frac{1}{2}(\varrho_1 + \varrho_2) . \tag{a}$$

Table 3.2. Elliptical coefficients according to Hertz [3.3, p. 100, 101]; [3.11]

cos τ	μ	ν	μν	$\frac{2K}{\pi\mu}$	cos τ	μ	ν	μν	$\frac{2K}{\pi\mu}$	cos τ	μ	ν	μν	$\frac{2K}{\pi\mu}$
0.9995	23.95	0.163	3.91	0.171	0.9820	6.19	0.321	1.99	0.447	0.959	4.47	0.380	1.70	0.550
0.9990	18.53	0.185	3.43	0.207	0.9815	6.12	0.323	1.98	0.450	0.958	4.42	0.382	1.69	0.553
0.9985	15.77	0.201	3.17	0.230	0.9810	6.06	0.325	2.97	0.453	0.957	4.38	0.384	1.68	0.556
0.9980	14.25	0.212	3.02	0.249	0.9805	6.00	0.327	1.96	0.456	0.956	4.34	0.386	1.67	0.559
0.9975	13.15	0.220	2.89	0.266	0.9800	5.94	0.328	1.95	0.459	0.995	4.30	0.388	1.67	0.562
0.9970	12.26	0.228	2.80	0.279	0.9795	5.89	0.330	1.94	0.462	0.954	4.26	0.390	1.66	0.565
0.9965	11.58	0.235	2.72	0.291	0.9790	5.83	0.332	1.93	0.465	0.953	4.22	0.391	1.66	0.568
0.9960	11.02	0.241	2.65	0.302	0.9785	5.78	0.333	1.92	0.468	0.952	4.19	0.393	1.65	0.571
0.9955	10.53	0.246	2.59	0.311	0.9780	5.72	0.335	1.92	0.470	0.951	4.15	0.394	1.65	0.574
0.9950	10.15	0.251	2.54	0.320	0.9775	5.67	0.336	1.91	0.473	0.950	4.12	0.396	1.64	0.577
0.9945	9.77	0.256	2.50	0.328	0.9770	5.63	0.338	1.90	0.476	0.948	4.05	0.399	1.63	0.583
0.9940	9.46	0.260	2.46	0.336	0.9765	5.58	0.339	1.89	0.478	0.946	3.99	0.403	1.62	0.588
0.9935	9.17	0.264	2.42	0.343	0.9760	5.53	0.340	1.88	0.481	0.944	3.94	0.406	1.61	0.593
0.9930	8.92	0.268	2.39	0.350	0.9755	5.49	0.342	1.88	0.483	0.942	3.88	0.409	1.60	0.598
0.9925	8.68	0.271	2.36	0.356	0.9750	5.44	0.343	1.87	0.486	0.940	3.83	0.412	1.59	0.603
0.9920	8.47	0.275	2.33	0.362	0.9745	5.39	0.345	1.86	0.489	0.938	3.78	0.415	1.58	0.608
0.9915	8.27	0.278	2.30	0.368	0.9740	5.35	0.346	1.85	0.491	0.936	3.73	0.418	1.57	0.613
0.9910	8.10	0.281	2.28	0.373	0.9735	5.32	0.347	1.85	0.493	0.934	3.68	0.420	1.56	0.618
0.9905	7.93	0.284	2.25	0.379	0.9730	5.28	0.349	1.84	0.495	0.932	3.63	0.423	1.55	0.622
0.9900	7.76	0.287	2.23	0.384	0.9725	5.24	0.350	1.83	0.498	0.930	3.59	0.426	1.54	0.626
0.9895	7.62	0.289	2.21	0.388	0.9720	5.20	0.351	1.83	0.500	0.928	3.55	0.428	1.53	0.630
0.9890	7.49	0.292	2.19	0.393	0.9715	5.16	0.353	1.82	0.502	0.926	3.51	0.431	1.52	0.634
0.9885	7.37	0.294	2.17	0.398	0.9710	5.13	0.354	1.81	0.505	0.024	3.47	0.433	1.51	0.638
0.9880	7.25	0.297	2.15	0.402	0.9705	5.09	0.355	1.81	0.507	0.922	3.43	0.436	1.50	0.642
0.9875	7.13	0.299	2.13	0.407	0.9700	5.05	0.357	1.80	0.509	0.920	3.40	0.438	1.50	0.646
0.9870	7.02	0.301	2.11	0.411	0.969	4.98	0.359	1.79	0.513	0.918	3.36	0.441	1.49	0.650
0.9865	6.93	0.303	2.10	0.416	0.968	4.92	0.361	1.78	0.518	0.916	3.33	0.443	1.48	0.653
0.9860	6.84	0.305	2.09	0.420	0.967	4.86	0.363	1.77	0.522	0.914	3.30	0.445	1.47	0.657
0.9855	6.74	0.307	2.07	0.423	0.966	4.81	0.365	1.76	0.526	0.912	3.27	0.448	1.46	0.660
0.9850	6.64	0.310	2.06	0.427	0.965	4.76	0.367	1.75	0.530	0.910	3.23	0.450	1.45	0.664
0.9845	6.55	0.312	2.04	0.430	0.964	4.70	0.369	1.74	0.533	0.908	3.20	0.452	1.45	0.667
0.9840	6.47	0.314	2.03	0.433	0.963	4.65	0.371	1.73	0.536	0.906	3.17	0.454	1.44	0.671
0.9835	6.40	0.316	2.02	0.437	0.962	4.61	0.374	1.72	0.540	0.904	3.15	0.456	1.44	0.674
0.9830	6.33	0.317	2.01	0.440	0.961	4.56	0.376	1.71	0.543	0.902	3.12	0.459	1.43	0.677
0.0825	6.26	0.319	2.00	0.444	0.960	4.51	0.378	1.70	0.546	0.900	3.09	0.461	1.42	0.680

cos τ	μ	ν	μν	$\frac{2K}{\pi\mu}$
0.895	3.03	0.466	1.41	0.688
0.890	2.97	0.471	1.40	0.695
0.885	2.92	0.476	1.39	0.702
0.880	2.86	0.481	1.38	0.709
0.875	2.82	0.485	1.37	0.715
0.870	2.77	0.490	1.36	0.721
0.865	2.72	0.494	1.35	0.727
0.860	2.68	0.498	1.34	0.733
0.855	2.64	0.502	1.33	0.739
0.850	2.60	0.507	1.32	0.745
0.84	2.53	0.515	1.30	0.755
0.83	2.46	0.523	1.29	0.765
0.82	2.40	0.530	1.27	0.774
0.81	2.35	0.537	1.26	0.783
0.80	2.30	0.544	1.25	0.792
0.75	2.07	0.577	1.20	0.829
0.70	1.91	0.607	1.16	0.859
0.65	1.77	0.637	1.13	0.884
0.60	1.66	0.664	1.10	0.904
0.55	1.57	0.690	1.08	0.922
0.50	1.48	0.718	1.06	0.938
0.45	1.41	0.745	1.05	0.951
0.40	1.35	0.771	1.04	0.962
0.35	1.29	0.796	1.03	0.971
0.30	1.24	0.824	1.02	0.979
0.25	1.19	0.850	1.01	0.986
0.20	1.15	0.879	1.01	0.991
0.15	1.11	0.908	1.01	0.994
0.10	1.07	0.938	1.00	0.997
0.05	1.03	0.969	1.00	0.999
0	1	1	1	1

3.9 Width of the Rectangular Contact Surface for Line Contact

Table 3.3. Hertzian elliptical coefficients for $k = 0.1$

$\cos \tau$	μ	ν	$\mu\nu$	$\dfrac{2K}{\pi\mu}$
0.9480	4.05	0.399	1.62	0.583
0.9467	4.01	0.401	1.61	0.587
0.9460	3.99	0.403	1.61	0.588

If the major axis of the ellipse tends to infinity the total compressive force Q must also do the same if the compressive force Q' is to remain finite.

The indefinite value $Q'/a = \infty/\infty$ is now put equal to an arbitrary constant p' which it will be shown corresponds to the force Q' per unit length of the cylinder. It follows that as $a \to \infty$, in (3.39) $A = 0$ and thus from Eq. (a)

$$B = 1/2\,(\varrho_1 + \varrho_2). \tag{b}$$

By comparing terms with the same coefficient in (3.39) and Eq. (b), one then obtains, as in Sect. 3.7

$$\frac{\vartheta_1 + \vartheta_2}{8}\, b p_0 \frac{J_3}{b^2} = \frac{1}{2}(\varrho_1 + \varrho_2). \tag{c}$$

As $a \to \infty$ we get $k'^2 = 1$ from (3.24), and from (3.37) we obtain $J_3 = 2E$ [1.10, p. 54]. $E = 1$ and hence $J_3 = 2$.

The relationship between the stress p_0 and the compressive force Q must now be established. As in Sect. 3.7 the maximum stress p_0 is obtained using Fig. 3.6 by making the volume of the semielliptical prism V_H over the contact surface equal to that of the rectangular prism

$$\frac{1}{2}\pi b p_0 l \overset{V_H}{=} 2 b l p_m \overset{V_R}{,}$$

$$p_0 = 4/\pi p_m = \frac{4}{\pi}\frac{Q}{A} = \frac{4}{\pi}\frac{Q}{2bl}. \tag{3.51}$$

Equation (3.51) substituted into (c) gives

$$\frac{\vartheta_1 + \vartheta_2}{8} b \frac{4}{\pi}\frac{Q}{2bl}\frac{2}{b^2} = \frac{1}{2}(\varrho_1 + \varrho_2)$$

and instead of l according to Fig. 3.6 with $l_w = l - qd$ we have

$$b = \sqrt{\frac{Q(\vartheta_1 + \vartheta_2)}{\pi l_w(\varrho_1 + \varrho_2)}}. \tag{3.52a}$$

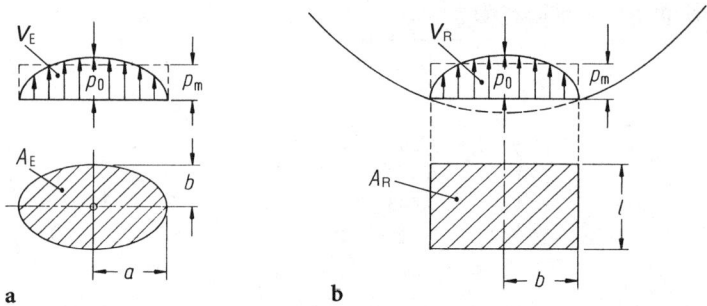

Fig. 3.6a, b. Semi-ellipsoidal distribution of the maximum surface stress p_0 over the contact surface. **a** Point contact, **b** line contact

Auxiliary calculation with $\vartheta = 4/E(1 - m^2)$, the constants $E = 2.08 \cdot 10^5$ and $m = 0.3$

$$2b = 2\sqrt{\frac{2 \cdot 4/E(1 - m^2)}{\pi}} \sqrt{\frac{Q}{l_w(\varrho_1 + \varrho_2)}} = \sqrt{\frac{2 \cdot 4 \cdot 0{,}91}{\pi \cdot 2{,}08 \cdot 10^5}} \sqrt{\frac{Q}{l_w(\varrho_1 + \varrho_2)}}$$

$$2b = \sqrt{11{,}140846 \cdot 10^{-6}} \sqrt{\frac{Q}{l_w(\varrho_1 + \varrho_2)}} = \frac{6{,}68}{10^3} \sqrt{\frac{Q}{l_w(\varrho_1 + \varrho_2)}}. \quad (3.52\text{b})$$

3.10 Deformation and Surface Stress at the Contact Face

It is now possible to calculate the deformation δ_0 and the surface stress p_0 at the contact surface using the results from Sects. 3.5, 3.7 and 3.9.

3.10.1 Point Contact

If the values for J_3 from (3.35) and a from (3.43) are put into (3.40) one gets

$$\delta_0 = \frac{3Q(\vartheta_1 + \vartheta_2)}{\pi \cdot a \cdot 2 \cdot 8} 2K = \sqrt[3]{\frac{8 \cdot 3^2}{(2 \cdot 8)^3}} (\vartheta_1 + \vartheta_2) \frac{2K}{\pi\mu} \sqrt[3]{Q^2 \Sigma \varrho}.$$

Auxiliary calculation with the values E and m as in (3.45)

$$\frac{8 \cdot 3^2}{(2 \cdot 8)^3} \left[2 \cdot \frac{4}{E}(1 - m^2) \right]^2 = \frac{0{,}2782}{10^3}.$$

The deformation thus becomes

$$\delta_0 = \frac{2{,}78}{10^4} \frac{2K}{\pi\mu} \sqrt[3]{Q^2 \Sigma \varrho}. \quad (3.53)$$

3.10 Deformation and Surface Stress at the Contact Face

To determine the Hertzian stress p_0 the semiaxis a from (3.43) and the semiaxis b from (3.44) are inserted in (3.39b)

$$p_0 = \frac{3Q}{\mu\nu \cdot 2\pi \sqrt[3]{\left[\frac{3Q(\vartheta_1 + \vartheta_2)}{8\Sigma\varrho}\right]^2}} = \sqrt[3]{\frac{3 \cdot 8^2}{2^3\pi^3(\vartheta_1 + \vartheta_2)^2}} \frac{1}{\mu\nu} \sqrt[3]{Q(\Sigma\varrho)^2}.$$

Auxiliary calculation with the values E and m as in (3.45)

$$\sqrt[3]{\frac{3 \cdot 8^2}{\pi^3 \cdot 2^3 \left[2\frac{4}{E}(1-m^2)\right]^2}} = 10^3 \cdot 0{,}85810\,.$$

$$p_0 = \frac{858}{\mu\nu} \sqrt[3]{Q(\Sigma\varrho)^2}. \tag{3.54}$$

3.10.2 Line Contact

Here the semiaxis $a \to \infty$ and thus $k^2 = 1$. It follows from [1.10, p. 54] that $K = \infty$ and $J_1 = 2K = 2\infty$. Substituting in (3.39) we get the following for the deformation in the contact surface (Fig. 3.6)

$$\delta_0 = \frac{\vartheta_1 + \vartheta_2}{8} b p_0 \cdot 2\infty\,. \tag{3.55}$$

In this case the deformation δ_0 can no longer be calculated using Hertz's method because the third condition, that the dimensions of the contact surface are very small compared with the radii of curvature, is not fulfilled since $a \to \infty$.

In 1926 Hellmuth Bochmann determined by empirical means that the deformation δ_0 is dependent on the first power of the load Q [3.5]. On the other hand Lundberg in 1939 [3.6] and Kunert in 1961 [3.7] found that

$$\delta_0 = \frac{4{,}05 \cdot P^{0.925}}{10^5 \cdot l_w^{0.85}} \text{ mm}\,. \tag{3.56}$$

Here the deformation is independent of the roller diameter. In 1963 Korrenn, Kirchner and Braune [3.8] confirmed Eq. (3.56) experimentally.

Since the width b of the rectangular contact surface is known, the mean stress is

$$p_m = \frac{Q}{2bl_w} \tag{a}$$

or, using (3.51), the maximum stress is

$$p_0 = \frac{4}{\pi} p_m = \frac{4}{\pi} \frac{Q}{2bl_w}\,. \tag{b}$$

To determine the maximum stress the term for the semiaxis b from (3.52) is put into (b) giving

$$p_0 = \frac{4}{\pi} \frac{Q}{2l_w} \frac{1}{\sqrt{\dfrac{Q}{l_w} \dfrac{\vartheta_1 + \vartheta_2}{\pi(\varrho_1 + \varrho_2)}}} = \sqrt{\frac{4Q(\varrho_1 + \varrho_2)}{\pi l_w (\vartheta_1 + \vartheta_2)}}.$$

Auxiliary calculation with the values E and m as in (3.45)

$$p_0 = \sqrt{\frac{4Q(\varrho_1 + \varrho_2)}{\pi l_w \left[2\dfrac{4}{E}(1 - m^2)\right]}} = \underbrace{\sqrt{\frac{2{,}08 \cdot 10^5}{\pi(1 - 0{.}3^2)}}}_{2.697 \cdot 10^2} \sqrt{\frac{Q}{l_w} \frac{(\varrho_1 + \varrho_2)}{2}}.$$

$$p_0 = 270 \sqrt{\frac{Q}{l_w} \frac{(\varrho_1 + \varrho_2)}{2}}. \tag{3.57}$$

This basic Eqs. (3.54) and (3.57) used in Chap. 4 for designing driveshafts are thus derived.

3.11 The validity of the Hertzian theory on ball joints

The size of the contact however is small relative to the size of the contacting bodies especially in the ball joints. This is often not valid. S. J. COLE researched in 1994 contact patch dimensions for ball-track contacts on track contact arcs up to $\theta = 50°$, as measured along the contact ellipse major-axis. The contact dimensions predicted by HERTZ are accurate to within 10%. He retains the assumption of elastic half-space behaviour. So, the difference between Hertz and the reality is:

HERTZ $P = \int p \cdot A \, dA$

Reality $P_v = \int (p \cdot A \cdot \cos \theta) \, dA$
whereby $P_v < P$

Exceeding $\theta = 50°$ the Hertzian stress should be corrected. In practice the normal range of conformity values used in ball joints is $x_Q = 1.002$ to 1.07. This high conformity brings ball-track contacts outside the scope of HERTZ' contact theory (Fig. 3.7). COLE's work predicts shortening of the contact length, increases in maximum contact pressure and a non-semi-ellipsoidal pressure distribution, compared with the Hertzian steel-on-steel value. His calculations show good agreement with experimental measurements of the contact patch dimensions.

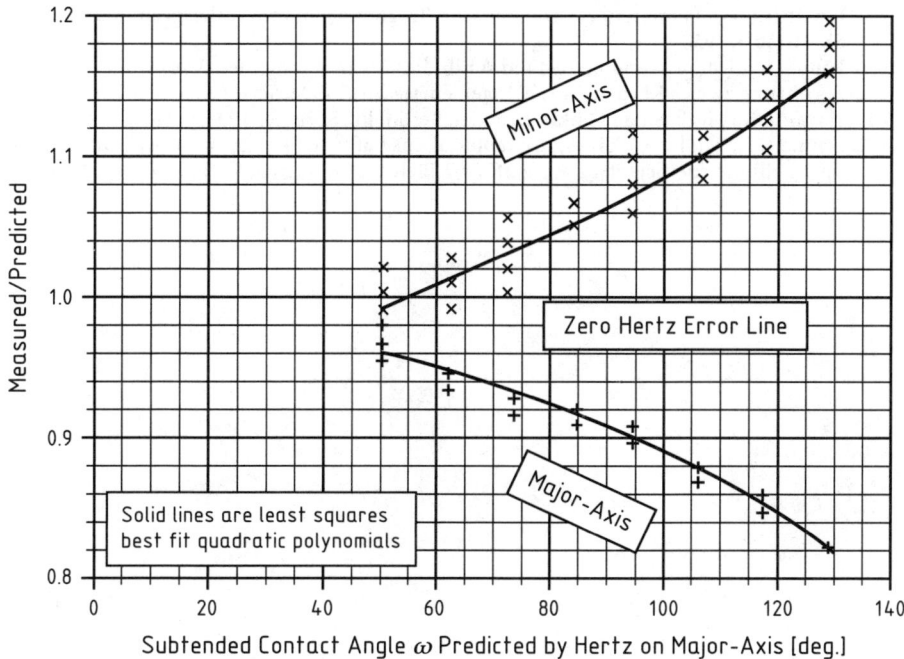

Fig. 3.7. Correction Factors for Elliptical Contact Patch Dimensions Predicted by Hertz Theory for CVJ Ball-Track Contacts exceeding 50° by S. J. COLE 1994

3.12 Literature to Chapter 3

3.1 Hertz, H. (1896) On the Contact of Solids (translated by Jones & Schott). Miscellaneous Papers, Macmillan, pp. 146–183
3.2 Mundt, R.: Über die Berührung fester elastischer Körper (Contact between solid elastic bodies). Schweinfurt: SKF Kugellagerfabriken GmbH, 1950
3.3 Boussinesq, J. V.: Application des potentiels à l'étude de l'équilibre et du mouvement des solides elastiques (Application of potentials to the study of equilibrium and movement of solid elastic bodies). Paris 1885
3.4 Eschmann, P.; Hasbargen, L.; Braendlein, J.: Ball and Roller Bearings. Munich/Chichester: Oldenbourg/John Wiley 1985, p. 140–144
3.5 Bochmann, H.: Die Abplattung von Stahlkugeln und Zylindern durch den Meßdruck (Flattening of steel balls and cylinders due to applied pressure). Diss. TH Dresden 1926. Erfurt: Deutsche Zeitschriften-Gesellschaft 1927, Z. Feinmechanik u. Präzision 35 (1927) No. 9, p. 95–100 and No. 11, p. 122–125
3.6 Lundberg, G.: Elastiche Berührung zweier Halbräume (Elastic contact between two semi-infinite bodies. Forsch. Ingenieurwes. 10 (1939) p. 201–211
3.7 Kunert, K. H.: Spannungsverteilung im Halbraum bei elliptischer Flachenpressungsverteilung über einer rechteckigen Drückfläche (Stress distribution in a semi-infinite body with elliptical surface stress distributiion over a rectangular stress area). Forsch. Ingenieurwes. 27 (1961) p. 165–174
3.8 Korrenn, H.; Kirchner, W.; Braune, G.: Die elastische Verformung einer ebenen Stahloberfläche unter linienförmiger Belastung (Elastic deformation of a flat steel surface under linear loading). Werkstattstechnik 53 (1963) p. 27–30

3.9 Lundberg, G.; Palmgren, A.: Dynamische Tragfähigkeit von Wälzlagern. Schweinfurt: SKF GmbH, 1950
3.10 Föppl, L.: Der Spannungszustand und die Anstrengung des Werkstoffes bei der Berührung zweier Körper. Forsch. Ingenieurwesen 7, 1936, Series A, p. 209–221
3.11 Hayashi, K.: Fünfstellige Funktionentafel. Berlin: Jul. Springer 1930, p. 127
3.12 Schmelz, F.; Müller, C.: Six-figure tables of $\cos \tau$ after H. Hertz, and the resulting four-figure elliptic coefficients μ, ν, $\mu\nu$, $2K'/\pi\mu$. Eichstaett/Bavaria: Polygon-Verlag, 2001

4 Designing Joints and Driveshafts

Demand within the automotive industry has led to strong competition among driveshaft manufacturers. For each application the strength and life must meet the specifications, while the weight and price must be kept as low as possible.

The useful life of a driveshaft is determined mainly by the joints. For this reason the other component parts will not be discussed here; they can in any case be designed using the general principles of strength of materials.

With good lubrication and sealing the life is limited by fatigue[1] of the material after millions of reversals. With bad sealing the life is limited by corrosion and, if there is contamination, by the resultant increase in wear. In order to calculate the life the first case is assumed.

4.1 Design Principles

The role of joints and driveshafts is to transmit torques between shafts which are not in line. These transmitted loads are limited by the capacity of the materials used. Hence in assessing the capacity of the joints it is most important to determine the pressure between the rolling bodies and their tracks. To this end (3.54) and (3.57) in Chap. 3 were derived using Hertz's theory. It must however be decided whether short or long term loading is involved:

- for short term loading the stress state is quasi-static. The joint is then designed so that excessive plastic deformation does not occur,
- for long term loading the stress state is dynamic. The joint must be designed for many millions of stress cycles.

The durability can only be determined by extensive rig tests, pure mathematical approaches are still inadequate. Only in the case of Hooke's joints is it possible to calculate the life using modified methods from roller bearing practice and the empirical Fischer equation (4.22). This cannot as yet be used for ball and pode joints.[1]

Joint manufacturers use formulae based on the empirical Palmgren Eq. (4.13) to convert the durability from rig tests into that corresponding to other torque, speed and angle conditions.

---

[1] The term originates from 1853. In 1929 Paul Ludwik described a "structural loosening". Wolfgang Richter spoke of a "shattering" of the material in 1960.

When designing machinery the designer does not know from the outset all the parameters, so he or she has to draw up the design from basic data, some of which are estimated. The designer then reassesses the joints with exact values for the static and dynamic stresses and can select a joint from manufacturers' catalogues: At this stage, all other factors such as the maximum articulation angle, plunge and installation possibilities must be taken into account.

4.1.1 Comparison of Theory and Practice by Franz Karas 1941

The problem of defining the joint of greatest material strain for two elastic bodies pressed together has not so far been solved. In 1941 Franz Karas [4.1] compared the five hypotheses of maximum stress:

- direct stress
- shear stress
- strain
- displacement
- change of shape

This confirmed what his predecessors had said [4.2–4.4], that the maximum stress is just behind the contact face and can be determined using a uniaxial reference stress p_0, see (3.54) or (3.57).

For the static stress of the joints' rolling body the tension hypothesis from OTTO MOHR 1882 is feasible. For pressure stress of ductile iron the main shear-strenght is decisive. The comparing tension for the three axled tension is:

$$\sigma_v = 2 \cdot \tau_{max} = (\sigma_z - \sigma_y)_{max}$$

Dubbel, 16. edition, C6 U [1.3.2].

The three families of Hooke's ball and pode joints give rise to different conditions for the transmission of torque because of their different types of construction. The permissible Hertzian stresses may be obtained from (3.54) or (3.57); by rearranging these equations one gets the compressive forces between the rolling bodies:

$$\text{For point contact} \quad P_{perm} = \left(\frac{p_0 \mu \nu}{858}\right)^3 \frac{1}{\Sigma \varrho^2} \tag{4.1}$$

and

$$\text{for line contact} \quad P_{perm} = \left(\frac{p_0}{270}\right)^2 \frac{2 l_w}{\varrho_1 + \varrho_2}. \tag{4.2}$$

In roller bearing practice it is usual to calculate using the unit loading $k = P/d^2$ or $k = P/dl_w$. (4.1) and (4.2) are therefore modified by d^2 or dl_w to give:

$$k = \frac{P_{perm}}{d^2} = \left(\frac{p_0 \mu \nu}{858}\right)^3 \frac{1}{(d\Sigma\varrho)^2} \tag{4.3}$$

4.1 Design Principles

and

$$k = \frac{P_{\text{perm}}}{dl_w} = \left(\frac{p_0}{270}\right)^2 \frac{2}{d(\varrho_1 + \varrho_2)}. \tag{4.4}$$

For abbreviation the coefficient of conformity c_p, determined from the dimensions of the joint using (3.15) and the coefficients μ and ν (Table 3.2), have been put into (4.1) and (4.2):

$$c_p = \frac{858}{\mu\nu} \sqrt[3]{(d \cdot \Sigma\varrho)^2} \ (\text{N/mm}^2)^{2/3} \tag{4.5}$$

or

$$c_p = 270 \sqrt{d/2(\varrho_1 + \varrho_2)} = 270 \sqrt{1 + d/D}. \tag{4.6}$$

The permissible compressive force on the roller body thus becomes:

- for point contact $\qquad P_{\text{perm}} = \left(\dfrac{p_0}{c_p}\right)^3 d^2 \tag{4.7a}$

  with specific loading $\qquad k = \left(\dfrac{p_0}{c_p}\right)^3 \Rightarrow p_0 = c_p \sqrt[3]{k}, \tag{4.7b}$

- for line contact $\qquad P_{\text{perm}} = \left(\dfrac{p_0}{c_p}\right)^2 dl_w \tag{4.8a}$

  with specific loading $\qquad k = \left(\dfrac{p_0}{c_p}\right)^2 \Rightarrow p_0 = c_p \sqrt{k}. \tag{4.8b}$

If the coefficient of conformity c_p is small, for the same Hertzian stress p_0 the unit loading k may be greater.

4.1.2 Static Stress

For the purposes of calculation one must decide whether the loading is static or dynamic. Here static loading refers to a compressive force arising from a torque on a body which is stationary or slightly oscillating. The torque can be constant or varying. The adjective "static" refers to the operating state, not the type of stress. In 1943 Palmgren, Lundberg and Bratt [1.27] carried out extensive tests with roller bearings to find empirical relationships between the permissible surface stress p_0 and the permanent plastic deformation δ_b given by Eschmann [3.3] as:

- for point contact $\qquad p_0 = 3300 \, c_p^{3/10} \left(\dfrac{\delta_b}{d}\right)^{1/5} \text{N/mm}^2, \tag{4.9}$

- for line contact $\qquad p_0 = 2690 \, c_p^{2/5} \left(\dfrac{\delta_b}{d}\right)^{1/5} \text{N/mm}^2. \tag{4.10}$

The permissible permanent deformation δ_b of the roller body diameter at the most heavily stressed point of contact must not exceed 0.01%. That means in the case of a ball 10 mm in diameter a combined plastic deformation of both bodies of 1 μm. According to the newest research Hooke's joints can be strained up to 0.04% without reducing the durability. Moreover experience has shown that these slight plastic deformations do not impair the smooth running. Since 1987 the static loading according to Hertz is set in with $p_0 = 4000$ N/mm².

4.1.3 Dynamic Stress and Durability

In the case of dynamic stress the two rolling bodies rotate relative to one another. As with static stress, the torque can be constant or varying.

Arvid Palmgren in 1924 [1.30] and Robert Mundt in 1919 [3.2] found that the relationship between the compressive force given in (4.7a) and (4.8a) and the durability is similar to the Woehler function for cyclically loaded structural parts (Fig. 4.1). Gough in 1926 and Moore and Kommers in 1927 surveyed this subject before and after Woehler [4.6; 4.7]. They said that there was a fatigue limit; a compressive force below this limit will give an infinite life. Palmgren's 1924 equation, somewhat rearranged, reads

$$L = \left(\frac{C}{k-u}\right)^p \quad (4.11)$$

where L is the life in millions of load cycles, C is a material constant, k is the specific load, u is the fatigue limit, and p is an exponent (about 3).

If the specific load is lowered to the fatigue limit ($k = u$), the life in (4.11) becomes infinite ($L = \infty$) which corresponds to the Woehler function. However attempts to determine the fatigue limit have met with unexpected difficulties. In 1943 and 1949 Niemann [4.8] determined the rolling fatigue limit to be:

- for point contact $p_0 \approx 0.44$ HB $= 0.44 \cdot 6965 = 3065$ N/mm²,
- for line contact $p_0 \approx 0.31$ HB $= 0.31 \cdot 6385 = 1980$ N/mm².

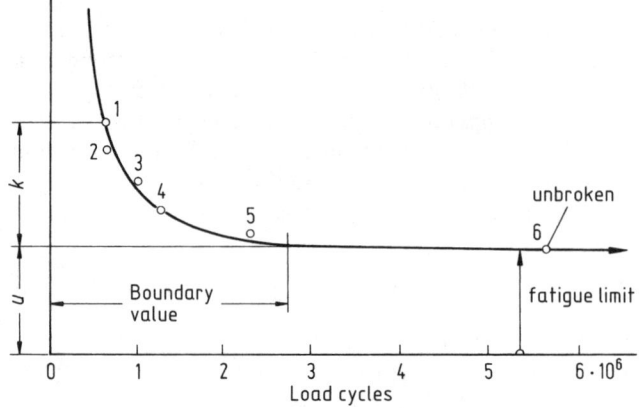

Fig. 4.1. Generalised Woehler (S/N) curve for steel [1.37]

4.1 Design Principles

Individual tests do not however provide reliable information because of the very large scatter, as demonstrated by the results of tests to find the durability of roller bearings (Fig. 4.2). Extensive tests by the roller bearing industry have shown that there is no fatigue limit, in the case of pure steel structure.

In 1936 Palmgren wrote on this subject [1.30]:
"Within the region examined a fairly low load always led to an increase in the number of revolutions which a bearing could sustain before fatigue sets in. Until this time it had been assumed that a low specific load would give a fatigue limit. It was established later through tests at low loads however that a fatigue limit, if one does in fact exist at all, is lower than all loads which occur, that is, in practice the life is always a function of the load." (Fig. 4.1 above).

After the research work of the ball bearing industry around 1966 the Woehler curve had been rediscovered. These revolutionary findings showed that heavy duty roller bearings can be run with unlimited durability if lubrication with highly filtered oil is provided [4.9, 4.10, 4.12].

It is now internationally accepted that:
"The service life of a sufficiently large quantity of apparently identical bearings is the number of revolutions (or the number of hours at constant speed) which 90% of this quantity of bearings withstand or exceed before the first signs of fatigue damage occur."

These tests necessarily take a long time because rollers need to be tested at low stresses giving very long running times. In SKF tests up to 1935 the longest life was $7 \cdot 10^9$ revolutions. Testing 24 hours a day at 1000 rpm gives a running time of

$$j = \frac{7 \cdot 10^9}{1000 \cdot 60 \cdot 24 \cdot 365} \approx 13 \tfrac{1}{3} \text{ years}$$

which is why durability tests are limited to medium high loads. The relationship between life L and compressive load P in the SKF service life tests was given by the equation

$$\frac{L_2}{L_1} = \left(\frac{P_1}{P_2}\right)^p. \tag{4.12}$$

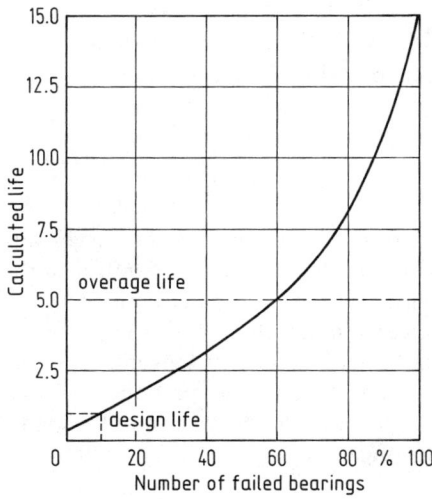

Fig. 4.2. Scatter of life for roller bearings [4.10]. For the design life only 10% of a large number of nominally identical bearings have failed

If one considers $P_1 = C$ as a reference load at which, with a constant speed, a rolling pair achieves the life $L_1 = 10^6$, the indices can be omitted and the equation written as

$$L = 10^6 \, (C/P)^p. \tag{4.13}$$

For a constant test speed the life L in (4.13) can be expressed in hours L_h

$$L_1 = L_{h1} n_1 \cdot 60 \,. \tag{4.14}$$

Comparing (4.14) with the Palmgren Eq. (4.13) gives

$$L_{h1} n_1 \cdot 60 = 10^6 \, (C/P_1)^p,$$

$$L_{h1} = \frac{10^6}{n_1 \cdot 60} \, (C/P_1)^p. \tag{4.15a}$$

If the durability of a joint is to be calculated for another compressive load P_2, it follows from (4.12) and (4.14) that:

$$\frac{L_2}{L_1} = \frac{L_{h2} n_2 \cdot 60}{L_{h1} n_1 \cdot 60} = \left(\frac{P_1}{P_2}\right)^p \Rightarrow L_{h2} = L_{h1} \left(\frac{P_1}{P_2}\right)^p. \tag{4.15b}$$

4.1.4 Universal Torque Equation for Joints

The torque transmitting capacity M of a joint with m elements is determined by the component P_x of the equivalent compressive load (Fig. 4.3)

$$P = \begin{pmatrix} P_x \\ P_y \\ P_z \end{pmatrix}.$$

It is tangential to the radius R and perpendicular to the axis of rotation r

$$M = m \, P_x R \Rightarrow P_x = \frac{M}{m \, R}. \tag{4.16}$$

The line of action p of the equivalent load P from the pressures at the curves surfaces of the drive body, transmitting element and driven body is identical to the surface normals n of the contact faces. Its general position is determined by the pressure angle α, the skew angle γ and the inclination or tilt angle ε. A joint has the greatest transmitting capacity when the compressive load P acts tangentially to the radius R and at right angles to the axis of rotation z because the components P_y and P_z are then zero. In order to understand the effect of the angles α, γ and ε the load P is rotated in space. The effect on the pressure angle α can be found by a rotation of $\vartheta = 90 - \alpha$ about the z-axis (Fig. 4.3).

4.1 Design Principles

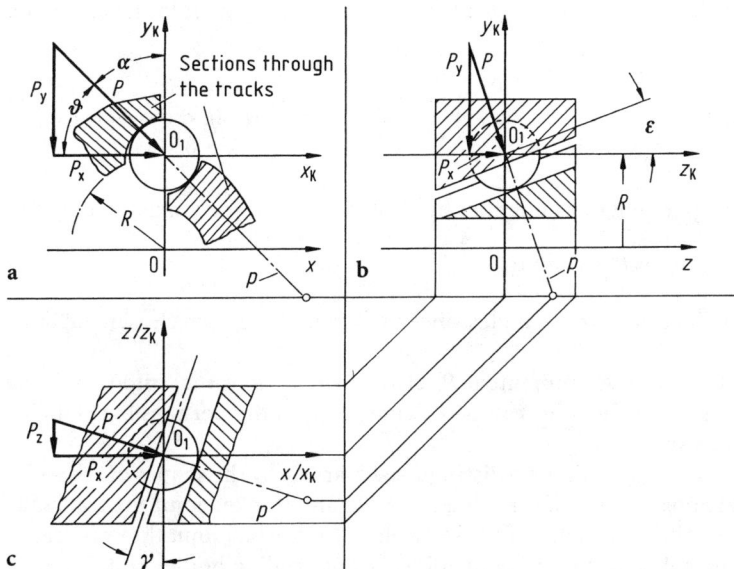

Fig. 4.3a–c. General position of the contact faces and line of action of the compressive loads relative to the axis of rotation z of the joint. The components of the equivalent compressive load are $P = \begin{pmatrix} P_x \\ P_y \\ P_z \end{pmatrix}$. **a** Front view, **b** side view, **c** plan view

$$P' = D_\vartheta P = \begin{pmatrix} \cos\vartheta & \sin\vartheta & 0 \\ -\sin\vartheta & \cos\vartheta & 0 \\ 0 & 0 & 1 \end{pmatrix} \begin{pmatrix} P \\ 0 \\ 0 \end{pmatrix} = \begin{pmatrix} P\cos\vartheta \\ -P\sin\vartheta \\ 0 \end{pmatrix} = \begin{pmatrix} P'_x \\ P'_y \\ 0 \end{pmatrix}.$$

The effective component is

$$P'_x = P\cos\vartheta = P\cos(90° - \alpha) = P\sin\alpha. \tag{4.16a}$$

If the load P is acting at a skew angle γ, its x-component is further reduced by a rotation about the y-axis at the angle γ

$$P'' = D_\gamma P' = \begin{pmatrix} \cos\gamma & 0 & \sin\gamma \\ 0 & 1 & 0 \\ -\sin\gamma & 0 & \cos\gamma \end{pmatrix} \begin{pmatrix} P\cos\gamma \\ -P\sin\vartheta \\ 0 \end{pmatrix} = \begin{pmatrix} P\cos\vartheta\cos\gamma \\ -P\cos\vartheta\sin\gamma \\ 0 \end{pmatrix} = \begin{pmatrix} P''_x \\ P''_y \\ 0 \end{pmatrix}.$$

The effective component is $P''_x = P\cos(90° - \alpha)\cos\gamma$,

$$P''_x = P\sin\alpha\cos\gamma. \tag{4.16b}$$

If P acts at the inclination angle ε, instead of the skew angle γ, by rotating about the x-axis we get:

$$P'' = D_\varepsilon P' = \begin{pmatrix} 1 & 0 & 0 \\ 0 & \cos\varepsilon & \sin\varepsilon \\ 0 & -\sin\varepsilon & \cos\varepsilon \end{pmatrix} \begin{pmatrix} P\cos\vartheta \\ -P\sin\vartheta \\ 0 \end{pmatrix} = \begin{pmatrix} P\cos\vartheta \\ -P\sin\vartheta\cos\varepsilon \\ P\sin\vartheta\sin\varepsilon \end{pmatrix} = \begin{pmatrix} P''_x \\ P''_y \\ P''_z \end{pmatrix}.$$

The effective x-component does not alter by a rotation about the x-axis; it remains

$$P''_x = P'_x = P\cos\vartheta = P\sin\alpha. \tag{4.16c}$$

If the load P is inclined at all three angles, once two angles have been chosen the third is defined.

By inserting the effective component P_x of Eq. (4.16a) to (4.16c) into (4.16), the transmitting capacity of the joint can be calculated from the permissible static or equivalent compressive load P_{perm}.

The three families of joints can be distinguished largely by the shapes of the roller bodies and by the position of their rolling faces. For the Hooke's joint the transmitting elements are roller bearings with many rollers; for the ball joint they are tracks each with a single ball; and for the pode joint they are rolling bodies running on a plain or a roller bearing.

These three families must be analysed separately because of the major differences in their configurations.

4.2 Hooke's Joints and Hooke's Jointed Driveshafts

The torque capacity of joints is determined by the position of the rolling surfaces with respect to the axis of rotation z. In the case of the Hooke's joint the load P acts at a pressure angle $\alpha = 90°$ to the y-axis (see Sect. 2.1) and at a skew angle $\gamma = 0°$ to

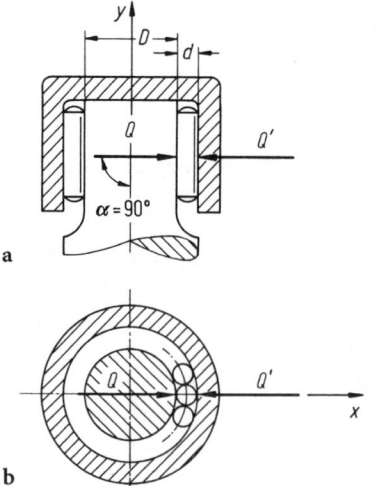

Fig. 4.4a, b. Position of the contact faces relative to the axis of rotation at the trunnion. a Side view, b plan view

the axis of rotation z. This is because the line of action p is perpendicular to z (Figs. 4.3 and 4.4). Hence we get from (4.16)

$$M = m \, P \underbrace{\sin 90°}_{1} \underbrace{\cos 0°}_{1} R \, .$$

Each transmitting element comprises a number of rolling bodies, the total capacity of which is Q_{total}. Thus

$$M = m \, Q_{\text{total}} \, R \, . \tag{4.17}$$

4.2.1 The Static Torque Capacity M_0

The needle or trunnion bearings can consist of i rows each containing z rollers. The pressure distribution on the individual rollers in the bearing was determined in 1901 by Richard Stribeck [1.22, p. 121] to be

$$Q_0 = \frac{s}{z} Q_{\text{total}} \cdot s_0 \tag{4.18}$$

where s is the distribution factor and s_0 the static safety factor for oscillating bearings on Hooke's joints, between 0.8 to 1.0.

For bearings with line contact, such as the needle or roller bearings of a trunnion, $s = 5$ [3.3, Sect. 2.4.2, p. 128], if one takes the play of the bearing into account. In (4.2), Q_0 is the maximum permissible compressive load on a rolling body. The static load capacity C_0 of a complete roller bearing is the total compressive load Q_{total} at which the maximum compressive load at the crown of the roller is equal to the permissible static compressive load Q_0.

$Q_{\text{total}} = C_0 = z/s \, Q_0$ follows from (4.18). Since the bearing has i rows of rollers it follows that the static load capacity C_0 of the trunnion bearing, with the permissible loading from (4.8), is

$$C_0 = i \frac{z}{s} \left(\frac{p_0}{c_p}\right)^2 dl_w = \underbrace{\frac{1}{s}\left(\frac{p_0}{c_p}\right)^2}_{f_0} izdl_w \tag{4.19a}$$

$$f_0 = 44 \, (1 - d/D_m) \text{ N/mm}^2, \text{ where } D_m = \frac{d}{2} + D \,^2 \tag{4.19b}$$

The procedure given in ISO 76/1987 [4.10] is to multiply the values obtained using $p_0 = 4000$ N/mm² and $s = 5$ by a factor which is a function of the diametric ratio d/D of Hooke's joints (0.15–0.16):

$$f_0 = \frac{1}{5}\left(\frac{4000}{270}\right)^2 \cdot \frac{1}{1+0.155} = \frac{1}{5} \cdot 219.4787 \cdot 0.863 = 38 \text{ N/mm}^2 \, .$$

---

[2] $D_m + d$ pitch line-∅, $d/D = 0{,}15$ to $0{,}16$ for cardan joint.

The following derives the factor 44 in (4.19b):

$$38 = x\,(1 - d/D_m) = x\left(1 - \frac{d}{D+d}\right) = x\left(1 - \frac{d}{\frac{d}{0.155} + d}\right)$$

$$38 = x\left(1 - \frac{0.155}{1.155}\right) = 0{,}863\, x$$

$$x = \frac{38}{0.863} = 44 \ \text{(rounded figure)}.$$

Substituting f_0 from (4.19b) we obtain

$$C_0 = 44\,(1 - d/D_m)\, izdl_w \approx 38\, izdl_w \ \ N. \tag{4.20}$$

The static torque capacity of the joint with $m = 2'$ elements in each yoke is obtained by incorporating $Q_{total} = C_0$ in (4.17) and introducing the static safety factor s_0 which is 0.8 to 1.0 for the oscillating bearings of Hooke's joints operating smoothly.

$$M_0 = \frac{2}{s_0}\, 38\, izdl_w\, R \ \ \text{Nm}. \tag{4.21a}$$

where i is the number of rows of rollers, z is the number of rollers in one row, d is the roller diameter, l_w is the effective length of the rollers[3], R is the effective joint radius in m, $F_0 = 38$ N/mm² the static load capacity coefficient for needle and roller bearings[4].

With $s_0 = 0.88$ from Figs. 4.10 and 5.11 we get

$$M_0 = \frac{2}{0.88}\, 38\, izdl_w\, R \approx 2{,}27\, C_0 R \tag{4.21b}$$

The static torque capacity M_0 has to be greater than or equal to the nominal torque M_N specified in manufacturer's catalogues.

4.2.2 Dynamic Torque Capacity M_d

The service life of a pair of rolling elements is limited by the fatigue damage to the rolling bodies[5] and the rolling surfaces after a large number of load cycles. The

---

[3] l_w is obtained from Fig. 4.6.
[4] $f_0 = 38$ is useful when the dimensions of the cross trunnion are still unknown.
[5] Here a roller is a rolling body which is ground to close tolerances to give flat ends, and guided (the length tolerance < 1/10 mm). A needle is a rolling body which, regardless of the profile of its ends, has a greater length tolerance (> 1/10 mm).

4.2 Hooke's Joints and Hooke's Jointed Driveshafts

higher the dynamic loading capacity C of these rolling elements, the greater is the life.

It is not possible to calculate this loading capacity C theoretically because the effect of individual parameters, such as the number of size (diameter, length) of the rolling bodies, the pressure angle, the conformity, the clearances, the material and its hardness, the rolling action, and the friction and lubrication conditions, cannot be established scientifically. It is however possible to develop empirical formulae, the coefficients of which come from tests on different types of bearings. These formulae must of course be updated regularly. The formula for roller bearings developed by Wilhelm Fischer 1959 [4.11; 4.12, p. 21] is

$$C = f_c \, (i \cdot l_w)^{7/9} z^{3/4} d^{29/27}. \tag{4.22}$$

In this formula f_c is a new value for the dynamic load capacity coefficient. It expresses the dependency of the load capacity on the material, the configuration, the proportions and the conformity between the rolling elements and the tracks. The value is not worked out using the mean value of a large number of tests but is obtained using a standardised method of calculation. The load capacity coefficient f_c can be derived simply from two factors f_1 and f_2 using

$$f_c = f_1 \, f_2.$$

The geometry coefficient f_1 is determined from Table 4.1 as a function of the value d/D_m. The bearing capacity coefficient f_2 is dependent on the type and guidance of the rolling bodies and is taken from Fig. 4.5. The effective length of the rolling bodies l_w is calculated from the table in Fig. 4.6.

Having obtained the dynamic load capacity C one could be tempted, as in (4.21), to derive the dynamic torque capacity using $M = 2\,C\,R$. This is not possible for two reasons:

- the cross trunnion bearing does not rotate but oscillates,
- the load on the bearing of the driven yoke is cyclic.

The actual stress must therefore the converted into an equivalent one.

4.2.3 Mean Equivalent Compressive Force P_m

In (4.15) P signifies a constant compressive load when the rolling bearing is rotating. Any oscillating movement must be converted into an equivalent rotary movement. The rocking movement occurs at twice shaft speed and through an angle equal to the articulation angle β of the joint. Figure 4.7 shows the path of a rolling body during one revolution of the joint.

When the angle of rotation $\varphi_1 = 90°$ and $270°$ the rolling bodies change their direction of movement. Since they cover this path twice per revolution the relationship between the oscillation angle β and the angle of rotation φ_1 is

$$\frac{4\beta°}{360°} = \frac{n'}{n} \Rightarrow n' \frac{\beta°}{90°} n.$$

Table 4.1. Geometry coefficient f_1 according to INA

d/D_m	f_1	d/D_m	f_1	d/D_m	f_1	d/D_m	f_1
0.001	43.29	0.051	103.14	0.102	117.11	0.202	122.92
0.002	50.50	0.052	103.56	0.104	117.44	0.204	122.87
0.003	55.27	0.053	103.96	0.106	117.77	0.206	122.81
0.004	58.92	0.054	104.36	0.108	118.08	0.208	122.76
0.005	61.92	0.055	104.75	0.110	118.37	0.210	122.69
0.006	64.49	0.056	105.13	0.112	118.66	0.212	122.62
0.007	66.74	0.057	105.50	0.114	118.94	0.214	122.55
0.008	68.75	0.058	105.87	0.116	119.20	0.216	122.47
0.009	70.58	0.059	106.23	0.118	119.46	0.218	122.39
0.010	72.25	0.060	106.59	0.120	119.70	0.220	122.31
0.011	73.80	0.061	106.94	0.122	119.94	0.222	122.22
0.012	75.24	0.062	107.28	0.124	120.16	0.224	122.12
0.013	76.59	0.063	107.62	0.126	120.38	0.226	122.02
0.014	77.86	0.064	107.95	0.128	120.59	0.228	121.92
0.015	79.07	0.065	108.27	0.130	120.78	0.230	121.82
0.016	80.21	0.066	108.59	0.132	120.97	0.232	121.71
0.017	81.29	0.067	108.91	0.134	121.15	0.234	121.59
0.018	82.33	0.068	109.22	0.136	121.32	0.236	121.48
0.019	83.32	0.069	109.52	0.138	121.48	0.238	121.36
0.020	84.27	0.070	109.82	0.140	121.63	0.240	121.23
0.021	85.19	0.071	110.11	0.142	121.78	0.242	121.11
0.022	86.06	0.072	110.40	0.144	121.91	0.244	120.98
0.023	86.91	0.073	110.68	0.146	122.04	0.246	120.84
0.024	87.73	0.074	110.96	0.148	122.16	0.248	120.70
0.025	88.52	0.075	111.24	0.150	122.27	0.250	120.56
0.026	89.29	0.076	111.51	0.152	122.38	0.252	120.42
0.027	90.03	0.077	111.77	0.154	122.48	0.254	120.27
0.028	90.75	0.078	112.03	0.156	122.57	0.256	120.12
0.029	91.45	0.079	112.29	0.158	122.65	0.258	119.97
0.030	92.12	0.080	112.54	0.160	122.73	0.260	119.81
0.031	92.78	0.081	112.79	0.162	122.80	0.262	119.65
0.032	93.43	0.082	113.03	0.164	122.86	0.264	119.49
0.033	94.05	0.083	113.27	0.166	122.92	0.266	119.32
0.034	94.66	0.084	113.50	0.168	122.97	0.268	119.15
0.035	95.26	0.085	113.73	0.170	123.01	0.270	118.98
0.036	95.84	0.086	113.96	0.172	123.05	0.272	118.81
0.037	96.40	0.087	114.19	0.174	123.08	0.274	118.63
0.038	96.95	0.088	114.40	0.176	123.10	0.276	118.45
0.039	97.49	0.089	114.62	0.178	123.12	0.278	118.27
0.040	98.02	0.090	114.83	0.180	123.14	0.280	118.09
0.041	98.54	0.091	115.04	0.182	123.14	0.282	117.90
0.042	99.04	0.092	115.25	0.184	123.14	0.284	117.71
0.043	99.54	0.093	115.45	0.186	123.14	0.286	117.52
0.044	100.02	0.094	115.65	0.188	123.13	0.288	117.32
0.045	100.49	0.095	115.84	0.190	123.12	0.290	117.13
0.046	100.96	0.096	116.03	0.192	123.10	0.292	116.93
0.047	101.41	0.097	116.22	0.194	123.07	0.294	116.73
0.048	101.86	0.098	116.40	0.196	123.04	0.296	116.52
0.049	102.30	0.099	116.58	0.198	123.01	0.298	116.32
0.050	102.72	0.100	116.76	0.200	122.97	0.300	116.11

$f_c = f_1 \cdot f_2$; $D_m = D + d$.

4.2 Hooke's Joints and Hooke's Jointed Driveshafts

Type of bearing (fully rolling)		f_2
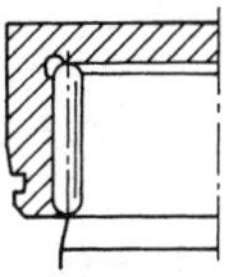	Unguided needle bearing	0.70
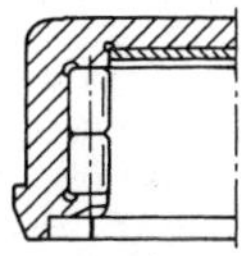	Multi-row roller bearing, with rigid guides	0.78
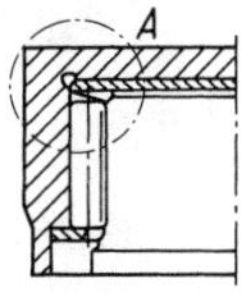	Single or multi-row roller bearing, with flexible guides and support disc, see Detail A	0.83
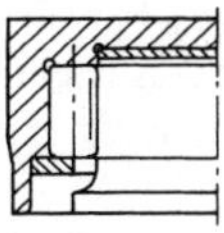	Single row roller bearing, with rigid guides $l_w/d \leq 2.5$	0.83

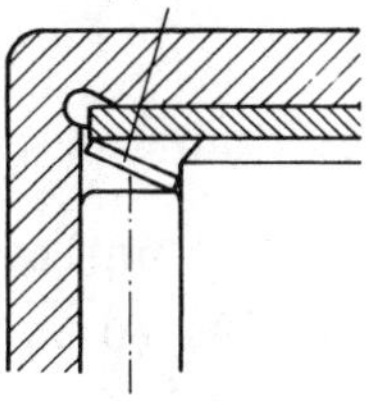

Detail A
Support disc

Fig. 4.5. Bearing capacity coefficient f_2 of rolling bearings

Rolling Body	Sketch	Length factor q
Cylindrical roller (flat end)	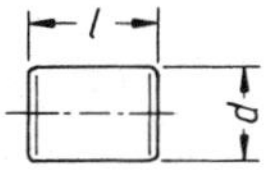	0.1
Needle (flat end)	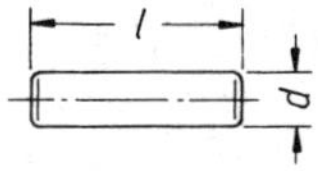	0.15
Needle (slightly curved end)		0.3
Needle (rounded end)	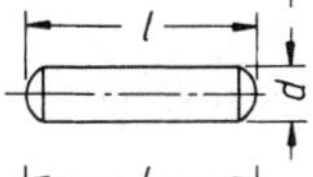	0.5
Needle (spherical end)	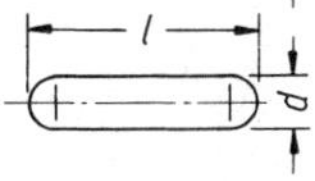	1.0

Fig. 4.6. Length factor q for calculating the effective length of the rolling bodies $l_w = l - qd$

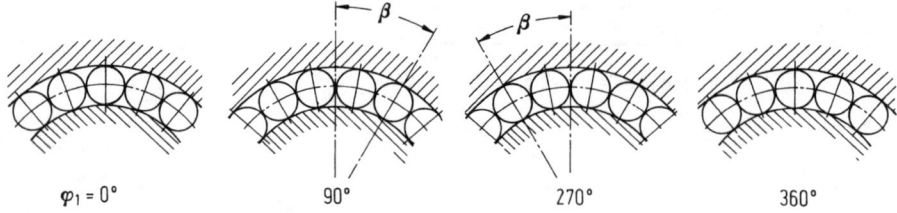

Fig. 4.7. Oscillation of the rolling bodies in a needle bearing. φ_1 is the angle of rotation of the joint, β is the oscillation and articulation angle

n' is the equivalent speed which is now inserted in (4.15):

$$L_h = \frac{10^6 \cdot 90°}{60 \cdot n\beta°}\left(\frac{C}{P}\right)^p = 1{,}5\,\frac{10^6}{n\beta°}\left(\frac{C}{P}\right)^p. \tag{4.23}$$

In (4.23) P is an idealised, constant compressive force and must be replaced. A mean, equivalent compressive force P_m is required which gives the same durability.

The Poncelet equation (1.2)

$$P_2 = P_1 \frac{1 - \sin^2\beta \sin^2\varphi_1}{\cos\beta}$$

states that the yoke force P_2 varies cyclically with the drive angle φ_1. The following equation applies as shown in Fig. 4.8

$$P_m = \frac{A_1^n}{\varphi_p} = \frac{1}{\varphi_p}\sum_{i=1}^{n}\Delta A = \frac{1}{\varphi_p}\sum_{i=1}^{n}P(\varphi)\Delta\varphi, \tag{4.24}$$

where A_1^n represents the area of all the strips.

As $\Delta\varphi \to 0$ we get

$$P_m = \frac{1}{\varphi_p}\int_0^{\varphi_p} P(\varphi)\,d\varphi. \tag{4.25}$$

The yoke force P_1 in (1.7) is related to a constant drive torque M as in (4.17), giving

$$P_2 = \frac{M}{2R}\frac{1 - \sin^2\beta \sin^2\varphi_1}{\cos\beta}.$$

The yoke force P_2, determined by the position of yoke 1 at angle φ_1, is designated by $P(\varphi)$ with the subscript 1 of φ_1 omitted for simplification. On then gets

$$P(\varphi) = \frac{M}{2R}\frac{1}{\cos\beta}(1 - \sin^2\beta \sin^2\varphi). \tag{4.26a}$$

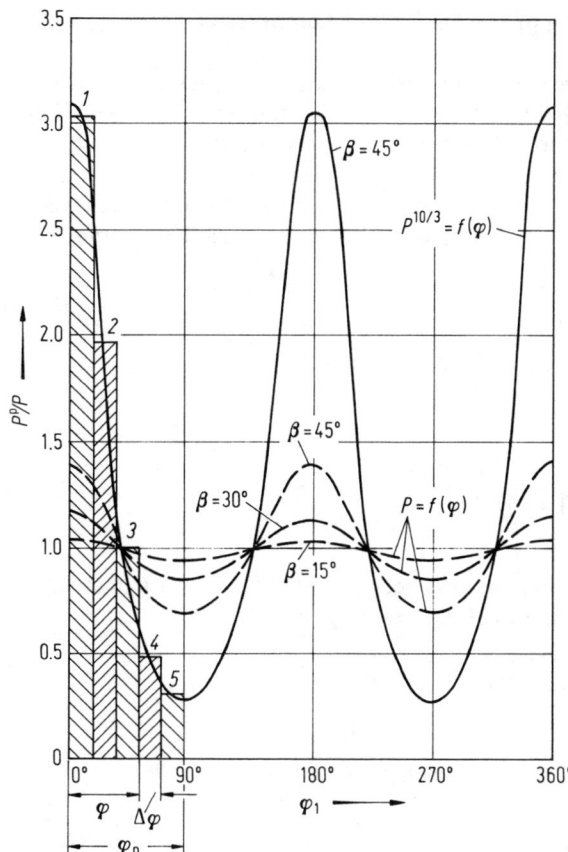

Fig. 4.8. Load curve $P = f(\varphi)$ and the curve raised to the p-*th* power $P^p = f(\varphi)$ to determine the equivalent compressive force P_m

If (4.26a) were put into (4.25) it would give an equation which would be difficult to integrate. It becomes even more difficult with $P^p(\varphi)$ in (4.23).

$$P_m^p = \left[\frac{1}{\varphi_p} \int_0^{\varphi_p} \frac{M}{2 \cdot R \cos \beta} (1 - \sin^2\beta \sin^2\varphi) \, d\varphi \right]^p. \tag{4.26b}$$

Conventional practice is to integrate graphically. One works out several ordinates $P^p(\varphi_p)$, plots the curve which has been raised to the p-th power and by planimetering or counting out the units of surface area, finds the surface area under the curve (Fig. 4.8). By dividing this surface area by the abscissa φ_p one obtains the value of which the p-th root is the desired equivalent compressive force P_m.

In order to get an idea of the effect of the equivalent compressive force P_m on the durability, an approximate value for P_m^p is determined mathematically in the next section.

4.2.4 Approximate Calculation of the Equivalent Compressive Force P_m

If the compressive force changes cyclically the equivalent compressive force P_m of one cycle is the same for all cycles. In the case in question the period is π. Since however the partial periods from 0 to $\pi/2$ and $\pi/2$ to π are symmetrical (Fig. 4.8). The calculation of P_m for a partial period $\varphi_p = \pi/2$ is sufficient.

The area A_1^n can be calculated to any desired accuracy from the ordinates given in (4.26a) and n strips of the same width $\Delta\varphi$ over the partial period $\varphi_p = n\Delta\varphi$. With (4.24) we then get

$$P_m = \frac{\Delta\varphi}{\varphi_p} \sum_1^n P_i(\varphi_i) = \frac{\Delta\varphi}{n\Delta\varphi} \sum_1^n P_i(\varphi_i),$$

or using (4.26a) again

$$P_m = \frac{1}{n} \sum_1^n \frac{M}{2R} \frac{1}{\cos\beta} (1 - \sin^2\beta \sin^2\varphi_i). \tag{a}$$

In Fig. 4.8 five values of the formula (a) have been raised to the power p in the interval from 0 to 90°. The mean ordinates of the five strips are:

No.	Interval	Mean Value
1	0 to 18°	9°
2	18 to 36°	27°
3	36 to 54°	45°
4	54 to 72°	63°
5	72 to 90°	81°

With (1.21) one gets

$$P_m^p = a_1 P_1^p + a_2 P_2^p + a_3 P_3^p + a_4 P_4^p + a_5 P_5^p$$

$$P_m^p = a_1 \left[\frac{M_d}{2R} \frac{1}{\cos\beta} (1 - \sin^2\beta \sin^2\varphi_1) \right]^p + a_2 \left[\frac{M_d}{2R} \frac{1}{\cos\beta} (1 - \sin^2\beta \sin^2\varphi_2) \right]^p \tag{b}$$

$$+ \ldots + a_5 \left[\frac{M_d}{2R} \frac{1}{\cos\beta} (1 - \sin^2\beta \sin^2\varphi_5) \right]^p.$$

The time intervals $a_1, a_2, \ldots, a_5$ all the same size and represent 1/5 of the durability L. k^2 is written as an abbreviation for $\sin^2\beta$

$$P_m^p = \left[\frac{M_d}{2R} \right]^p \left[\frac{1}{\cos\beta} \right]^p \frac{1}{5} \{ (1 - k^2 \sin^2\varphi_1)^p + (1 - k^2 \sin^2\varphi_2)^p$$

$$+ \ldots + (1 - k^2 \sin^2\varphi_5)^p \}. \tag{c}$$

4.2 Hooke's Joints and Hooke's Jointed Driveshafts

It follows from the binomial theorem for fractional exponents [1.10, p. 100], if the negative sign in the bracket is also taken into account, that

$$(1-x)^{10/3} = 1 - \frac{10}{3}x + \frac{70}{18}x^2 - \frac{280}{162}x^3 + \frac{1120}{1944}x^4 - \frac{2240}{29160}x^5$$
$$+ \frac{11200}{524880}x^6 - \frac{56000}{9447840}x^7 + \ldots . \tag{d}$$

The term

$$\frac{1}{5}\left\{\sum_{i=1}^{5}(1-k^2\sin^2\varphi_i)^p\right\}$$

from (c) is now calculated for $p = 10/3$.

The equivalent compressive force P_m is introduced into (4.23)

$$L_h = 1{,}5 \frac{10^6}{n\beta^\circ}\left(\frac{2CR}{M_d}\right)^{10/3} \underbrace{\left[\frac{\cos\beta}{\frac{1}{2}\sum_{i=1}^{n}(1-\sin^2\beta\sin^2\varphi_i)}\right]^{10/3}}_{k_\omega} \tag{e}$$

$$L_h = 1{,}5 \frac{10^6}{n\beta^\circ}\left(\frac{2CR}{M_d}\right)^{10/3} k_\omega . \tag{4.27}$$

The *equivalence factor* k_ω calculated for the cyclic compressive force is shown in Fig. 4.9. It can be disregarded if $\beta < 15°$.

The calculation is much simpler for point contact. The cubic mean value P_m^3 is required and the relevant term

$$\frac{1}{n}\sum_{i=1}^{5}(1-\sin^2\beta\sin^2\varphi_i)^3$$

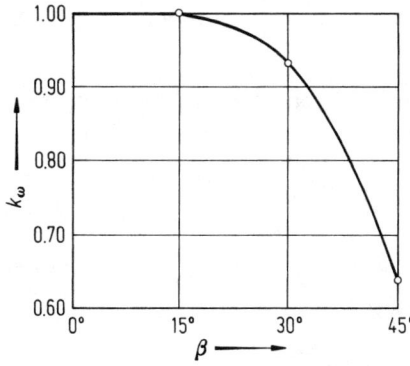

Fig. 4.9. Dependence of the equivalence factor k_ω on the articulation angle β

does not give an infinite series but, since $(1-x)^3 = 1 - 3x + 3x^2 - x^3$, gives an expression with only four terms [1.10, p. 70].

The bearing factor k_t is needed next. It accounts for the reduced life arising because the needles of the cross trunnion bearing are more heavily loaded externally and less internally. In GWB-type Hooke's joints k_t is in the range 1.00 to 1.66:

Bearing factor	k_t
heavy shaft with small tolerances and rigid yoke, low stress	1.00–1.33
light shaft with normal tolerances, high stress	1.33–1.66

Where an operation involves shocks, the effect of this loading is taken into account by a special shock or *operating factor* k_s. It depends both on the prime mover and also on the operating equipment connected to it (Table 4.2).

4.2.5 Dynamic Transmission Parameter 2 CR

Equation (4.27) can be re-arranged to give

$$2CR = \frac{M_d k_t}{k_\omega} \sqrt[10/3]{\frac{L_h n \beta}{1.5 \cdot 10^6}}.$$

If the product $L_h n \beta$ is selected to give $5000 \cdot 100 \cdot 3 = 1.5 \cdot 10^6$, then $2\,CR$ represents the torque at which, at a speed of 100 rpm and an articulation angle of 3°, the joint has a life of 5000 hours:

$$2CR = \frac{M_d k_t}{k_\omega} \sqrt[10/3]{\frac{1.5 \cdot 10^6}{1.5 \cdot 10^6}} = \frac{M_d k_t}{k_\omega} \sqrt[10/3]{1} = \frac{M_d k_t}{k_\omega} \cdot 1 = M_d k_t \,.\,^6$$

2CR is a reference value for universal joints, and as such is included in the last row of Fig. 4.10. It has been adopted by manufacturers and is called the transmission parameter 2CR corresponding to the loading capacity C. It is based on the assumption that

$$L_h n \beta = 5000 \cdot 100 \cdot 3 = 1{,}5 \cdot 10^6 \,.$$

For other operating conditions, in terms of the life L_x, speed n_x and the articulation angle β_x, the values must be converted accordingly

$$\left(\frac{2CR}{M_x k_t}\right)^{10/3} = \frac{L_{hx} n_x \beta_x}{5000 \cdot 100 \cdot 3} \Rightarrow L_{hx} = \frac{1.5 \cdot 10^6}{n_x \beta_x}\left(\frac{2CR}{k_t M_x}\right)^{10/3} \tag{4.28}$$

---

[6] From Fig. 4.9 $k_\omega \approx 1$ for $\beta < 15°$, see Fig. 4.9.

4.2 Hooke's Joints and Hooke's Jointed Driveshafts

Table 4.2. Shock or operating factor f_{ST}

Type of loading	Type of machinery	Shock factor k_s
	Prime movers	
	Electric motors	1.0
	Spark ignition engines	1.0…1.15
	Diesel engines	1.1…1.2
	Motor vehicles	
	Passenger cars	1.0…1.2
	Lorries	1.2…1.6
	Construction vehicles	1.5…5.0
	Operating Equipment	
constant	Generators and conveyors, centrifugal pumps, light fans	1.0…1.1
light shock	Generators and conveyors (with non-uniform loading), centrifugal pumps, medium fans, machine tools, printing, wood-working, light duty paper and textile machines	1.0…1.5
medium shock	Multi-cylinder piston pumps and compressors, large fans, ships' drives, calenders, roller tables, rod mills, light drive rollers and tube rolling mills, primary drives for locomotives, heavy paper and textile machinery	1.5…2.0
heavy shock	Single cylinder compressors and piston pumps, mixers, crane final drives, excavators, bending machines, presses, rotary drills, locomotive drives, reversing roller tables (continuous operation), medium and heavy and rolling mills and continuous tube mills	2.0…3.0
very heavy shock	Rolling mill stands, feed rollers of wide strip coilers, reversing roller tables and heavy rolling mills, vibrating conveyors, crushers	3.0…6.0
extreamely heavy shock	Heavy rolling mills, gate shears, jaw braker, swing conveyor	6…10

or

$$2CR = M_x k_t^{10/3} \sqrt{\frac{L_{hx} n_x \beta_x}{1{,}5 \cdot 10^6}}. \tag{4.29}$$

Using (4.29) it is possible to calculate the required transmission parameter $2CR$ for the given conditions. A suitable driveshaft can then be selected from Fig. 4.10 or from catalogues.

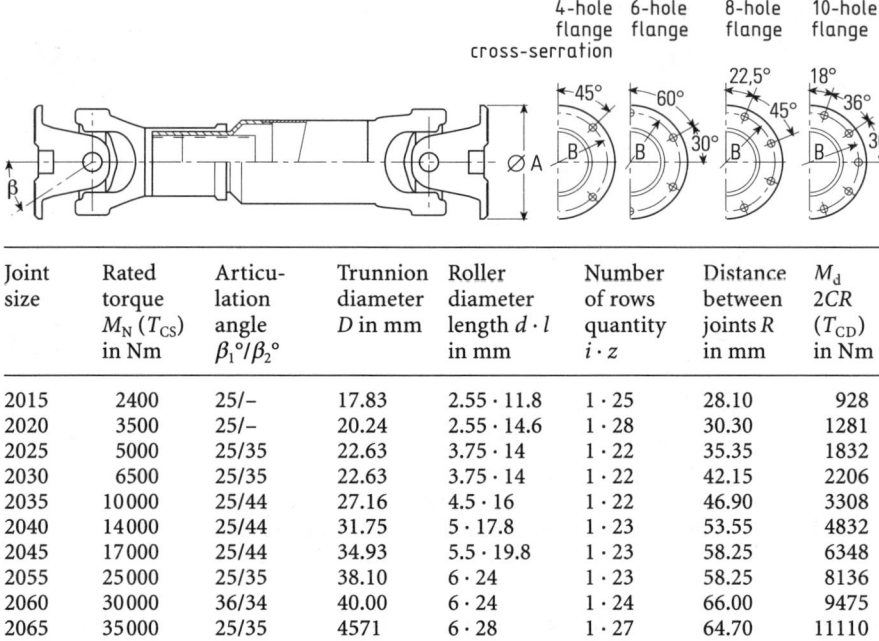

Joint size	Rated torque M_N (T_{CS}) in Nm	Articulation angle $\beta_1°/\beta_2°$	Trunnion diameter D in mm	Roller diameter length $d \cdot l$ in mm	Number of rows quantity $i \cdot z$	Distance between joints R in mm	M_d $2CR$ (T_{CD}) in Nm
2015	2400	25/–	17.83	2.55 · 11.8	1 · 25	28.10	928
2020	3500	25/–	20.24	2.55 · 14.6	1 · 28	30.30	1281
2025	5000	25/35	22.63	3.75 · 14	1 · 22	35.35	1832
2030	6500	25/35	22.63	3.75 · 14	1 · 22	42.15	2206
2035	10000	25/44	27.16	4.5 · 16	1 · 22	46.90	3308
2040	14000	25/44	31.75	5 · 17.8	1 · 23	53.55	4832
2045	17000	25/44	34.93	5.5 · 19.8	1 · 23	58.25	6348
2055	25000	25/35	38.10	6 · 24	1 · 23	58.25	8136
2060	30000	36/34	40.00	6 · 24	1 · 24	66.00	9475
2065	35000	25/35	4571	6 · 28	1 · 27	64.70	11110

Fig. 4.10. Principal and cross dimensions of a Hooke's jointed driveshaft for light loading, GWB design. $2CR$ value of the dynamic transmission parameter; d roller diameter in mm; D trunnion diameter; l length of a roller in mm; i number of rows of rollers; R effective joint radius or distance between joints in mm; z number of rollers in a row; M_d dynamic torque in Nm; β articulation angle

4.2.5.1 Example of Specifying Hooke's Jointed Driveshafts in Stationary Applications

This example is for a Hooke's jointed driveshaft transmitting the torque (M_d) of 400 Nm in a geared electric motor. At a speed of 200 rpm and an articulation angle β of 4° the target life (L_h) is 50,000 hrs. The following must be calculated:

a) the transmission parameter $2CR$,
b) the Hertzian pressure p_0 in the needle bearing of the selected joint size,
c) the width of the contact face $2b$,
d) the total elastic deformation δ_0 between the needles and the cross trunnion.

Solution: a) The dynamic transmission parameter $2CR$ (Fig. 4.11) is calculated from (4.29) as

$$2CR = 2 \cdot 400 \sqrt[\frac{10}{3}]{\frac{50000 \cdot 200 \cdot 4}{1.5 \cdot 10^6}} = 2 \cdot 400 \sqrt[\frac{10}{3}]{\frac{40}{15}} = 2 \cdot 400 \sqrt[\frac{10}{3}]{2.6667} = 2 \cdot 400 \cdot 1.3421$$
$$2CR = 2142 \text{ Nm}$$

The joint size GWB 2033, selected from Fig. 4.10 or from the manufacturer's catalogue, has $2CR = 1281$.

4.2 Hooke's Joints and Hooke's Jointed Driveshafts

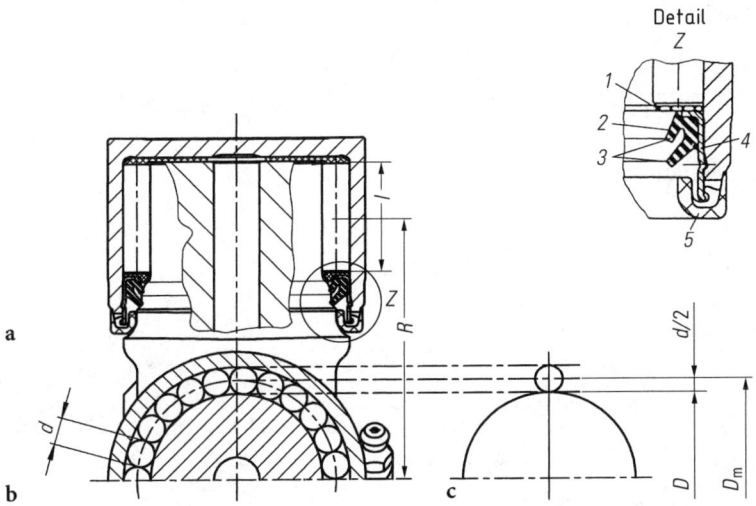

Fig. 4.11 a, b. Section through the Hooke's joint in the example, **a** Side view; **b** plan view; **c** sketch showing the mean diameter of the needle bearing D_m. $D = 22.24$ mm; $d = 2.5$ mm; $l = 15.8$ mm; number of rollers $z = 31$; number of rows of rollers $i = 1$; $R = 37$ mm; $D_m = D + d = 24.74$ mm, $l_w = l - q \cdot d = 15.8 - 2.5 = 13.3$; Detail Z: *1* guiding disc for roller, *2* sealing lip against guiding disc, *3* sealing lips against oil loss, *4* guiding jack in metal plate, *5* sealing ring with two sealing spots and one lip

The characteristic data of the journal cross fittings are:

$$i \cdot z = 1 \cdot 28 \qquad\qquad d/D_m = 2.55/22.79 = 0.1119$$
$$d \cdot l = 2.55 \cdot 14.6 \qquad\qquad \text{from Table 4.10 } f_1 = 118.66$$
$$l_w = 14.6 - 0.15 \cdot 2.55 = 14.22 \qquad \text{from Fig. 4.5 } f_2 = 0.7$$
$$D = 20.24 \qquad\qquad f_c = f_1 \cdot f_2 = 118.66 \cdot 0.7 = 83.062$$
$$D_m = 20.24 + 2.55 = 22.79$$

a) From (4.22) it follows

$$C = 83.062 \, (1 \cdot 14.22)^{7/9} \cdot 28^{3/4} \cdot 2.55^{29/27} = 21\,782 \Rightarrow 2\,CR$$
$$\qquad\qquad\qquad\downarrow\qquad\quad\downarrow\qquad\quad\downarrow$$
$$\qquad\qquad\qquad 7.883\quad 12.172\quad 2.733$$
$$C = 2 \cdot 21\,782 \cdot 0.03030 = 1320 \text{ Nm}$$

b) from (3.57) it follows

$$p_0 = 270 \cdot \sqrt{\frac{Q_0}{l_w} \cdot \frac{\varrho_1 + \varrho_2}{2}}$$

$$\varrho_1 + \varrho_2 = 1/r_1 + 1/r_2 = 2/20.24 + 2/2.55 = 0.0988 + 0.7843 = 0.8831$$

From (4.16) it follows

$$Q_{total} = \frac{400}{2 \cdot 0.03030} = 6601 \text{ N}.$$

After Stribeck's equation (4.18) it is

$$Q_0 = \frac{s}{z} Q_{ges} = \frac{5}{28} 6601 = 1179 \text{ N}$$

and again from (3.57) it follows

$$p_0 = 270 \cdot \sqrt{\frac{1179}{14.22} \cdot \frac{0.8831}{2}} = 270 \cdot \sqrt{36.610} = 1634 \text{ N/mm}^2$$

c) from (3.52b) it follows

$$2b = \frac{6.68}{10^3} \cdot \sqrt{\frac{1179}{14.22 \cdot 0.8831}} = \frac{6.68}{10^3} \cdot \sqrt{93.887} = 0.0647 \text{ mm}$$

d) after (3.56) it follows

$$\delta_0 = \frac{4.05}{10^5} \cdot \frac{1179^{0.925}}{14.22^{0.85}} = \frac{4.05}{10^5} \cdot \frac{639.7}{9.549} = \frac{267.64}{10^5} = 0.00271 \text{ mm} \approx 3 \text{ μm}$$

This counts for constant torque and rpm.

4.2.6 Motor Vehicle Driveshafts

Driveshafts in vehicles are subjected to two types of torque:

- the starting torque M_A. The portion of the engine torque M_M acting on each shaft is given by the factor ε_V or ε_H. It is shared between the u driveshafts and is multiplied by the overall gear ratio i_A between the engine and the shaft

$$M_A = f_{ST} \frac{\varepsilon_{V,H}}{u} M_M i_A \cdot e_{TM}^{n1}; \quad \frac{1}{0.8u} \approx 0.6 \text{ für } u = 2 \quad (4.30)$$

- the adhesion torque M_H. For commercial vehicles the dynamic load transfer between axles is considered into a lump through the shock-factor $f = 1.2 - 4.0$, the static axle load G, the axle ratio i_H before the shaft and the static rolling radius R_{stat}

$$M_H = f_{ST} \cdot \mu \frac{G \cdot g}{i_H} R_{stat} \cdot \frac{1}{e_{AD}^{n2}} \quad (4.31)$$

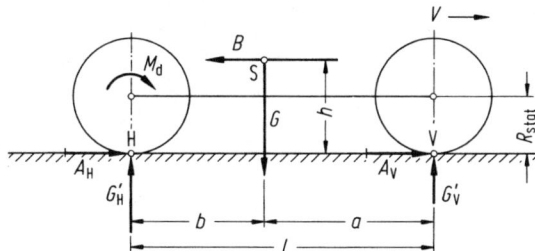

Fig. 4.12. Adhesion torque for a motor car moving away from rest

For passenger cars the starting torque is much higher and hence the dynamic load transfer between the axles must be taken into account.

As shown in Fig. 4.12 the following applies on the flat

- for a stationary passenger car, using the equilibrium condition $\sum M = 0$ and pivot in

$$H: G_V l - Gb = 0 \Rightarrow b = \frac{G_V}{G} l, \tag{a}$$

$$V: G_h l - Ga = 0 \Rightarrow a = \frac{G_H}{G} l, \tag{b}$$

- when accelerating there is an inertia force B. Taking moments about

$$V: -G'_H l + Ga + Bh = 0, \tag{c}$$

$$H: G'_V l - Gb + Bh = 0. \tag{d}$$

Without rolling and air resistance – taken to be negligible – the values are

Four wheel drive $\quad B = A_V + A_H = \mu(\overbrace{G'_V + G'_H}^{G}),$ (e)

Front wheel drive $\quad B = A_V = \mu G'_V,$ (f)

Rear wheel drive $\quad B = A_H = \mu G'_H.$ (g)

The values G'_V and G'_H, with primes, are, in contrast to the static G_V and G_H, the dynamic forces of axle loads.

Hence the loads for *four wheel drive*: equation (e), substituted in (c), becomes

$$G'_V l - Gb + \mu Gh = 0$$

$$G'_V \cdot l = G(b - \mu h)$$

$$G'_V = \frac{G}{l}(b - \mu h) \tag{h}$$

and equation (e), substituted in (d)

$$-G'_H l + Ga + \mu Gh = 0,$$
$$G'_H l = G(a + \mu h),$$
$$G'_H = \frac{G}{l}(\alpha + \mu h). \tag{i}$$

From (h) and (i), [4.11], the ratio of the traction forces A_V and A_H when all wheels are fully utilised for acceleration, is

$$\frac{G'_V}{G'_H} = \frac{b - \mu h}{a + \mu h} = \frac{A_V}{A_H} = \frac{M_V}{M_H} = \frac{\varepsilon_V}{\varepsilon_H} = \varepsilon. \tag{k}$$

Ideally the drive $\varepsilon_V/\varepsilon_H$ should vary as the ratio of the dynamic loads G'_V/G'_H. Since the gear ratios between the engine and the drive axles are usually constant, one has to decide on the best value for ε. It is found by measuring the dynamic axle loads, e.g. $G'_V/G'_H = \varepsilon = 0.56$. The distribution ε_V and ε_H to the drive axles can be calculated from

$$\varepsilon_V M_A + \varepsilon_H M_A = M_A,$$
$$\varepsilon_V + \varepsilon_H = 1. \tag{l}$$

and

$$\varepsilon_V = \varepsilon \cdot \varepsilon_H, \tag{m}$$
$$\varepsilon_H = \frac{1}{\varepsilon} \varepsilon_V. \tag{n}$$

From equations (l) to (n) one gets

$$\varepsilon_V + \frac{1}{\varepsilon}\varepsilon_V = 1.$$

$$\varepsilon_V\left(1 + \frac{1}{\varepsilon}\right) = 1 \Rightarrow \varepsilon_V = \frac{\varepsilon}{1+\varepsilon} = \frac{0.56}{1.56} = 0.36 \tag{4.32a}$$

and

$$\varepsilon_H \cdot \varepsilon + \varepsilon_H = 1,$$

$$\varepsilon_H(\varepsilon + 1) = 1 \Rightarrow \varepsilon_H = \frac{1}{1+\varepsilon} = \frac{1}{1.56} = 0.64. \tag{4.32b}$$

If one takes into account the dynamic transfer of axle loads for *front wheel drive*, using Eq. (f), substituted in (d),

$$G'_V l - Gb + \mu G'_V h = 0,$$
$$G'_V l(l + \mu h) = Gb \Rightarrow G'_V = G\frac{b}{l + \mu h}$$

4.2 Hooke's Joints and Hooke's Jointed Driveshafts

From (b)

$$G'_V = G \frac{G_V}{G} \frac{l}{l + \mu h}.$$

For each driveshaft

$$M_{HV} = k_s \mu \frac{G_V}{2} \frac{l}{l + \mu h} R_{stat}. \qquad (4.33\,\text{a})$$

For *rear wheel drive* substituting Eq. (g) in (c)

$$-G'_H l + Ga + \mu G'_H h = 0,$$
$$G'_H (l - \mu h) = Ga,$$
$$G'_H = G \frac{a}{l - \mu h}.$$

From (a)

$$G'_H = G \frac{G_H}{G} \frac{l}{l - \mu h}.$$

For each driveshaft

$$M_{HH} = k_s \mu \frac{G_H}{2} \frac{l}{l - \mu h} R_{stat}. \qquad (4.33\,\text{b})$$

In *commercial vehicles* Hooke's jointed driveshafts are used in axle and auxiliary drives (Fig.4.13). They should be designed according to the lower torque, which occurs in the driveline

$$M_N \text{ selected} > M_A \text{ or } M_H$$

The rated torque M_N of the driveshaft should not be exceeded.

In a vehivle with varying torque M_d and varying speed n, the equivalend values for M_d and n must be taken from service load spectra. It can be assumed that the factors used in the service life prediction of roller bearings have the same effect on the durability of the journal cross bearings.

Damage can result in the shaft if the chosen size is too small or there is an additional clutch in the *assembly*. In any case, the rated torque M_N must not be exceeded.

4.2.7 GWB's Design Methodology for Hooke's joints for Vehicles

In the Gelenkwellenbau GmbH Essen (GWB) "compact 2000" series the rated torque M_N of the driveshaft is determined with a 5% safety margin against plastic deformation. The shock effect of a "dropped" clutch during starting must therefore be

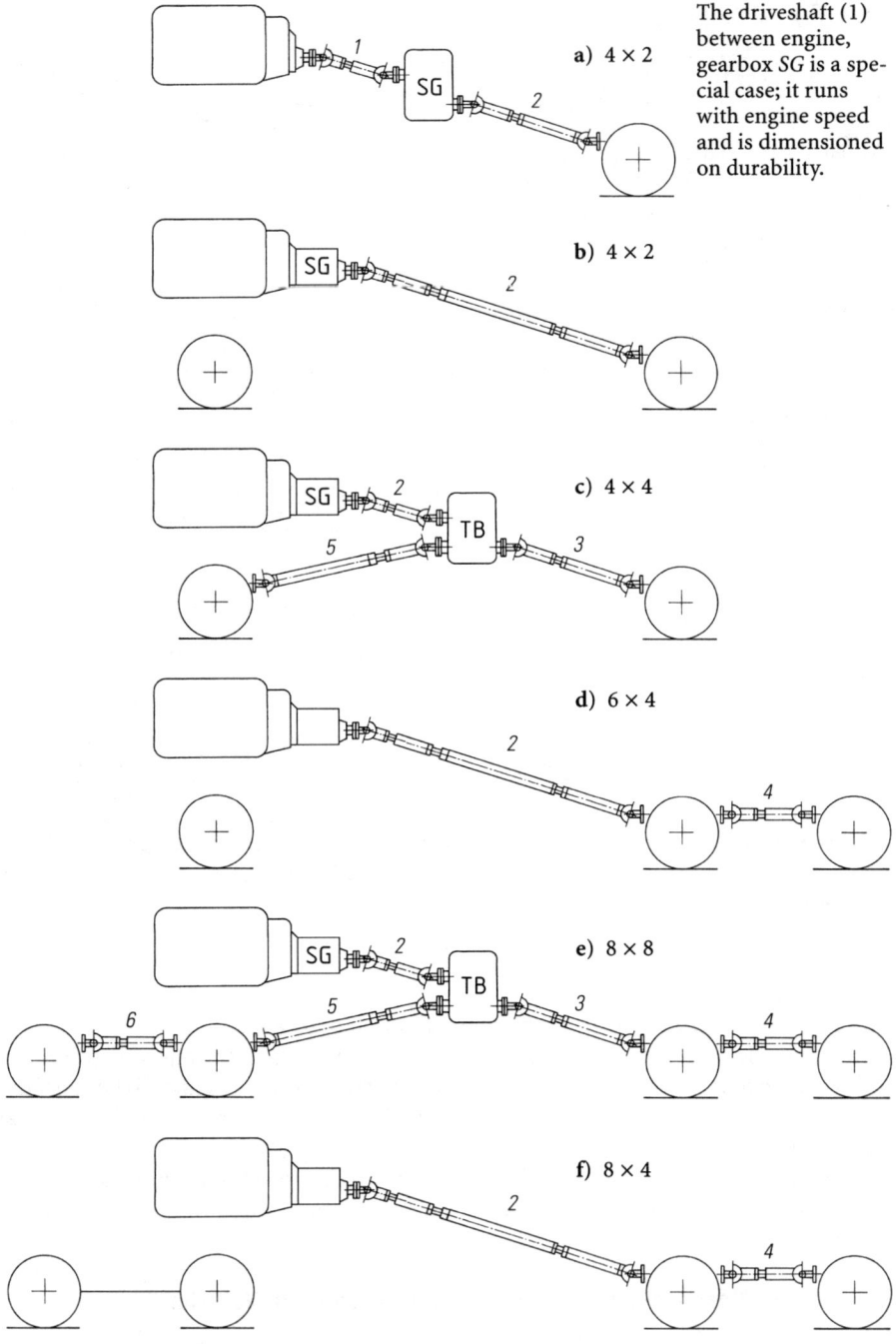

Fig. 4.13 a–f. Driveshafts between engine, gear boxes and driven axles. *TB* = transfer box, *SG* = standard gear box (flanged), 2+3 rear axle drive & coupling shaft, 4+6 coupling shaft for double axle, 5 front axle drive shaft

considered. The starting torque M_A and the adhesion torque M_H depend on the following factors:

- moment of inertia of the engine
- starting speed
- engine torque
- closing time of the clutch
- gear ratio and efficiency
- stiffness of the whole driveline including halfshafts and tyres
- moment of inertia of wheels
- running resistance
- mass of the vehicle and position of its centre of gravity
- coefficient of friction (μ) between tyres and road surface (0.9 on road, 1.2 off-road)

During starting the kinetic energy of the rotating parts acts before the transfer box. The torque arising from this can exceed the nominal engine torque M_M if one's foot slips off the pedal. Since the data above are not generally known, factors determined from simulation calculations and measurements have been used in simplified equations (4.30) and (4.31).

Specification
The driveshaft is specified from the starting torque and the adhesion torque. If the starting torque is greater than the adhesion torque, the wheels slip. Therefore, only the smaller torque is taken into account.

Driving torque
manual transmission automatic transmission

$$M_A = M_M \cdot f_{ST} \cdot i_A \cdot \varepsilon_{V,H} \cdot e_{TM}^{n1} \qquad M_A = M_M \cdot i_C \cdot i_A \cdot \varepsilon_{V,H} \cdot e_{TM}^{n1} \qquad (4.34)$$

where: i_A = highest (numerical) gear ratio between the engine and the shaft being specified

i_A = 10 on-highway, 12.5 off-road

f_{ST} = 1.6 dropped clutch factor for shaft 2 after the gear box – 1.2 for all shafts after the transfer box

e_{TM} = 0.95^{n1} efficiency of a gear stage

n2 = number of gear stages between engine and shaft from Tab. 4.3

$\varepsilon_{V,H}$ = torque distribution in the transfer box, front/rear

i_c = multiplying factor for automatic transmissions

n1 = Efficiency of a gear stage

Adhesion torque

$$M_{H\,shaft2} = G_{whole\ vehicle} \cdot g \cdot R_{ST} \cdot f_{TL} \cdot \mu \cdot \frac{1}{i_H \cdot e_{AD}^{n2}} \qquad (4.35)$$

where: $\mu_{roadway}$ = 0.9 on-highway and 1.2 for off-road as friction constant

e_{AD} = 0.95^{n2} efficiency between all gearwheels

n2 = number of gear stages between wheel and driveshaft

$g = 9.806$ m/sec^2
R_{ST} = static tyre radius
f_{TL} = factor for dynamic axial displacement
   1.0 for shafts 2 and 4; 1.2 for shafts 3 and 5
i_H = ratio between wheel and shaft that is being calculated

For vehicles with four-wheel drive differing driveshaft configurations are required, depending on the number of axles. In Figure 4.13a, driveshaft 1 is connected directly to the engine; it turns at the same (high) speed as the engine.

4.2.7.1 Example of Specifying Hooke's Jointed Driveshafts for Commercial Vehicles

For a three-axle, 26t all-wheel drive Iveco-Magirus tipper truck (Fig. 4.13e), with front mounted engine, the required sizes of the driveshafts 2 to 5 should be determined from the starting torque M_A and the adhesion torque M_H for the different gear stages.

Engine	Gear box 4 speed	Transfer box box 2 speed	Axle gear box	Tyres
$N_M = 199$ kW	$i_{smax} = 10{,}25$	$i_{vmax} = 1{,}75$	$i_R = 3{,}478$	$R_{ST} = 0{,}515$ m 12.00 R 20
$M_M = 1100$ Nm at 1300 rpm	$i_{smin} = 1{,}0$	$i_{vmin} = 1{,}0$	$i_{diff} = 1{,}647$	
$n_{max} = 2200$ rpm	$i_{rear} = 9{,}8$	diff. lock	diff. lock	
transfer box:	engine torque distribution $\varepsilon_V/\varepsilon_H = 0.29/0.71$			

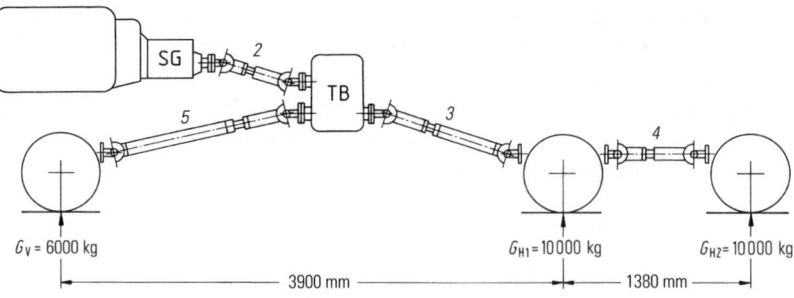

Fig. 4.14. Driveline layout and data for a three-axle 26t 6 × 6 Iveco-Magirus truck; total weight = 420,000 N and GVW = 260,000 N. Average articulation angle β 7°, 12° on-highway, 20° off-highway. *2* coupling shaft, *3* propeller shaft, *4* connecting shaft for double axle, *5* front axle drive shaft

4.2 Hooke's Joints and Hooke's Jointed Driveshafts

Shaft	Starting torque M_A acc. to (Chapter 4.3.4) in Nm	Adhesion torque M_H acc. to (Chapter 4.3.5) in Nm	Required joint size
	$M_A = M_M \cdot f_{ST} \cdot i_A \cdot \varepsilon_{V,H} \cdot e_{TM}^{n_1}$ $i_A = i_S \, i_V = 10{,}25 \cdot 1{,}75 = 17{,}938$	$M_H = f_{TL}\, \mu\, G \cdot g \cdot R_{ST}\, \dfrac{1}{i_{TH} \cdot e_{AD}^{n_2}}$ $i_H = i_R i_{Diff} = 3{,}478 \cdot 1{,}647 = 5{,}728$	
2	$M_A = 1{,}6 \, \dfrac{1}{1 \cdot 1} \, 1100 \cdot 10{,}25 \cdot 0{,}95^1 = 17\,138$	$M_H = 1{,}0 \cdot 1{,}2 \, \dfrac{26000 \cdot 9{,}81}{1 \cdot 5{,}728} \, 0{,}515 \cdot \dfrac{1}{0{,}95^{2,25}} = 30\,885$	2045
3	$M_A = 1{,}2 \cdot 0{,}71/1 \cdot 1100 \cdot 17{,}938 \cdot 0{,}95^2 = 15\,172$	$M_H = 1{,}2 \cdot 1{,}2 \, \dfrac{20000 \cdot 9{,}81}{1 \cdot 5{,}728} \, 0{,}515 \cdot \dfrac{1}{0{,}95^{1,5}} = 27\,433$	2045
4	$M_A = \dfrac{f_{ST}}{2} \cdot M_{A3} = \dfrac{1{,}2}{2} \cdot 15172 = 9103$	$M_H = \dfrac{1}{2} \cdot M_{H3} = \dfrac{1}{2} \cdot 27433 = 13717$	2035
5	$M_A = 1{,}2 \cdot 0{,}29/1 \cdot 1100 \cdot 17{,}938 \cdot 0{,}95^2 = 6197$	$M_H = 1{,}2 \cdot 1{,}2 \, \dfrac{6000 \cdot 9{,}81}{1 \cdot 5{,}728} \, 0{,}515 \cdot \dfrac{1}{0{,}95} = 8022$	2030

where $f_{ST} = 1.6$ only for shaft 2 because of the kinetic energy of the rotating masses in front of the transfer box
$f_{ST} = 1.2$ for shafts 3,4,5 behind the transfer box
ε_V = the index V stands for the share of the engine torque for FRONT
ε_H = the index H stands for the share of the engine torque for REAR
f_{TL} = factor in considering the dynamic axle load displacement
1.2 for shafts 3 and 5
1.0 for shafts 2 and 4
n_1, n_2 = from Table 4.3.
* = $1/1.666 = 1/(0.888 \cdot 2) = 1/(c \cdot 2) = 0.6$ (manufacturer's empirical value)

Table 4.3. Values for the exponents n_1 and n_2 after GWB

Shaft	2					3	4	5			6	
Vehicle model	4 × 2 6 × 2 8 × 2	6 × 4 8 × 4 4 × 4	4 × 4	6 × 6	8 × 8	4 × 4	6 × 6 8 × 8	6 × 4 6 × 6 8 × 8	4 × 4 6 × 6	8 × 8	8 × 8	
n_1	1	1	1	1	1	2	3	*	2	2	*	
n_2	1	1.5	2	2.25	2.5	1	1.5	*	1	1.5	*	

(Example: vehicle model 4 × 2 means tha the vehicle has 4 wheels, two of which are driven).
* The specification of shaft 4 is developed from the specification for shaft 2 of 6 × 4 or shaft 3 for 6 × 6 vehicles. Shaft 6 from the specification of shaft 5 of 4 × 4, 6 × 6 or 8 × 8 vehicles. No n_1 or n_2 values are needed for this.

The *broken* exponents for n_2 result from the torque distribution in the tandem-axle.

4.2.8 Maximum Values for Speed and Articulation Angle

It follows from Sect. 1.2 that a single Hooke's joint can only be used where the non-uniformity of its torque transmission is negligible.

Two Hooke's joints with a telescopic sliding piece are generally described as a propshaft. A Z- or W-configuration of the individual joints results in constant velocity conditions (Fig. 1.9). But even here the middle part rotates in a non-uniform way, see (1.1a) and (1.2). The first and second derivatives with respect to time from (1.1b) give the angular velocity ω_2 and the angular acceleration α_2 both of which are cyclic.

The inertia torque comes from the moment of inertia J_2 and the angular acceleration α_2

$$M_\alpha = J_2 \alpha_2 \tag{a}$$

The resulting torque in the centre section is then

$$M_{\text{res}} = M_2 + M_\alpha. \tag{b}$$

With (1.2) and (1.3) one obtains from (b)

$$M_{\text{res}} = M_d \omega_1 \frac{1 - \sin^2\beta \sin^2\varphi_1}{\cos\beta} + J_2 \omega_1^2 \frac{\cos\beta \sin^2\beta \sin 2\varphi_1}{(1 - \sin^2\beta \sin^2\varphi_1)^2}. \tag{4.36}$$

The alternating inertia torque M_α can bring about torsional vibrations and rattling noises the causes of which are often not correctly identified. These noises occur when the smallest value from (4.34) becomes nought or negative [4.12].

Theory and experience show that in order to achieve smooth running of driveshafts certain levels of inertia torque M_α must not be exceeded. Table 4.4 gives the maximum permissible values for the product $n\beta$ in the service life equation (4.28).

4.2 Hooke's Joints and Hooke's Jointed Driveshafts

Table 4.4. Maximum speeds and maximum permissible values of $n\beta$ arising from the moment of inertia of the connecting parts (length of driveshaft 1500 mm)

GWB-Joint size	n_{max} [1/min.]	$(n\beta_{perm})$ [°/min.]
2015	6000	27000
2020	6000	26000
2025	6000	24000
2030	6000	22000
2035	6000	21000
2040	5600	20000
2045	5200	19000
2050	5000	19000
2055	5000	18000
2060	4600	17000
2065	4300	17000

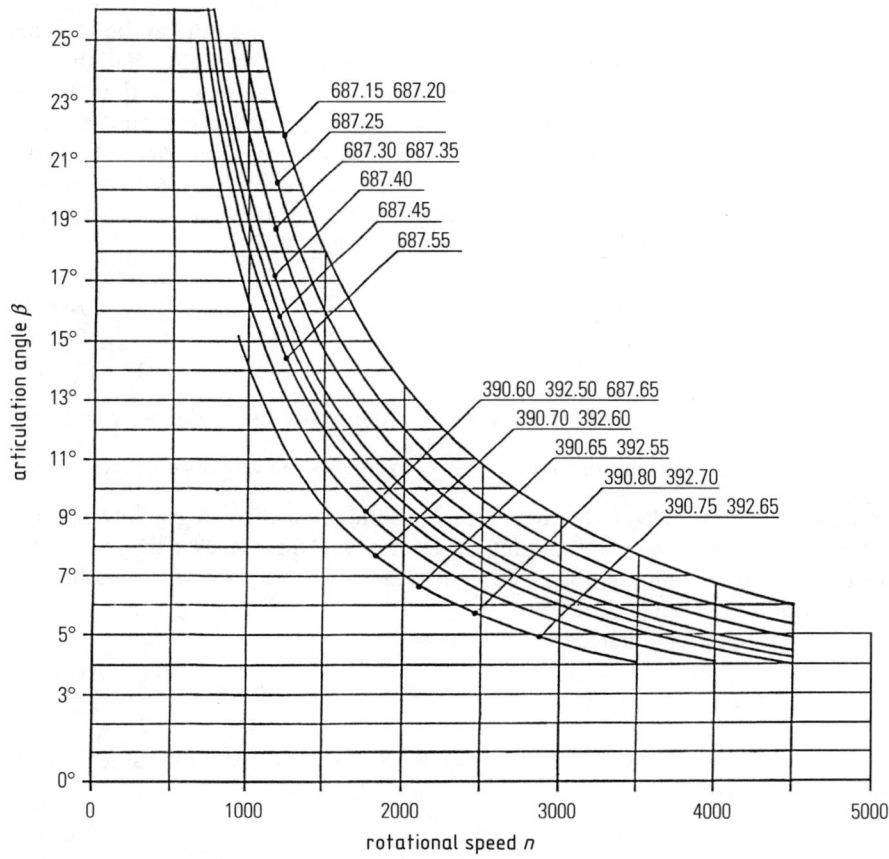

Fig. 4.15. $n\beta$ – values for Hooke's driveshafts, heavy series

These values can only be considered as guidelines because smooth running also depends largely on the way the joints are connected. In 1974 Hans-Joachim Kleinschmidt recommended that, for smooth running the acceleration torque M_α in (4.32) should not exceed 8% of the continuous torque [4.13]. M_d can be taken; to a first approximation, as 25% of the rated torque M_N in manufacturers' catalogues (Fig. 4.10) [4.15].

4.2.9 Critical Speed and Shaft Bending Vibration

The speeds of the shafts in machines and drives increased considerably towards the end of the 19th century, by which time electric motors and internal combustion engines achieved 800 to 1600 rpm. However compared with these Laval's steam turbine which operated at 10 000 to 30 000 rpm was a particularly bold step.

In 1887 Carl Gustav de Laval found that a rotor on a resilient shaft does not always deflect more and move from its axis of rotation as the speed increases: after passing through a "critical" speed it runs more smoothly as the speed increases. This critical speed must be passed through quickly so that no damage occurs.

Contrary to earlier opinions about the causes of vibration it can now be said that if the critical speed were only a resonance phenomenon between the rotational frequency ω_r and the inherent bending frequency ω_b of the rotor shaft, then a rotor with a vertical axis [7], where the overall centre of gravity S lies exactly on the axis of rotation (Fig. 4.16), would also resonate. In fact such an ideal rotor does not get into a critical state [4.14–4.16]. If the vertically standing rotor on its resilient shaft is struck, even when stationary it will vibrate transversely. These vibrations can be calculated from the well-known equation [1.37], p. 37, Eq. (30)]:

$$\omega_e = \sqrt{\frac{c}{m}} = \sqrt{\frac{48\,EJ}{ml^3}}, \qquad (4.37)$$

where c is the spring constant in N/m, E is the modulus of elasticity $= 2.08 \times 10^{11}$ N/m², J is the moment of inertia $= \dfrac{\pi d^4}{64}$ in m⁴, l is the length of the rotor shaft in m, m is the mass in kg. If the rotor is turning and if its centre of gravity S lies outside the axis of rotation by the amount e, the centrifugal force leads to the shaft bending outwards by an amount f. Since the centrifugal force and force are in equilibrium there is nothing to cause vibration.

In 1894 Dunkerley equated the centrifugal force

$$C = m(e + f)\,\omega^2$$

with the aligning force of the rotor $F = cf$. From this he derived the value f for the bending outwards during rotation [4.14]

---

[7] The weight is taken by the lower bearing so it does not act on the rotor.

$$m(e+f)\omega_r^2 = cf$$

$$me\omega_r^2 + mf\omega_r^2 - cf = 0$$

$$f(m\omega_r^2 - c) = -em\omega_r^2$$

$$f = \frac{e \cdot m\omega^2}{c - m\omega^2} = \frac{e \cdot \omega^2}{c/m - \omega^2}.$$

Since $\sqrt{c/m} = \omega_b$ is the inherent bending frequency of the rotor, the following equation applies

$$f = e\frac{\omega^2}{\omega_e^2 - \omega^2}. \tag{4.38}$$

For $\omega = \omega_e = \omega_k$ the denominator becomes $= 0$ and $f = \infty$, there is resonance. It can also be recognised that for $\omega < \omega_e$ f is positive and therefore $r > e$ (Fig. 4.16a) and for $\omega > \omega_e$ f is negative and therefore $r < e$ (Fig. 4.16b), i.e. after the critical speed ω_k has been exceeded the shaft centres itself automatically and thus runs more smoothly. This effect can however only be used for very rigid shafts, e.g. turbine rotors, that can be balanced very exactly. Also, the critical speed must be passed through very quickly so that no damage occurs.

A result similar to that given in (4.36) can also be derived for shafts with uniformly distributed mass [4.18]. The characteristic frequencies are obtained from the general solution for bars with their own mass, without a point mass [1.37, p. 37, Eq. (30)]:

$$\omega_i = \frac{\beta_i^2}{l^2}\sqrt{\frac{EJ}{\varrho A}} \tag{a}$$

where density $\varrho = 7.8 \cdot 10^3$ (kg/m³) and the cross-sectional area of the shaft $A = \pi/4d^2$ in m^2, with the rest as in (4.35).

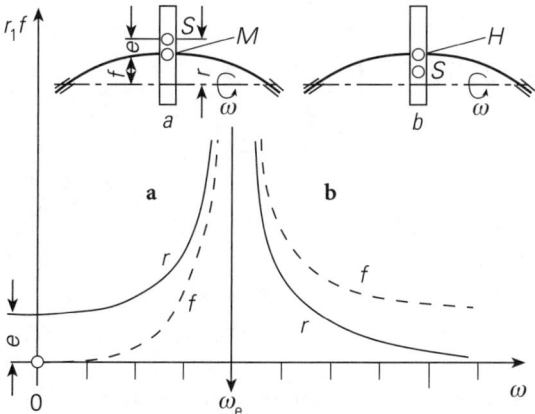

Fig. 4.16. Resonance graph for a rotating shaft with a point mass, **a** sub-critical range, **b** super-critical range. e = offset of centre of mass, f = elastic deflection under rotation, r = offset of centre of mass from the axis under rotation [4.18], M = geometrical centre of disc

Using these values and $\beta_1 = \pi$ from [1.37, p. 36, Table 4] and substituting in (4.35) gives

$$\omega_1 = \frac{\pi^2}{l^2}\sqrt{\frac{2.08 \cdot 10^{11} \cdot \pi d^4 \cdot 4}{7.8 \cdot 10^3 \cdot 64 \cdot \pi d^2}} = \frac{\pi^2 d}{l^2}\underbrace{\sqrt{1.667 \cdot 10^8}}_{0.129 \cdot 10^4}$$

$$\omega_1 = \frac{\pi^2}{l^2} 10^4 \cdot 0.129 \cdot d.$$

If one needs the 1st critical bending speed ω_1 instead of $n_{1\,\text{crit}}$, if follows from $\omega = \pi n/30$ that

$$n_1 = \frac{30}{\pi}\omega_1 = \frac{30}{\pi}\frac{\pi^2}{l^2} \cdot 1.29 \cdot 10^3 \cdot d.$$

If the dimensions are put into mm, "10^8" must be written instead of "10^5". We then have:

$$n_{kI} = 1.22 \cdot 10^8 \cdot \frac{d}{l^2} \text{ rpm} \qquad \text{for a solid shaft} \qquad (4.39a)$$

$$n_{kI} = 1.22 \cdot 10^8 \frac{\sqrt{D^2 + d^2}}{l^2} \text{ rpm} \qquad \text{for a fixed tubular shaft} \qquad (4.39b)$$

$$n_{kI} = (0.7 \to 0.9) \cdot 1.22 \cdot 10^8 \frac{\sqrt{D^2 + d^2}}{l^2} \text{ rpm} \quad \begin{array}{l}\text{for a plunging}\\ \text{tubular shaft}\end{array} \qquad (4.39c)$$

(0.7 – short tube; 0.9 – long tube. Measured between joint centres)
The maximum permissible speed

$$n_{\max} = 0.8 \cdot n_{kI} \text{ rpm} \qquad (4.39d)$$

For a fixed tubular shaft

$$J = \frac{\pi}{64}(D^4 - d^4) = \frac{\pi}{64}(D^2 - d^2)(D^2 + d^2)$$

$$A = \frac{\pi}{4}(D^2 - d^2).$$

Manufacturers have found that in general practice the critical bending speed of the first mode is very important (Fig. 4.15).
In order to avoid vibrations a safety margin must be maintained

$$n_{\max.\,\text{permissible}} = A \cdot n_{kI} \text{ rpm}$$

$$A = 0.75 \text{ to } 0.85 \text{ depending on balancing quality}$$

4.2 Hooke's Joints and Hooke's Jointed Driveshafts

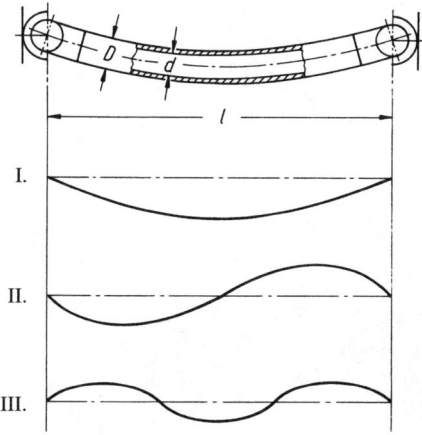

Fig. 4.17. 1st, 2nd and 3rd modes of vibration

Modes of shaft vibrations

Apart from the 1st mode (at the critical speed n_{kI}) other higher modes occur. The first three are shown in Figure 4.17.

For Hooke's jointed driveshafts higher modes are not of interest because it is not possible to pass through the critical speed n_{kI}. This is due to the play of the joints and the splines of the plunging part which do not allow exact balancing.

Modes of critical revolutions

The mode indicates the impulse coefficient of the excitation force per revolution. Dangerous vibrations only occur if there is resonance between the mode coefficient ω_e and the excitation impulse coefficient. The critical speed thus becomes:

$$n_{\text{kin}} = \frac{\text{mode coefficient } n_{ki}}{\text{impulse coefficient } n/\text{revolution}}$$

Two cases, as a rule, relate to the Hooke's jointed driveshaft (Fig. 4.18):

1. Imbalance excitation
   This occurs with an impulse coefficient of 1;
   thus $n_{kI1} = \dfrac{n_{kI}}{1} = n_{kI}$ [rpm]

2. Kinematic excitation
   As a result of the Hooke's joint kinematics under articulation, the excitation force F_K occures twice per revolution, that is, with the impulse coefficient 2;
   thus $n_{kI2} = \dfrac{n_{kI}}{2}$ [rpm]

Dangerous vibrations can however only occur here if certain limit values for the articulation angle β and torque M_1 and therefore M_b are exceeded. The manufacturers provide information about this.

Even higher orders can be stimulated by connected units such as piston engines.

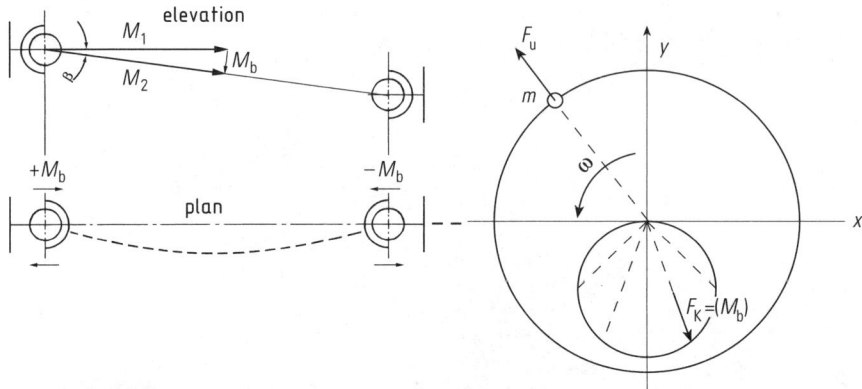

Fig. 4.18. Excitation force ratios on the Hooke's jointed driveshaft.
M_1, M_2 = torques on the driveshaft
M_b = bending reaction torque, F_u = imbalance excitation force
$F_k = f(M_b)$ = kinematic excitation force, β = articulation angle

4.2.10 Double Hooke's Joints

The double Hooke's joint comes about if the middle part between two single Hooke's joints is rigid and very short. Constant velocity conditions are almost fulfilled in the quasi-homokinetic joint (Fig. 1.11b) and completely fulfilled in the strictly homokinetic joint (Fig. 1.11c) because of their intrinsic centring. Uncentred joints (Fig. 1.11a) can only be used with external centring, as in the steer axle of a commercial vehicle (Fig. 4.20). The centring ensures that the axes of the input and output shafts always intersect and that the difference between the individual articulation angles $\Delta\beta = \beta_1 - \beta_2$ does not become too great.

In Fig. 4.21 shaft *1* is firmly mounted in the steering knuckle. The centre point G of the joint coincides with the point of rotation D which lies on the kingpin axis. If the output shafts rotate in the direction of the arrow through the angle β from the inline position AD towards A'D, a progressively increasing difference $\Delta\beta$ develops. At the same time the point B moves by the amount x to B'. The designer must know the amount of displacement x needed and make provision for it on the movable shaft *2*. It can be calculated using Fig. 4.21 with the following equations:

triangle A'DF

$$\sin\beta = A'F/c \Rightarrow A'F = c\sin\beta, \tag{a}$$

$$\cos\beta = DF/c \Rightarrow DF = c\cos\beta. \tag{b}$$

Triangle A'B'F

$$\sin\beta_2 = A'F/2c \Rightarrow A'F = 2c\sin\beta_2. \tag{c}$$

4.2 Hooke's Joints and Hooke's Jointed Driveshafts

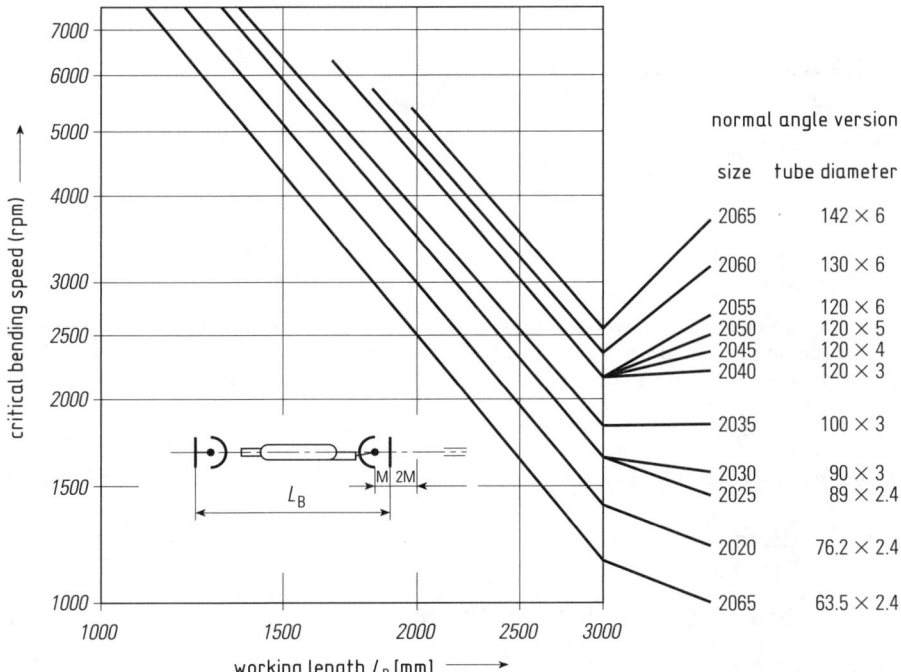

Fig. 4.19. Critical bending speeds for vehicle driveshafts, as a function of the working length L_B

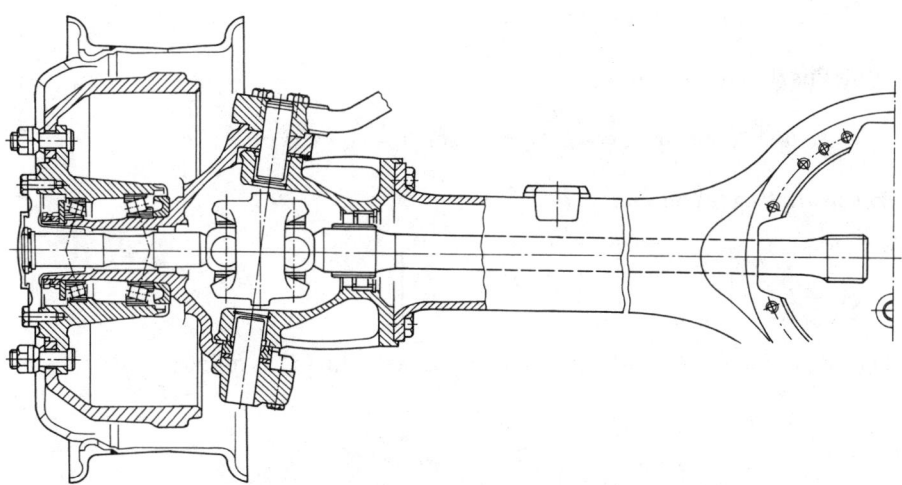

Fig. 4.20. Externally centred double Hooke's joint with quasi-homokinetic characteristics because it is guided by the steer axle of a lorry. GWB design

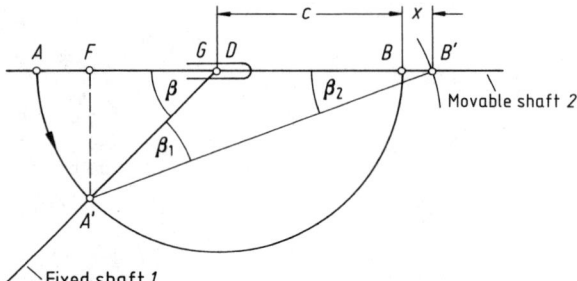

Fig. 4.21. Kinematics of the quasi-homokinetic double Hooke's joint. The centre of the joint G and the point of rotation D coincide. $\beta_{1,2}$ individual articulation angles; x displacement, A and B cross centres; c distance of cross centres A and B from the joint centre G

From (a) and (c) it follows that:

$$2c \sin\beta_2 = c \sin\beta \Rightarrow \sin\beta_2 = 1/2 \cdot \sin\beta \tag{d}$$

$$\cos\beta_2 = \sqrt{1 - \frac{1}{4}\sin^2\beta}. \quad [1.10, \text{Sec. } 3.1.2.9, \text{S. } 85]. \tag{e}$$

Triangle $A'B'F$

$$\cos\beta_2 = \frac{B'F}{2c} = \frac{DF + (c+x)}{2c}$$

and with (b)

$$(c+x) + c\cos\beta = 2c\cos\beta_2.$$

From this the compensating value x

$$x = 2c\cos\beta_2 - c\cos\beta - c = c[2\cos\beta_2 - (1 + \cos\beta)].$$

By substituting (e) one gets

$$x = c\left[2\sqrt{1 - \frac{1}{4}\sin^2\beta} - (1 - \cos\beta)\right]. \tag{4.40}$$

The angular difference $\Delta\beta = \beta_1 - \beta_2$ is calculated from (d) as follows

$$\sin\beta_2 = 1/2 \cdot \sin\beta$$

$$\beta_2 = \arcsin\left(\frac{1}{2} \cdot \sin\beta\right)$$

$$\beta_1 + \beta_2 = \beta$$

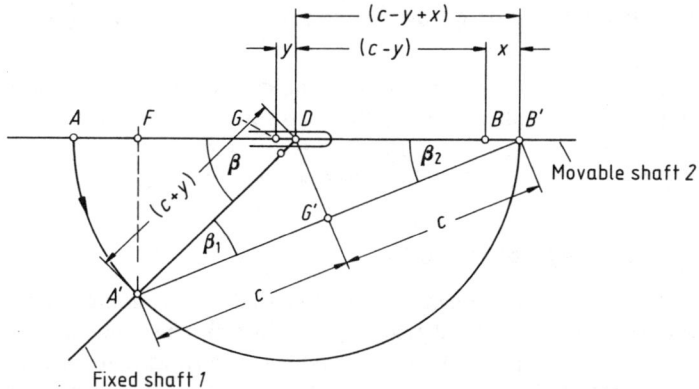

Fig. 4.22. Kinematics of the quasi-homokinetic double Hooke's joint with its joint centre G offset from the axis of rotation D by the compensating value y

$$\beta_2 = \beta - \beta_1 = \beta - \arcsin\left(\frac{1}{2} \cdot \sin\beta\right)$$

$$\beta_2 - \beta_1 = \beta - \arcsin\left(\frac{1}{2} \cdot \sin\beta\right) - \arcsin\left(\frac{1}{2} \cdot \sin\beta\right)$$

$$\Delta\beta = \beta - 2\arcsin\left(\frac{1}{2} \cdot \sin\beta\right). \quad (4.41)$$

The angular error can be reduced if the double joint centre G is moved from the pivot point D by a compensating value y to the fixed side. The value of y depends on the distance c and the angle β. This gives rise to constant velocity operation, for one angle β_y. β_y must be selected so that the smallest differences of the individual angles β_1 and β_2 occur for the intended operating range.

In Fig. 4.22 $A'B'D$ then becomes an isosceles triangle, i.e. $\beta_1 = \beta_2 = \beta_y/2$. Dropping a perpendicular from D to G' it follows that, for the angle β_1,

$$\cos\beta_1 = \cos(\beta_y/2) = \frac{c}{c+y}.$$

Compensating value y

$$y = \left(\frac{c}{\cos\beta_y/2}\right) - c = c\left(\frac{1}{\cos\beta_y/2} - 1\right). \quad (4.42)$$

The value of y depends on the angle β_y at which constant velocity operation is required and the distance between the joint centres.

4.3 Forces on the Support Bearings of Hooke's Jointed Driveshafts

Apart from the torsional and bending vibrations in the centre part of a Hooke's jointed driveshaft, the two joints also cause cyclic forces and torques which are transmitted to their support bearings. These forces are analysed by examining the interaction of forces within the individual joints [4.19; 4.20].

4.3.1 Interaction of Forces in Hooke's Joints

In Fig. 4.23a the yoke *l* lies in the plane formed by the two shafts *1* and *2* and rotates in an anti-clockwise direction (see Sect. 1.2.1, in-phase position). The angle of rotation φ_1 of shaft *1* is measured from this position. Two torques now have to be considered: the input torque M_1 and the output torque M_2. They load the cross with forces P_1 and P_2 respectively and give rise to couples $P_1 h$ and $P_2 h$. Equilibrium must however prevail on the cross. Consequently additional forces Z must also act; together with the forces P_2 on yoke 2, these form a couple Rh in the plane of the cross since it can only transmit a torque in this plane. In the position shown in Fig. 4.23 additional forces are only possible in the plane of yoke 2 and together with the forces P_2 form the couple Rh. In Fig. 4.23 is

$$P_2 = R \cos \beta, \tag{a}$$

$$Z = R \sin \beta. \tag{b}$$

If (a) is substituted, one obtains the output torque

$$M_2 = P_2 h = Rh \cos \beta. \tag{c}$$

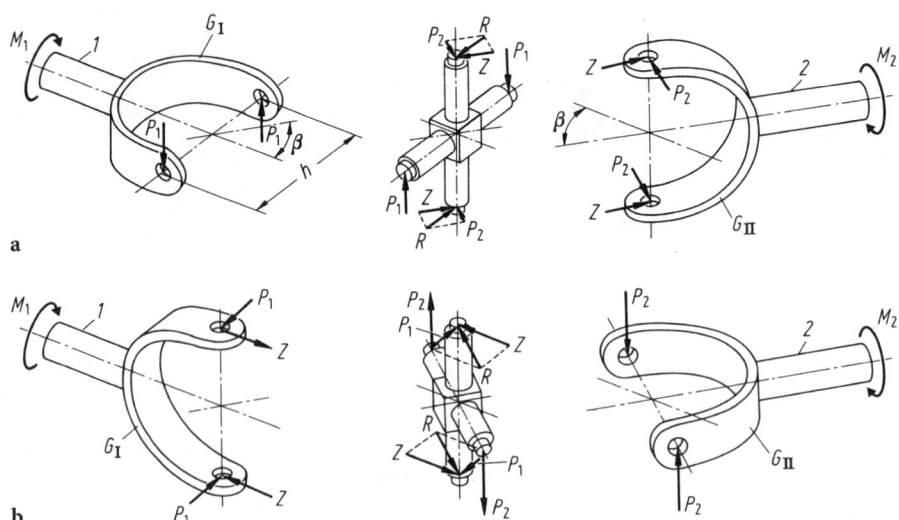

Fig. 4.23a, b. Interaction of forces in the Hooke's joint for the two main positions. **a** Starting position $\varphi_1 = 0$, **b** $\varphi_1 = \pi/2$

4.3 Forces on the Support Bearings of Hooke's Jointed Driveshafts

If (b) is substituted, one obtains the secondary couple

$$M_Z = Zh = Rh \sin \beta. \tag{d}$$

It can be seen from Fig. 23a, that

$$Rh = P_2 h = M_1. \tag{e}$$

It follows then from (c) and (d) with (e) for the in-phase position $\varphi_1 = 0$ with $M_d = M_1$

$$M_2 = M_d \cos \beta, \tag{4.43}$$

$$M_Z = M_d \sin \beta. \tag{4.44}$$

In Fig. 4.23b the yoke 2 lies in the plane formed by shafts 1 and 2. Yoke 1 has rotated by the angle $\varphi_1 = 90°$. The input torque M_1 acts again to give the two forces P_1 on the trunnion cross. The reaction torque M_2 generates the two forces P_2 on the cross which, without additional forces, is not in equilibrium. Additional forces are now possible only in the plane of yoke 1 and together with the forces P_1 give the couple $R \cdot h$.

In Fig. 4.23b is

$$P_1 = R \cos \beta, \tag{a}$$

$$Z = R \sin \beta. \tag{b}$$

If (a) is substituted, the output torque is

$$M_2 = Rh = \frac{P_1}{\cos \beta} h, \tag{c}$$

and since $P_1 h = M_1 = M_d$

$$M_2 = \frac{M_d}{\cos \beta}. \tag{4.45}$$

If (b) is substituted the secondary couple is

$$M_2 = Zh = Rh \sin \beta,$$

and with $R = P_1/\cos \beta$ from (c) we get

$$M_Z = \frac{P_1}{\cos \beta} h \sin \beta = M_d \tan \beta. \tag{4.46}$$

Equations (4.45) and (4.46) apply for the orthogonal position $\varphi_1 = 90°$ of the input yoke 1.

In the case of the double Hooke's joint in the W-configuration the loading is mirrored by the output joint 2. The forces acting are all the same size but, except for the additional forces Z, have opposite signs. Since the double Hooke's joint has a very short, rigid middle part, analysis of the second joint is not necessary.

Cyclic bearing forces, the maximum values of which have yet to be determined, do however develop in the support bearings from the secondary couples M_{Z1} and M_{Z2} from joints *1* and *2*.

4.3.2 Forces on the Support Bearings of a Driveshaft in the W-Configuration

Calculation of the bearing forces *A* and *B* [4.13] starts in the orthogonal position of the yoke 2 (Fig. 4.24a) with the angle of rotation $\varphi_1 = 0°$ on drive yoke *1*. The secondary couple M_Z which acts as a bending moment on the intermediate part (Fig. 4.24b) must be supported by the force $Q = M_Z/l$ in the other joint and brings about an equal reaction force Q' in its own joint. In this way the forces $(Q' + Q)$ act on each cross.

Taking moments about B_1 the following applies for the input joint

$$A_1 a_1 = (Q_1' + Q_2) b_1 \Rightarrow A_1 = (Q_1' + Q_2)\frac{b_1}{a_1}, \qquad \text{(a)}$$

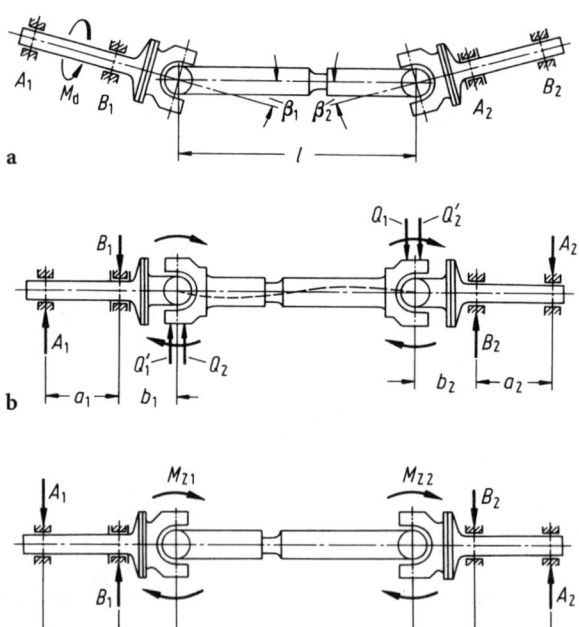

Fig. 4.24a–c. Forces on the support bearings of a Hooke's jointed driveshaft in a W-configuration. **a** Side view, **b** plan view with yoke angle $\varphi_1 = 0°$, **c** plan view with yoke angle $\varphi_1 = 90°$

4.3 Forces on the Support Bearings of Hooke's Jointed Driveshafts

and, by taking moments about A_1

$$B_1 a_1 = (Q_1' + Q_2)(a_1 + b_1) \Rightarrow B_1 = (Q_1' + Q_2)\frac{a_1 + b_1}{a_1}. \tag{b}$$

From (4.43) we know that the torque in the intermediate part is

$$M_2 = M_d \cos\beta_1.$$

In order to calculate the bearing forces the joints must be considered, from the point of view of the intermediate part, as being loaded with the torque M_2. Therefore, from (4.46), the following applies for the orthogonal position of yoke 2:

$$Q_1 = Q_1' = \frac{M_{Z1}}{l} = \frac{M_2 \tan\beta_1}{l} = \frac{M_d \cos\beta_1 \tan\beta_1}{l}, \tag{c}$$

$$Q_2 = Q_2' = \frac{M_{Z2}}{l} = \frac{M_2 \tan\beta_2}{l} = \frac{M_d \cos\beta_1 \tan\beta_2}{l}. \tag{d}$$

If (c) and (d) are substituted in (a) and (b) the resultant bearing forces on the input joint are

$$A_1 = \frac{M_d \cos\beta_1}{l} \frac{b_1}{a_1}(\tan\beta_1 + \tan\beta_2), \tag{4.47a}$$

$$B_1 = \frac{M_d \cos\beta_1}{l} \frac{a_1 + b_1}{a_1}(\tan\beta_1 + \tan\beta_2). \tag{4.47b}$$

The following apply for the output joint:

$$A_2 = \frac{M_d \cos\beta_1}{l} \frac{b_2}{a_2}(\tan\beta_1 + \tan\beta_2), \tag{4.47c}$$

$$B_2 = \frac{M_d \cos\beta_1}{l} \frac{a_2 + b_2}{a_2}(\tan\beta_1 + \tan\beta_2). \tag{4.47d}$$

The in-phase position of yoke 2 corresponds to an angle of rotation $\varphi_1 = 90°$ of the input yoke 1 (Fig. 4.24c). In this position the cross cannot transmit any additional torques to the intermediate part.

For the *input* joint, by using (4.47) and by taking moments about B_1, we get

$$A_1 a_1 = M_{Z1} = M_d \tan\beta_1 \rightarrow A_1 = \frac{M_d \tan\beta_1}{a_1},$$

and by taking moments about A_1, we get

$$B_1 a_1 = M_{Z1} = M_d \tan\beta_1 \Rightarrow B_1 = \frac{M_d \tan\beta_1}{a_1},$$

that is

$$A_1 = B_1 = \frac{M_d \tan \beta_1}{a_1}. \tag{4.48}$$

Figure 4.24c shows that even when the articulation angle $\beta_1 = 90°$ the whole torque M_d still loads the support bearing.

For the *output* joint we get, by using (4.44) and $M_2 = M_d/\cos \beta_1$ from (4.45), by taking moments about B_2

$$A_2 a_2 = M_{Z2} = M_2 \sin \beta_2 = \frac{M_d}{\cos \beta_1} \sin \beta_2 \Rightarrow A_2 = \frac{M_d \sin \beta_2}{a_2 \cos \beta_1},$$

by taking moments about A_2

$$B_2 a_2 = M_{Z2} = M_2 \sin \beta_2 = \frac{M_d}{\cos \beta_1} \sin \beta_2 \Rightarrow B_2 = \frac{M_d \sin \beta_2}{a_2 \cos \beta_1},$$

that is

$$A_2 = B_2 = \frac{M_d \sin \beta_2}{a_2 \cos \beta_1}. \tag{4.49}$$

4.3.3 Forces on Support Bearings of a Driveshaft in the Z-Configuration

The forces on the support bearings when the individual joints are in a Z-configuration can be calculated using (4.44) to (4.47) as for the W-configuration in Sect. 4.3.2.

For the angle of rotation $\varphi_1 = 0°$ (Fig. 4.25a and b)

$$A_1 = M_d \frac{\cos \beta_1}{l} \frac{b_1}{a_1} (\tan \beta_1 - \tan \beta_2), \tag{4.50a}$$

$$B_1 = M_d \frac{\cos \beta_1}{l} \frac{a_1 + b_1}{a_1} (\tan \beta_1 - \tan \beta_2), \tag{4.50b}$$

$$A_2 = M_d \frac{\cos \beta_1}{l} \frac{b_2}{a_2} (\tan \beta_1 - \tan \beta_2), \tag{4.50c}$$

$$B_2 = M_d \frac{\cos \beta_1}{l} \frac{a_2 + b_2}{a_2} (\tan \beta_1 - \tan \beta_2). \tag{4.50d}$$

The following applies in Fig. 4.25c for the angle of rotation $\varphi_1 = 90°$:

$$A_1 = B_1 = A_2 = B_2 = M_d \frac{\tan \beta_1}{a_1}. \tag{4.51}$$

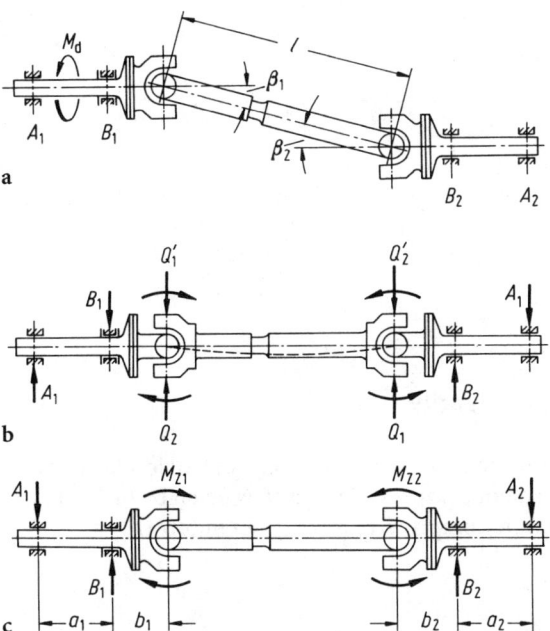

Fig. 4.25 a–c. Forces on the support bearings of a Hooke's jointed driveshaft in a Z-configuration. **a** Side view. **b** plan view with yoke angle $\varphi_1 = 0°$. **c** plan view with yoke angle $\varphi_1 = 90°$

4.4 Ball Joints

The transmitting elements in ball joints are balls which run independently along guide tracks. It is assumed that the compressive force P is evenly distributed among all m-active balls. This number of balls m depends on whether the joint is divided according to the Rzeppa principle (concentrically) or the Weiss principle (radially), see Fig. 4.26.

The dividing planes between the input and output members are then either concentric with (Fig. 4.26a) or radial to the longitudinal axis of the joint (Fig. 4.26b). In the Rzeppa arrangement, all the balls act for each direction of rotation whereas with the Weiss arrangement only half the balls act. In the universal torque equation for joints (4.16), for the Rzeppa arrangement $m = z$, and for the Weiss arrangement $m = z/2$.

In addition to the way in which the planes run relative to the longitudinal axis of the joint, the shape of the tracks in this direction is also critical; this is the "effective geometry" in Sects. 2.3.1 and 2.3.2.

Where there are curved tracks the joint can only articulate, there can be no plunging. Straight and helical tracks allow articulation *and* plunge of the joint. Ball joints can therefore be further categorised as fixed joints and plunging joints.

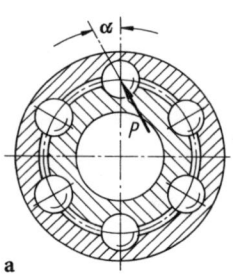
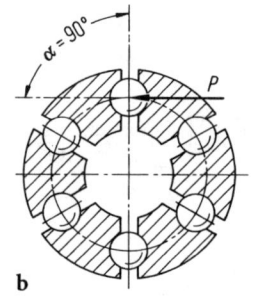

Fig. 4.26a, b. Number of active balls m in joints with z balls.
a Rzeppa type $m = z$,
b Weiss type $m = z/2$

4.4.1 Static and Dynamic Torque Capacity

As with Hooke's joints, the dependence of the torque capacity of ball joints on the effective component P_x of the equivalent compressive force P is given by (4.16a to c).

The permissible static or dynamic compressive force given in (4.7a)

$$P_{\text{perm}} = \left(\frac{p_0}{c_p}\right)^3 d^2 = k d^2$$

is inserted in (4.16b) to give

$$P_x = \left(\frac{p_0}{c_p}\right)^3 d^2 \sin\alpha \cos\gamma = k d^2 \sin\alpha \cos\gamma.$$

For Rzeppa-joints with meridian races ($\gamma = 0$) and the working number of balls $m = z$, the static and dynamic torque capacity is transformed to:

$$M_0 = k_0 z d^2 R \sin\alpha \tag{4.52a}$$

$$M_d = k_d z d^2 R \sin\alpha \tag{4.52b}$$

The coefficient of conformity c_p in this equation is calculated from the principal curvatures using the Hertzian coefficient (3.15), from the principal curvatures of the surfaces of the joint elements and balls which are pressed together.

$$\cos\tau = \frac{(\varrho_{11} - \varrho_{12}) + (\varrho_{21} - \varrho_{22})}{\Sigma\varrho}.$$

The two curvatures ϱ_{11} and ϱ_{12} of the balls are the same and are equal to the reciprocal of the ball radius r_1

$$\varrho_{11} = \varrho_{12} = \frac{1}{r_1} = \frac{1}{d/2} = \frac{2}{d}.$$

4.4 Ball Joints

The curvatures of the track in the longitudinal direction, ϱ_{21} and ϱ'_{21} [8], are determined as shown in Fig. 4.27 from the pressure angle α, angle of tilt ε, opening angle $\delta = 2\varepsilon$, joint radius R and distance apart b.

$b = 0$: the generating centres E and E' of the tracks lie on the joint axis,
$b > 0$: barrel-shaped joints according to I. R. Phillips/H. Winter [4.24],
$b = \infty$: Straight tracks.

The following apply in Fig. 4.27.

$$\cos \varepsilon = \frac{MF}{ME} \Rightarrow ME = \frac{MF}{\cos \varepsilon}. \tag{a}$$

$$\cos \alpha = \frac{ME}{MD}.$$

Equation (a) substitution gives

$$\cos \alpha = \frac{MF}{MD \cos \varepsilon} \Rightarrow MD = \frac{MF}{\cos \alpha \cos \varepsilon}, \tag{b}$$

$$MF = (R + b). \tag{c}$$

$$r_{21} = MD - d/2 = \frac{MF}{\cos \alpha \cos \varepsilon} - \frac{d}{2} = \frac{R + b}{\cos \alpha \cos \varepsilon} - \frac{d}{2}, \tag{d}$$

$$-r'_{21} = MD - d/2 = \frac{MF}{\cos \alpha \cos \varepsilon} + \frac{d}{2} = \frac{R + b}{\cos \alpha \cos \varepsilon} + \frac{d}{2}, \tag{e}$$

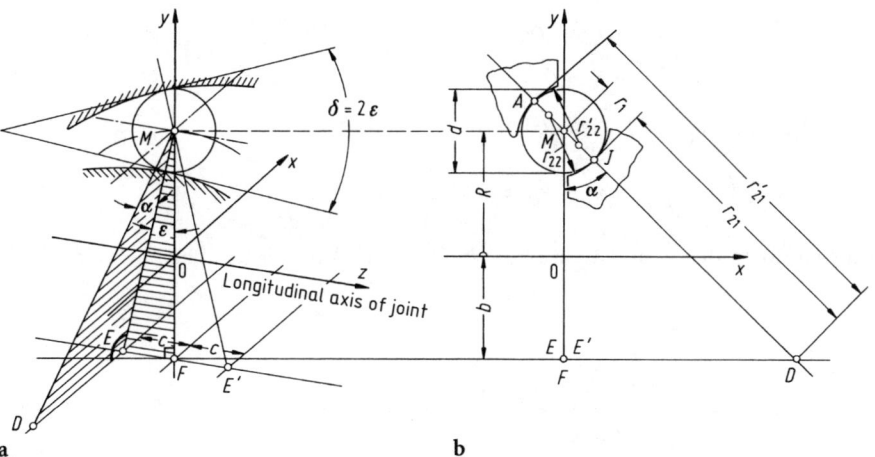

Fig. 4.27 a, b. General position of the generating centres of the tracks. **a** 3-D view. E and E' generating centres of the tracks, c offset; **b** cross-section. Enlargement of the radius of curvature r_{21} of the tracks for a pressure angle α and the angle of tilt ε. A outer contact point; J inner contact point: r_{22} radius of curvature of the tracks

---

[8] Values having a prime suffix such as ϱ'_{21}, related to the outer contact point A in Fig. 4.27b.

In Sect 1.5.2 the conformity χ_L is defined as the ratio r_1/r_{21} so that the radius r_{21} and the ball diameter can be put into the equation

$$\frac{r_1}{r_{21}} = \frac{d/2}{r_{21}} = \chi_L \Rightarrow \frac{1}{r_{21}} = \frac{2\chi_L}{d} = \varrho_{21}. \tag{f}$$

If (d) is substituted in (f) the conformity χ_L in the longitudinal direction can be worked out as

$$\frac{2\chi_L}{d} = \frac{1}{r_{21}} = \frac{1}{\dfrac{R+b}{\cos\alpha\cos\varepsilon} \mp \dfrac{d}{2}}$$

$$\chi_L = \frac{d}{2\left(\dfrac{R+b}{\cos\alpha\cos\varepsilon} \mp \dfrac{d}{2}\right)} = \frac{d}{\dfrac{2(R+b)}{\cos\alpha\cos\varepsilon} \pm d}. \quad ^9 \tag{4.53}$$

The curvatures of the track in the transverse direction ϱ_{22} and ϱ'_{22} are generally the same. The radii r_{22} and r'_{22} are made larger than the ball radius by ψ, the reciprocal of the conformity to give $r_{22} = \psi d/2$. One can get a good idea of the value for the conformity from (1.16)

$$\chi_Q = \frac{r_K}{-r_R} = \frac{r_K}{-\psi\, r_K} = \frac{1}{-\psi} \tag{4.54}$$

By taking the sign of the curvature into account, one gets for example a conformity $\chi_Q = 1/-1.03 = -0.971 = 97.1\%$ (see Sect. 1.5.2).

$\Sigma\varrho$ in the denominator of (3.15) can now be worked out as

$$\Sigma\varrho = \varrho_{11} + \varrho_{12} + \varrho_{21} + \varrho_{22} = 2/d + 2/d + \frac{2\chi_L}{d} + \frac{2\chi_Q}{d}$$
$$= 2/d\,(2 + \chi_L + \chi_Q). \tag{4.55}$$

In addition the value $d\,\Sigma\varrho$ is needed in (4.3) and (4.5)

$$d\,\Sigma\varrho = d\,\frac{2}{d}\,(2 + \chi_L + \chi_Q) = 2\,(2 + \chi_L + \chi_Q). \tag{4.56}$$

If the conformities χ_L and χ_Q are known, it follows from (3.15) that

$$\cos\tau = \frac{\overbrace{\left(\dfrac{2}{d} - \dfrac{2}{d}\right)}^{0} + \left(\dfrac{2\chi_L}{d} - \dfrac{2\chi_Q}{d}\right)}{\dfrac{2}{d}(2 + \chi_L + \chi_Q)} = \frac{\chi_L - \chi_Q}{2 + \chi_L + \chi_Q}. \tag{4.57}$$

The signs of the conformities χ_L and χ_Q which are governed by the Hertzian convention for the signs of the curvatures ϱ_{21} and ϱ_{22} should be noted here. [10]

---

[9] $\cos\varepsilon \approx \cos 9° = 0.9877$; the minus sign applies for the inner contact point J, as well as for the radius r_{21} of curvature. χ_L alongside χ_Q transverse curvature of the track.

4.4 Ball Joints

Table 4.5. Radial bearing connection forces of Constant Velocity Joints (CVJ) with shafts in one plane, from Werner KRUDE

in Z-articulation   in W-articulation

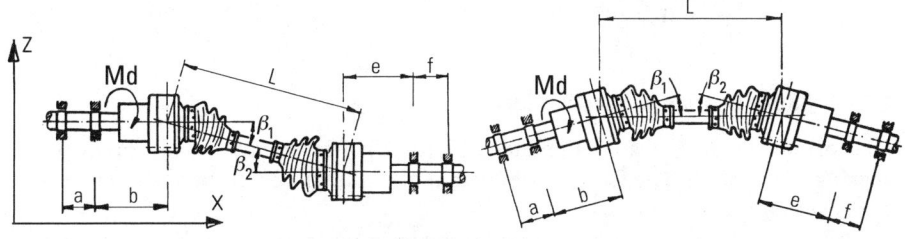

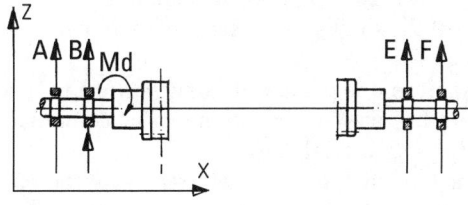

Z-articulation

$$A = \frac{M_d}{a}\left[\tan\frac{\beta_1}{2} + \frac{b}{L}\left(\tan\frac{\beta_2}{2} - \tan\frac{\beta_1}{2}\right)\right]$$

$$B = \frac{M_d}{a}\left[\tan\frac{\beta_1}{2} + \frac{a+b}{L}\left(\tan\frac{\beta_2}{2} - \tan\frac{\beta_1}{2}\right)\right]$$

$$F = \frac{M_d}{f}\left[\tan\frac{\beta_2}{2} - \frac{e}{L}\left(\tan\frac{\beta_2}{2} - \tan\frac{\beta_1}{2}\right)\right]$$

$$E = \frac{M_d}{f}\left[\tan\frac{\beta_2}{2} - \frac{e+f}{L}\left(\tan\frac{\beta_2}{2} - \tan\frac{\beta_1}{2}\right)\right]$$

W-articulation

$$A = \frac{M_d}{a}\left[\tan\frac{\beta_1}{2} - \frac{b}{L}\left(\tan\frac{\beta_2}{2} + \tan\frac{\beta_1}{2}\right)\right]$$

$$B = \frac{M_d}{a}\left[\tan\frac{\beta_1}{2} - \frac{a+b}{L}\left(\tan\frac{\beta_2}{2} + \tan\frac{\beta_1}{2}\right)\right]$$

$$F = \frac{M_d}{f}\left[\tan\frac{\beta_2}{2} - \frac{e}{L}\left(\tan\frac{\beta_2}{2} + \tan\frac{\beta_1}{2}\right)\right]$$

$$E = \frac{M_d}{f}\left[\tan\frac{\beta_2}{2} - \frac{e+f}{L}\left(\tan\frac{\beta_2}{2} + \tan\frac{\beta_1}{2}\right)\right]$$

Driveshaft guidance with the same articulation angles: $\beta_1 = \beta_2$

Z-articulation

$$B = \frac{M_d}{a} \times \tan\frac{\beta}{2} = -A$$

$$E = \frac{M_d}{f} \times \tan\frac{\beta}{2} = -F$$

W-articulation

$$A = -\frac{M_d}{a} \times \tan\frac{\beta}{2}\left[1 - \frac{2b}{L}\right]$$

$$B = \frac{M_d}{a} \times \tan\frac{\beta}{2}\left[1 - 2\frac{a+b}{L}\right]$$

$$F = \frac{M_d}{f} \times \tan\frac{\beta}{2}\left[1 - \frac{2e}{L}\right]$$

$$E = -\frac{M_d}{f} \times \tan\frac{\beta}{2}\left[1 - 2\frac{f+e}{L}\right]$$

The Hertzian coefficient cos τ is the key to the elliptical coefficients μ, ν, $\mu\nu$ and $2k/\pi\mu$, see Sect. 3.8, Table 3.2.

The coefficient of conformity can be worked out from (4.5)

$$c_p = \frac{858}{\mu\nu} \sqrt[3]{(d\Sigma\varrho)^2} \quad (\text{N/mm}^2)^{2/3}. \tag{4.58}$$

4.4.1.1 Radial bearing connection forces

The torque transfer of an articulated CV Joint generates additional forces in the bearing of the shaft. The forces are constant over the full revolution, also at a three-dimensional order.

The size and the direction of stress at the bearings depend on the torque, the articulation, the length of the main drive shaft L and the bearing distances. There is further to consider the rotation direction of the shaft and the signs of their articulation angles.

The bearing forces of a jointed driveshaft in one plane work only vertical to this plane. The forces of this driveshaft (Table 4.5) in W-articulation can be computed with the formulae for Z-articulation, if $\beta_2 = -\beta_2$ is related to.

If the permissible specific loading k is known[11], the Hertzian stress can be obtained from (4.7b)

$$p_0 = c_p \sqrt[3]{k}.$$

In the coefficient of conformity c_p from (4.5) has been found, the permissible static surface stress can also be calculated from (4.9) using ball bearing theory and compared with the values of J. W. Macielinski[12] (Table 4.5). With (4.9) it is possible to compare the torque capacity of all non-articulated joints provided that the permanent deformation $\delta_b/d \leq 0.0001 = 10^{-4}$.

4.4.2 The ball-joint from the perspective of rolling and sliding bearings

The ball-joint a in Fig. 4.28 could be thought of as a combination of the 4-point ball bearing b and the double radial bearing c. The 4-point ball bearing transmits the compressive force P through the balls via two opposing points. The innercentring proposed by Rzeppa is undertaken by the ball cage. The cage is centred with its two spherical surfaces (inner and outer) on the two spherical surfaces of the inner and outer joint bodies. The cage can then be seen as a double radial bearing.

---

[10] If the centre of curvature M is then according to HERTZ, Fig. 5.74 [1.21]

$\varrho = \dfrac{1}{+r} = (+)$ $\varrho = \dfrac{1}{-r} = (-)$

inside the body   outside the body

[11] k is the specific load value.
[12] GKN Transmissions Ltd.

4.4 Ball Joints

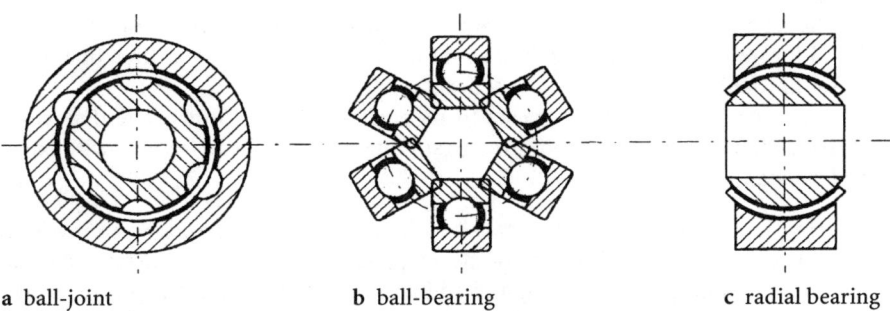

a ball-joint b ball-bearing c radial bearing

Fig. 4.28. Theoretical formation of a ball-joint from ball and radial bearings

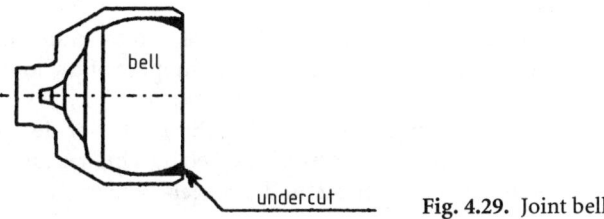

Fig. 4.29. Joint bell

Since, however, the outer and inner tracks cross over one another during the rotational articulation a second centring of the inner joint body occurs here via the balls. However the double radial bearing is already undertaking this task. The classic case of redundant centring is the result. This joint can only work properly with a fixed, common, precise joint centre. Under rotation and torque the two centrings work against each other and this can lead to considerable internal distortion. Eccentricities, wobbling and out-of-roundness of the bell and the inner race can cause radial stresses that constantly alter the position of the cage between these two joint parts. There are several patented solutions to avoid radial centring.

Radial centring through the cage can be avoided. The outer surface of the ball cage must be kept with clearance from the outer body (Fig. 4.62). Then, only the balls in the intersecting grooves centre the joint. The ball cage sliding on the inner race must be able to follow this ball centring.

In addition to the radial centring, axial centring is however also necessary. Because the ball cage has *no undercut* it can only absorb tensile forces.

4.4.3 A mutual, accurate joint centre

The two guidance centres don't lie in a common point owing to tolerances in manufacturing (excentrism, staggering, ovalism, distortion after hardening, indexing error and the required radial clearance). There are three ways to reach the accurate common centre of the joint:

1. hard grinding or a combination of hard milling (tracks) and hard turning (spheres).

2. get a separate undercut bell. To minimize machining, forged components are used for inner and outer race with some stock. This is removed as final operation after hardening, to eliminate distortion.
3. the forging process for the tracks to its ends shape (UF joint). The distortion through hardening is adjusted through sorting out a selected assembly with equal tolerances.

During history of the constant velocity joint, various proposals were made to overcome the generic disadvantage of the Rzeppa principle: the over-determination.

Figure 4.30 shows a proposal of a fixed ball joint, having an undercut free shape also in the cage sphere of the disk type joint outer race. For full guidance of the cage a cup shape component is added. By a bold connection of such undercut free outer race to the stem element, a full axial determination of the joint components one against the other can be achieved.

A further evolution of this concept is shown in Fig. 4.31. Here also an undercut free layout of the cage could be achieved, enabled by an indirect support of the joint inner race to the stem.

Though functional improvements in terms of overcoming the joint over-determination are achieved, additional manufacturing costs are caused by a higher number of components and a necessary connection method between outer race and stem.

4.4.3.1 Constant Velocity Ball Joints based on Rzeppa principle

The main application of Constant Velocity Ball Joints based on Rzeppa principle is the wheel side or outboard joint of a front wheel driven vehicle. The development of this joind type seemed to be completed by A. Rzeppa and B. Stuber (Fig. 1.25/26, 1.28). Nevertheless still today there an ongoing development need for cost reduction, higher durability, larger articulation angles and for a better noise behaviour.

Especially to achieve a higher articulation angle of the Rzeppa type joint, but also to allow manufacturing methods with a reduction of machining time, concepts were proposed with an undercut free shape of the joint components.

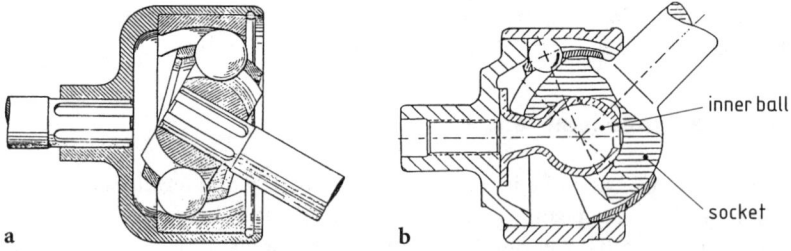

Fig. 4.30. Comparison of two ball joint shapes. **a** Earlier version by Rzeppa 1927, **b** Joint with undercut-free cage after Werner Jacob 1992 (German Patent 4208786); the driveshaft forms a socket for the inner ball and guides the cage, which is open at one end and undercut, on an outer ball surface

4.4 Ball Joints

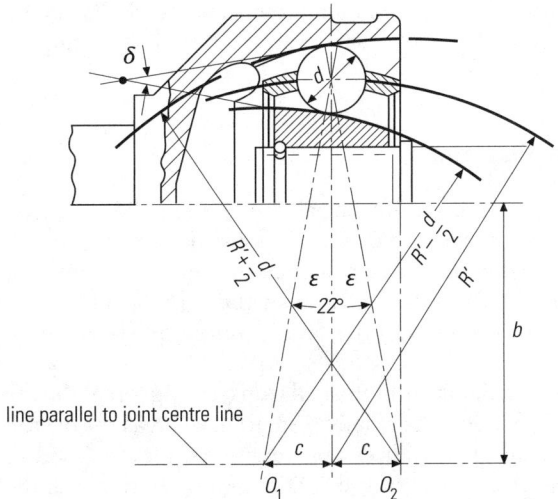

Fig. 4.31. Eccentric arrangement of the generating centres O_1 and O_2 with continuous tracks as far as the front surface of the bell, after H.-H. Welschof/E. Aucktor (German Patent 2252827/1972)

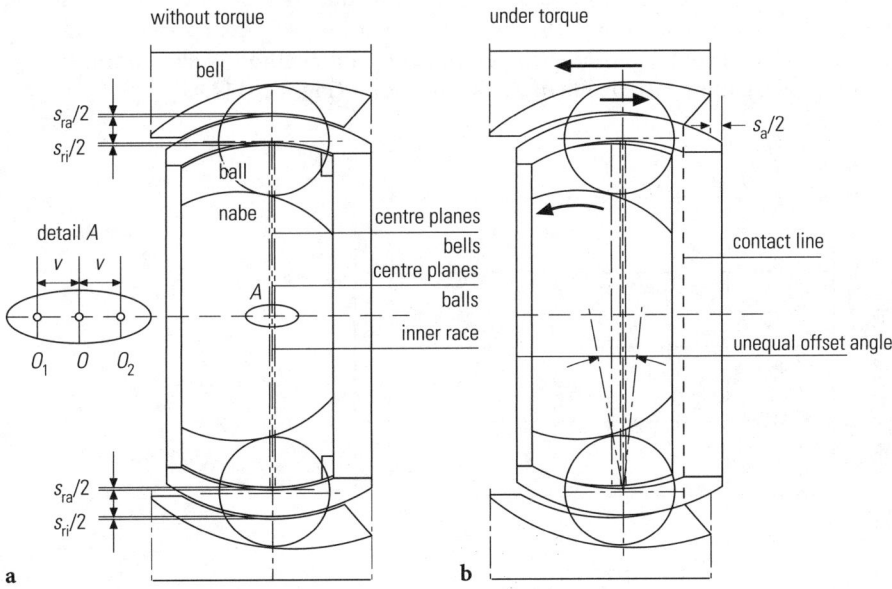

Fig. 4.32. Effect of the torque on the movement of the cage according to Werner Jacob.
a Position of the centres of the three joint parts without torque,
b Joint under applied torque stiffened by the axial force resulting arising from the axial play s_a.

H.-H. Welschof and E. Aucktor proposed an undercut free track layout by introduction of a radial offset b of the ball track centres O_1 and O_2, compare Fig. 4.31.

4.4.4 Internal centering of the ball joint

The internal centering of the ball joint is done via the ball cage with its sliding spherical surfaces. The inner race is guided by its outer sphere inside the inner diameter of the cage, the cage however is held with its outer diameter in the inner sphere of the outer race (bell). Under ideal conditions all centres of the components (inner race, cage and outer race) should be on one and the same position inside the joint, see Fig. 4.32b.

It must be considered that by the axial force of the ball transmitting forces during torque transmission, the joint components are axially displaced, whereas the overall axial play is a result of manufacturing tolerances and functional plays s_{ra} and s_{ri} in the spheres. Figure 4.33 shows as an effect of these axial movements, that in the assembled and loaded joint, the inner race is pushed into the joint, whereas the cage is pushed in the direction of the track opening. As a result of these movements, the resulting working offsets and the offset angles become unequal.

To overcome this effect and any potential impact on joint durability, K. Taniyama et al. proposed to introduce an unequal offset of inner race and outer race [4.41].

Set the outer race offset at a slightly larger value than that of the inner race. So you obtain the same value for both the outer and inner race apparent offsets. In this way you consider the axial gap of each component. This method is also useful for extending the spalling fatigue life of the Rzeppa joint (Fig. 4.33) [4.41].

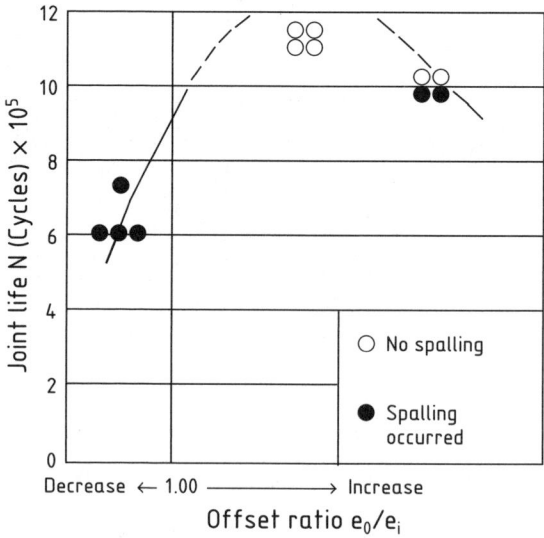

Fig. 4.33. Relation between Offset Ratio and Rzeppa Joint Life [4.41]

4.4 Ball Joints

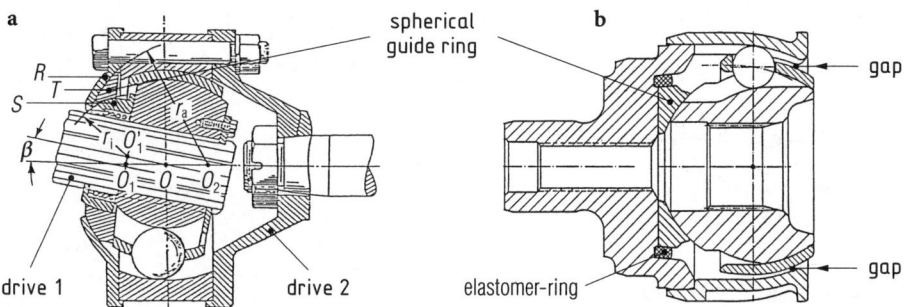

Fig. 4.33. Spherical guide rings to support the cage. **a** A proposal by Alfred Hans Rzeppa 1933 (US-PS 2010899, DRP 624463): the inner race S is supported on T. **b** A proposal by Werner Jacob 1989 (German PS 4317606, 3904655): the joint with its spherical shaped inner race is supported on a spherical guide ring. The ball cage guided on the inner race is open towards the supported side and can therefore be designed without undercut

4.4.4.1 The axial play s_a

Assuming purely spherical shapes for the principle of Fig. 4.33a, for the play

between outer race (bell) and cage $s_{ra}/2 = 0.025 +/- 0.009$ mm
between cage and inner race $s_{ri}/2 = 0.020 +/- 0.009$ mm

the total radial play is

$$s_r = s_{ra} + s_{ri} = 2 \cdot 0.025 + 2 \cdot 0.020 = 0.09 \text{ mm}$$

Such radial play leads to an axial displacement in the non-articulated position of the joint, which – as explained in section 4.4.4 – has an impact of the steering behaviour, i.e. offset of the joint assembly.

The axial displacement v in the non-articulated position arises from the geometrical relationships given by the radial play s_r. If one imagines the whole radial play s_r in the cross section divided into the four gaps of the joint (Fig. 4.33a), the following relationship results for the outer and inner gaps (Fig. 4.34).

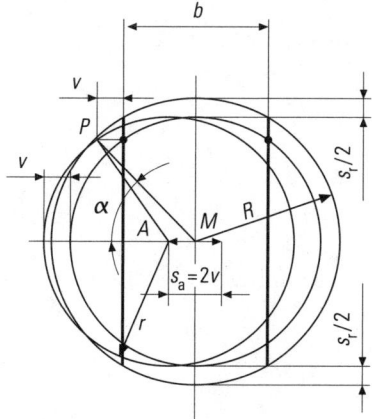

Fig. 4.34. Geometrical relationship between radial play $s_r/2$ and axial displacement v; this is half the radial play s_a

Figure 4.34 follows from the cosine law in the triangle AMP:

$$r^2 = R^2 + v^2 - 2Rv \cdot \cos \alpha \tag{a}$$

$$\cos \alpha = \frac{b/2 + v}{R} \tag{b}$$

$$R = r + \frac{s}{2} \tag{c}$$

Inserting (b) into (a)

$$r^2 = R^2 + v^2 - 2Rv\,\frac{b/2 + v}{R}$$

$$r^2 = R^2 + v^2 - 2v\frac{b}{2} - 2v^2$$

$$v^2 + bv = R^2 - r^2$$

$$v = -\frac{b}{2} \pm \sqrt{R^2 - r^2 + \left(\frac{b}{2}\right)^2} \tag{4.59}$$

The joint parts can also be displaced in the opposite direction by the same amount. The total axial play thus becomes

$$s_a = 2\,(v_i + v_a) \tag{4.60}$$

4.4.4.2 Three examples for calculating the axial play s_a

1. I. W. Macielinski (1970), AC 100 joint

Inner gap	$r_i = 32$ mm	$s_{ri}/2 = 0.035$ mm
Cage	$b = 36$ mm	$s = 5$ mm
Outer gap	$r_a = 37$ mm	$s_{ra}/2 = 0.040$ mm

From equation (4.59) it follows

$$v_i = -\frac{36}{2} + \sqrt{32.035^2 - 32^2 + \left(\frac{36}{2}\right)^2} = -18 + \sqrt{2.2412 + 324}$$

$$= -18 + 18.06215 = 0.06215 \text{ mm}$$

$$v_a = -\frac{36}{2} + \sqrt{37.040^2 - 37^2 + \left(\frac{36}{2}\right)^2} = -18 + \sqrt{2.9616 + 324}$$

$$= -18 + 18.0821 = 0.0821 \text{ mm}$$

$$v = 2\,(v_i + v_a) = 2\,(0.06215 + 0.0821) = 0.2885 \text{ mm}$$

$$s_a = 2v = 2 \cdot 0.2885 = 0.577 \text{ mm}$$

Macielinski's empirical value is about 0.60 mm

4.4 Ball Joints

2. E. Aucktor (1972), AC 95 joint

Inner gap	$r_i = 30$ mm	$s_{ri}/2 = 0.022$ mm
Cage	$b = 34$ mm	$s = 5$ mm
Outer gap	$r_a = 35$ mm	$s_{ra}/2 = 0.024$ mm

From equation (4.59) it follows

$$v_i = -\frac{34}{2} + \sqrt{30.022^2 - 30^2 + \left(\frac{34}{2}\right)^2} = -17 + \sqrt{1.32048 + 289}$$

$$= -17 + 17.03879 = 0.03879 \text{ mm}$$

$$v_a = -\frac{34}{2} + \sqrt{35.024^2 - 35^2 + \left(\frac{34}{2}\right)^2} = -17 + \sqrt{1.68076 + 289}$$

$$= -17 + 17.04935 = 0.04935 \text{ mm}$$

$$v = 2(v_i + v_a) = 2(0.03879 + 0.04935) = 0.1763 \text{ mm}$$

$$s_a = 2v = 2 \cdot 0.1763 = 0.3526 \text{ mm}$$

3. W. Jacob (1987), AC 125 joint

Inner gap	$r_i = 40$ mm	$s_{ri}/2 = 0.020$ mm
Cage	$b = 46$ mm	$s = 6$ mm
Outer gap	$r_a = 47$ mm	$s_{ra}/2 = 0.025$ mm

From equation (4.59) it follows

$$v_i = -\frac{46}{2} + \sqrt{40.020^2 - 40^2 + \left(\frac{46}{2}\right)^2} = -23 + \sqrt{1.6004 + 529}$$

$$= -23 + 23.03476 = 0.03476 \text{ mm}$$

$$v_a = -\frac{46}{2} + \sqrt{46.025^2 - 46^2 + \left(\frac{46}{2}\right)^2} = -23 + \sqrt{2.3006 + 529}$$

$$= -23 + 23.04996 = 0.04996 \text{ mm}$$

W. Krude's empirical value is about 0.35 mm, because

$$v = 2(v_i + v_a) = 2(0.03476 + 0.04996) = 0.16944 \text{ mm}$$

$$s_a = 2v = 2 \cdot 0.16944 = 0.3389 \text{ mm}$$

The impact of axial lash on the joint can be summarized as follows:

It was shown, that by an axial movement of the joint components relative to the other, there is a direct influence of the axial play on joint offset in assembled and loaded conditions. Countermeasures like introducing unequal component offsets can be considered to minimize any potential impact on joint durability.

Furthermore the axial movement of the joint components can lead to an increase of the lash between the balls and its corresponding tracks. As will be shown in chapter 4.4.5.2, a potential negative effect on contact conditions and pressure distribution between ball and track flank needs to be considered during the selection of the ball track shape.

Considering state-of-the-art manufacturing tolerances and functional plays of current Rzeppa type joints in the spheres and tracks, the typical maximum value of the axial lash is ~ 0.65 mm. Only by a modification of the joint concept, as e.g. introduced in Fig. 4.30, a significant reduction of the maximum value can be achieved. W. Krude reported of a maximum value of ~ 0,35 mm for a modified joint concept and improved manufacturing tolerances.

4.4.4.3 The forced offset of the centres

The forced centre offset is generated by the articulation angle β. As a result of functional plays and manufacturing tolerances, the offsets in the assembled conditions can become unequal. This has some effect on the position of the inner race and cage centre inside the outer race, when the joint is articulated.

Figure 4.35 introduces a radial displacement of the centres, when the joint is articulated, which can lead to an increased articulation effort.

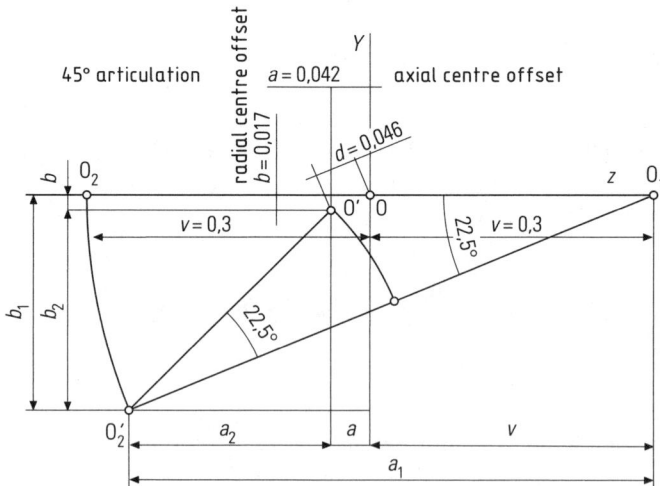

Fig. 4.35. Calculation of the forced centre offset d as a function of the articulation angle β. This is one of the causes of jamming of the joint under large articulation angles

4.4 Ball Joints

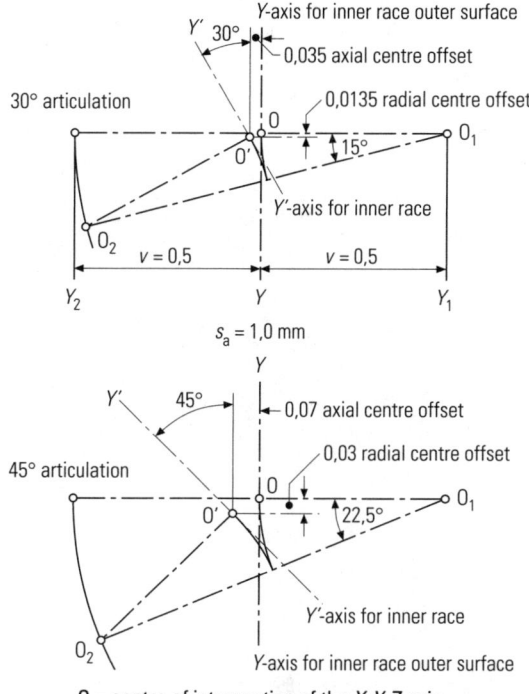

Fig. 4.36. The forced centre offset increases with articulation angle:
$\beta = 30°$ $\beta = 45°$
from $s_a = 0.035$ to $s_a = 0.07$
from $s_r = 0.0135$ to $s_r = 0.03$

From Fig. 4.35:

$$a = -a_1 + a_2 + v = -2v \cos \beta/2 + v \cos \beta + v$$

$$b = -b_1 + b_2 = -2v \sin \beta/2 + v \sin \beta$$

$$d^2 = a^2 + b^2 \rightarrow d = \sqrt{a^2 + b^2} \tag{4.61}$$

For Macielinski's limit value of $s_a = 2v = 0.60$ mm and $\beta = 45°$ we get:

$$a = -2 \cdot 0.3 \cdot 0.9239 + 0.3 \cdot 0.7071 + 0.3 = -0.0422 \text{ mm}$$

$$b = -2 \cdot 0.3 \cdot 0.3827 + 0.3 \cdot 0.7071 \qquad = -0.0175 \text{ mm}$$

Figure 4.37 shows the positions of a ball with 360° rotation of the joint.

The forced centre offset d drives the ball into 0/360° and 180° positions. 4-point contact of the balls results and the cage is jammed. In the 270° position wedging occurs and in the 90° position, the tracks are opened, the balls no longer have contact with the tracks.

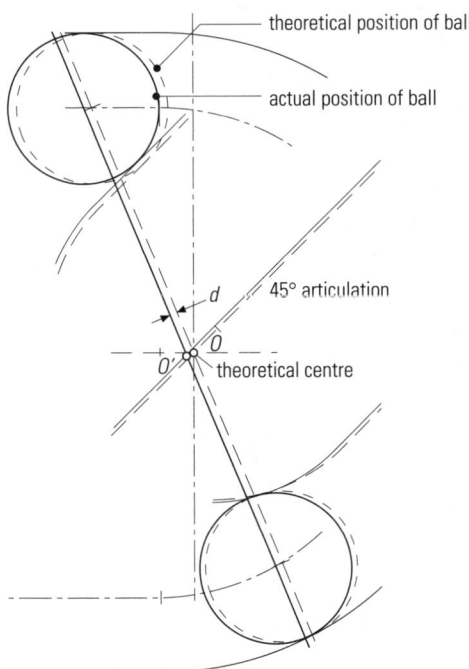

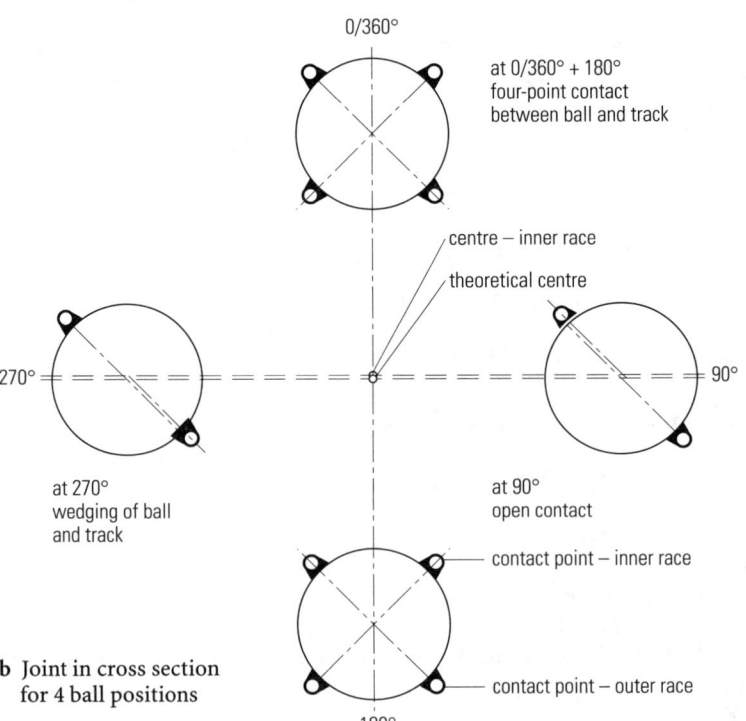

Fig. 4.37. Forced centre offset d according to Werner Jacob 1992.
a Longitudinal section of the joint, forced centre for $\beta = 45°$
b Cross-section of the joint with 5 possible positions of the ball

4.4 Ball Joints

Figure 4.37 shows the local ball track loading conditions for a joint having a forced centre offset. Assuming that the cage is held in a half-angle plane, the local loading of the ball under 270° would increase whereas the ball under 90° would be free of load.

Due to the fact that the cage has the possibility to centre itself in load balanced angular position between the outer race and inner race, an angular movement of the cage outside of the drawing plane along a vertical axis will appear (i.e. by which peak loads in the balls during rotation can be reduced).

Such movement of the cage outside of the half-angle plane has no measurable effect on constant velocity behaviour.

4.4.4.4 Design of the spherical contact areas

In Rzeppa type joints, the ball cage is guided by its outer and inner sphere between inner and outer race. The main load direction between the sphere contacts are in axial direction, but – as explained in chapter 4.4.4.3 – also some push in radial direction can be possible.

Lashes are required to ensure full function under all working conditions. Under high power transmission heat generation inside the joint functional surfaces together with cooling by surrounding air can lead to an unequal heat distribution, i.e. the inner components have a higher temperature than the outer race. This can lead to an unequal dimensional growth and a risk of undetermined contacts in the spheres. Therefore some functional lashes are required and special sphere profiles are introduced.

One solution is to use spherical surfaces, but introduce a defined sphere profile to avoid any contact in or close to the ball plane. Typical angles for such sphere profile which can be applied in both inner and outer spheres are +/– 15°, compare Fig. 4.38.

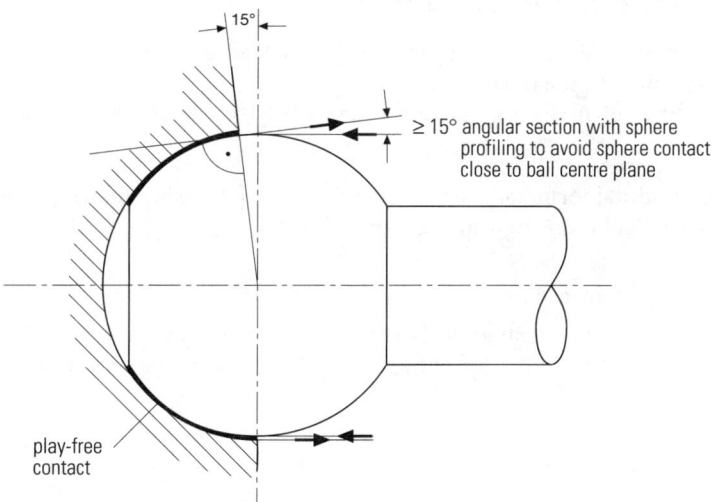

Fig. 4.38. Significance of sphere profiling: by introduction of a sphere profile, a sphere contact close to the ball centre plane can be avoided

An alternative solution is to use barrelled sphere shapes. Here all outer spheres have a radius with offset to the joint centre line, which is larger than half the outer diameter. All inner spheres have a radius which is smaller than half of the inner diameter. Such geometry modification leads to well defined contact areas with small functional lashes in the end region of the spheres to allow a good support of the axial loads generated by the tracks, but introduces a functional lash in the region of the ball centre plane.

4.4.5 The geometry of the tracks

The nature of the pairs of elements and the geometry of their tracks on which they are guided determine the torque capacity and the design shapes of the articulating ball joint (Fig. 2.16). Rzeppa type fixed joints (AC type) have circular shaped tracks in longitudinal direction, which – in combination with a spherical guidance between inner race, cage and outer race – does not allow any plunge movement.

4.4.5.1 Longitudinal sections of the tracks

The various longitudinal track patterns possible in constant velocity ball joints thereby determining the effective geometry) are shown in Section 2.1 (Figs 2.9–2.11). Although according to (2.8) and (2.9) all longitudinal grooves that are mirror symmetrical in pairs in the outer and inner bodies are possible for constant velocity joints, for production reasons straight, circular or helical shaped grooves are preferred. Since the Weiss joints are rarely used these days, the main aspects of the longitudinal track shape are dealt with for Rzeppa ball joints which are manufactured in large quantities.

Some problems for the offset let us take into account:

- the decrease in the track depth with increasing articulation angle β (Fig. 1.26a)
- the axial reinforcement of the outer cage
- reinforcement of the hub in the cage, adversed stress and friction against the edge of the track.

Some possible longitudinal forms of the tracks and their relationship between the offset angle and the articulation angle are shown in Fig. 4.41.

The simple offset-meridian track I (Fig. 4.41)

In AC and UF joints, the tracks run as arcs of a circle, the generating centres of which lie on the axis of rotation of the joint, offset by the angle $\varepsilon \approx 9°$ from the centre of the joint.

4.4 Ball Joints

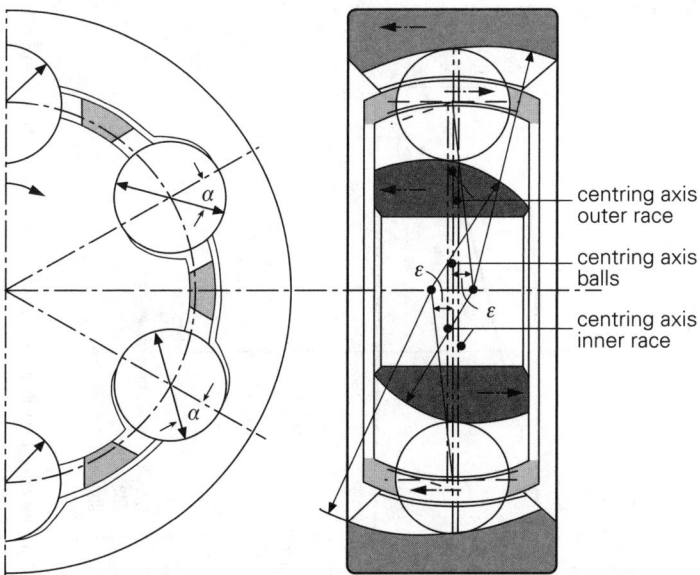

Fig. 4.39. Fixed joint with grooves and alternating offset to the longitudinal axis. It is centred solely by the pre-loaded balls. The cage holds the balls in the constant velocity plane (opposing track joint)

The alternating offset-track I/Ia (counter track joint, Fig. 4.39)

If the AC and UF joints with the simple offset meridian tracks are altered such that three offset tracks extend towards one half and three tracks extend towards the other half of the joint centre on the axis of rotation, this is called the "alternating configuration". See continuous line I and the continuous line Ia lying under the β-axis.

A joint with improved properties results:
- A move away from the disruptive cage centring in favour of purely ball centring. The two groups of balls mutually centre themselves and allow any ball pre-tensioning. Under torque the joint adjusts itself to equal offsets to the left and right.
- Compensation of the axial forces through mutual support of the balls on the cage windows. Cage friction therefore does not apply, only the rolling and boring friction of the balls remains.
- A saving of 4 locations because of the lack of cage centring.

With increasing articulation angle β the offset causes *a decrease in the track depth* (Fig. 1.26a) which causes the ball to be supported increasingly less by the *flanks* of the track: the joint loading is then limited.

If the joint has to work at a large articulation angle, it is advisable to choose the eccentrically offset II/Ia, instead of the track pattern I/Ia, see Fig. 4.41. Greater track depths and shorter radial ball movements are thus achieved. The joint then has similar improved properties to the I/Ia.

172   4 Designing Joints and Driveshafts

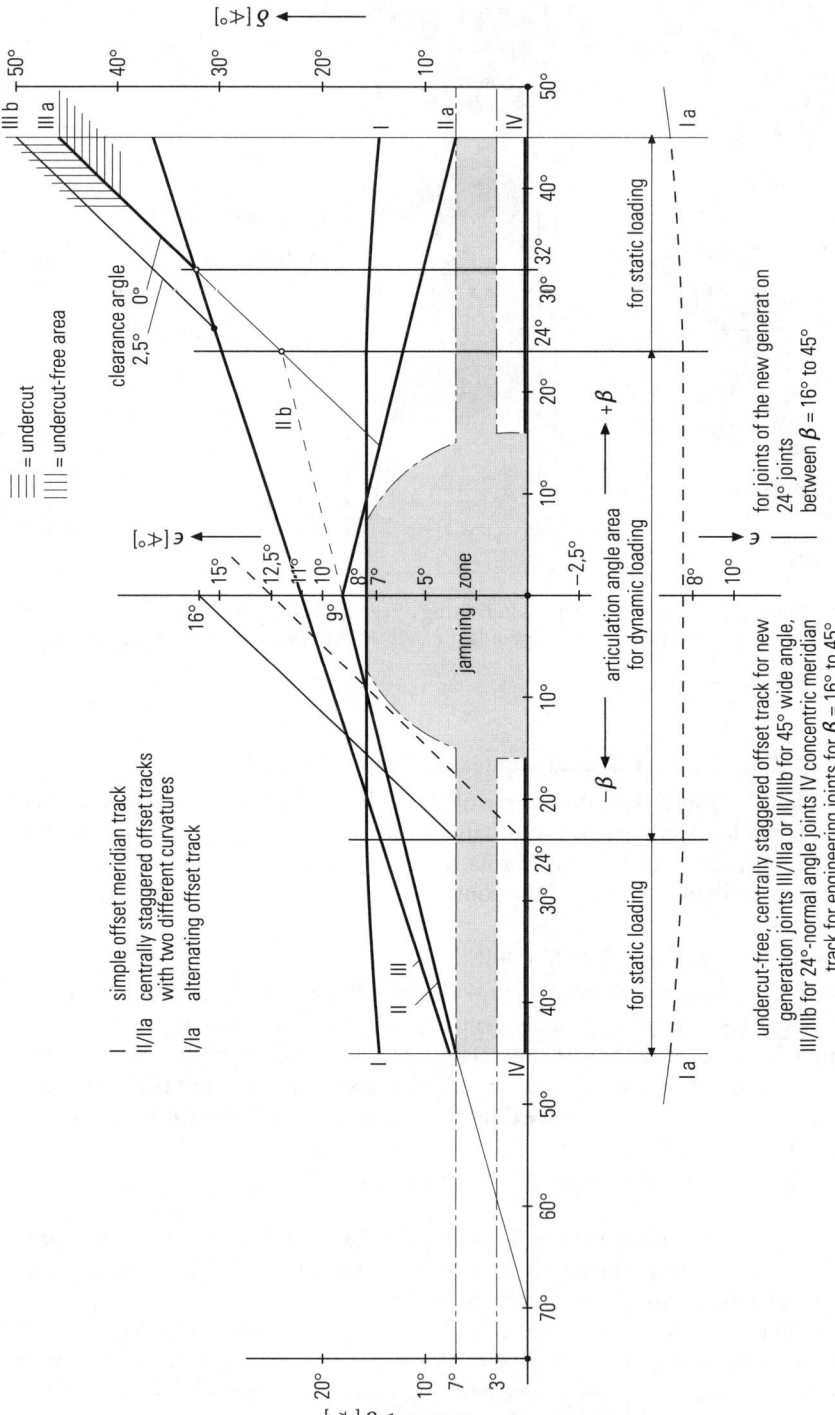

Fig. 4.41. Possible track patterns in the longitudinal section of constant velocity rotary ball joints and their relationships to the wedge angle δ and offset angle ε, as a function of the articulation angle β on the running ball, according to Erich Aucktor

4.4 Ball Joints

The centrally staggered offset-track II/IIa with two different curvatures (Fig. 4.41)

If the normal, dynamic working area of $\beta \leq 15°$ is exceeded, a high cage friction results with the simple track I that can be reduced with the centrally offset track. The cage friction is generated by the axial plunging force. It is derived from the torque to be transmitted and the wedge angle $\delta = 2\varepsilon$ acting on the balls. Using the example of an articulation angle $\beta = 45°$, track I has a value $\delta = 16°$ whereas track II has a value $\delta = 7°$; the cage friction is thus halved for the same torque. The track with the two centrally offset, differing curvatures also reduces the radial ball movement with increasing articulation angle β. Because of this the cage wall can be designed thinner and the track deeper. Thus the pressure ellipse becomes longer with $\beta = 45°$ and allows higher joint loading.

The undercut-free offset-track III/IIIa

The tracks pattern follows in principle an eccentric, constant curvature. The jamming zone must remain untouched here. With articulated joint it defines the smallest wedge angle $\delta \approx 7°$, and with an extended joint the semicircular shaped area of the smallest offset angle ε according to III/IIIa (Fig. 4.41).

The articulation angle

$\beta = 0 \ldots 15°$ corresponds to normal driving (operation angle)
$\beta = 0 \ldots 50°$ corresponds to car ranking or parking (wide angle)

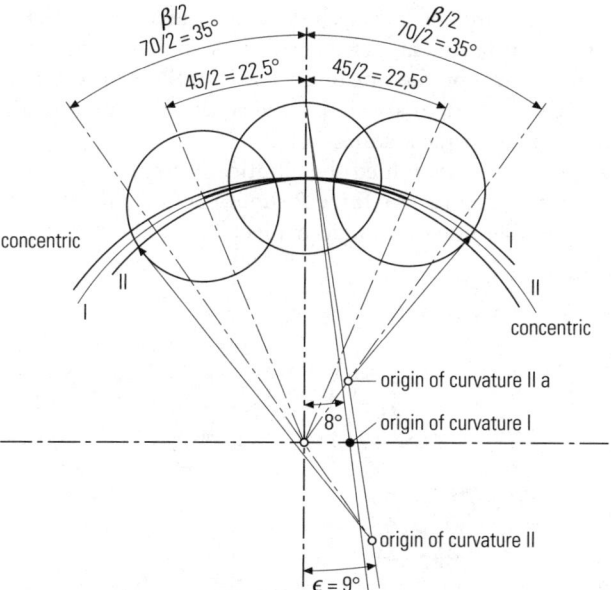

Fig. 4.42. Design of the eccentric offset track with two centrally offset, differing curvatures II/IIa compared to the simple offset meridian track II. They reduce the cage friction to less than half the value

Table 4.6. Most favourable track patterns for the two groups of joints

	track pattern from Fig. 4.41	
	III/III a and b	II/IIb
Articulation angle/wedge angle		
$-\beta/+\delta$	45°/7°	24°/7°
$-\beta/+\delta$	45°/45°	24°/24°
offset angle ε	≈ 11°	≈ 9°

δ = wedge angle, jamming angle or steering angle, $+\beta$ = angle of articulation towards the outside, $-\beta$ = angle of articulation towards the inside.

4.4.5.2 Shape of the Tracks

The tracks in the input and output bodies of the joint can be elliptical, circular or gothic arch in shape (Fig. 4.43). If the ball is to move at all, the radius of the side of the track must be greater than the ball radius.

The circular track cross-section (4.43b) can easily be checked with a ball gauge. If the radial play between the inner and outer tracks is 1/2 to 1/3 of the tangential play, with inverse values for the conformity of ψ_Q = 1.03 to 1.06, pressure angles α = 40° to 45° arise. The pressure angle α can be calculated or estimated only approximately here (Figs. 4.43 to 4.44).

The use of "lemon play" gives the circular tracks a high torque capacity due to the high conformity. The greater the torque, the higher the ball rises up as a result of the elastic deformation of the contact faces on the side of the track, until its pressure ellipse uses up the whole flank. All track edges are chamfered so that they do not jam because of distortion under the high torque loading.

William Cull (British patent 637718) examined the elliptical track cross section (Fig. 4.45) in 1946. The equivalent compressive force P should act on the effective joint radius R at the largest possible pressure angle α (Sect. 4.1.4, Fig. 4.3). On a

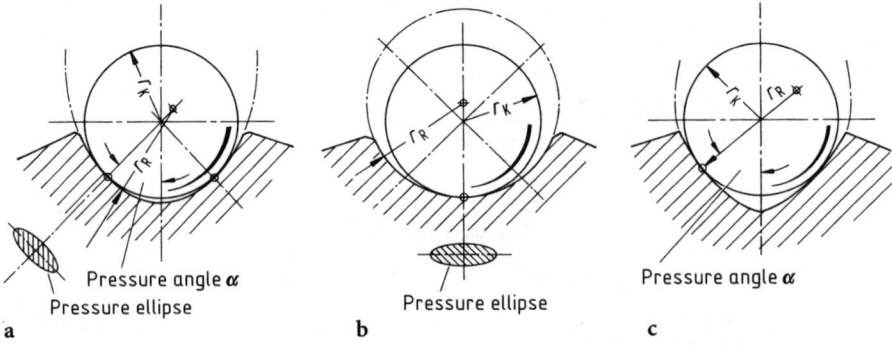

Fig. 4.43a–c. Track cross section. a Elliptical, b circular, c gothic arch (ogival)

4.4 Ball Joints

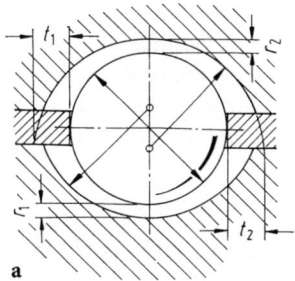 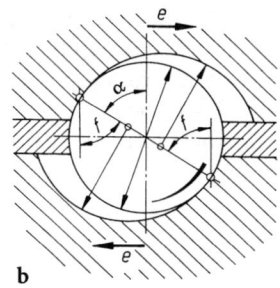

Fig. 4.44 a, b. Circular track cross-section by Willimek/Aucktor 1964 (German patent 1169727). a Centre position ("lemon play"), b joint under torque

finished track (Fig. 4.45), the distance A between the two contact points can be measured and the shape of the ellipse can be calculated from the measurement DF between the ball gauge and the base of the ellipse. For a ball diameter $d = 16.104$ mm, the difference between the values of DF for inverse conformity values $\psi = 1.03$ and 1.05 is only 14 μm (from Fig. 4.45). This difference between the desired shapes of the two ellipses is so small that little can be said about the actual track shape.

The gothic arch or ogival cross-section can be chosen (Fig. 4.43c) instead of the elliptical cross-section. At each point where the ball contacts the track flanks there is constant conformity. The bottom of the track can be radiussed or left as formed.

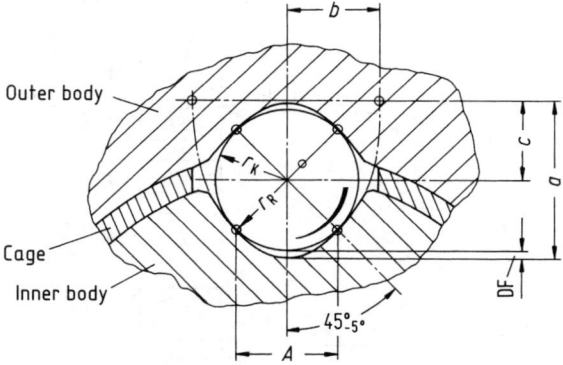

Ball-ϕ	Coeff. of conformity	Dimensions in mm				
$2R_K$	$\psi = r_R/r_K$	a	b	c	DF	A
15,875	1,03 1,04 1,05	8,305 8,432 8,562	8,059 8,101 8,144	0,347 0,468 0,590	0,020 0,027 0,033	11,225
16,104	1,03 1,04 1,05	8,425 8,554 8,685	8,176 8,218 8,261	0,352 0,474 0,599	0,020 0,027 0,034	11,387

Fig. 4.45. Tracks with elliptical cross section (ellipse greatly magnified to show it clearly)

In practice, the difference between the last two track cross-sections (Figs. 4.43b and c) is slight; the choice between them depends on the cost of tooling.

In addition to the shape of the track cross-section, the locus of the centre point of the ball is important. The track flanks govern the steering and also determine the pressure angle α and indeed the form of the joints. The parameters shown in Figs. 2.13 and 4.27, lead to a

Family tree of ball joints

(Table 4.7). In practice, straight, helical and circular tracks are used, although in theory any geometry is possible.

Table 4.7. Ball joint family tree

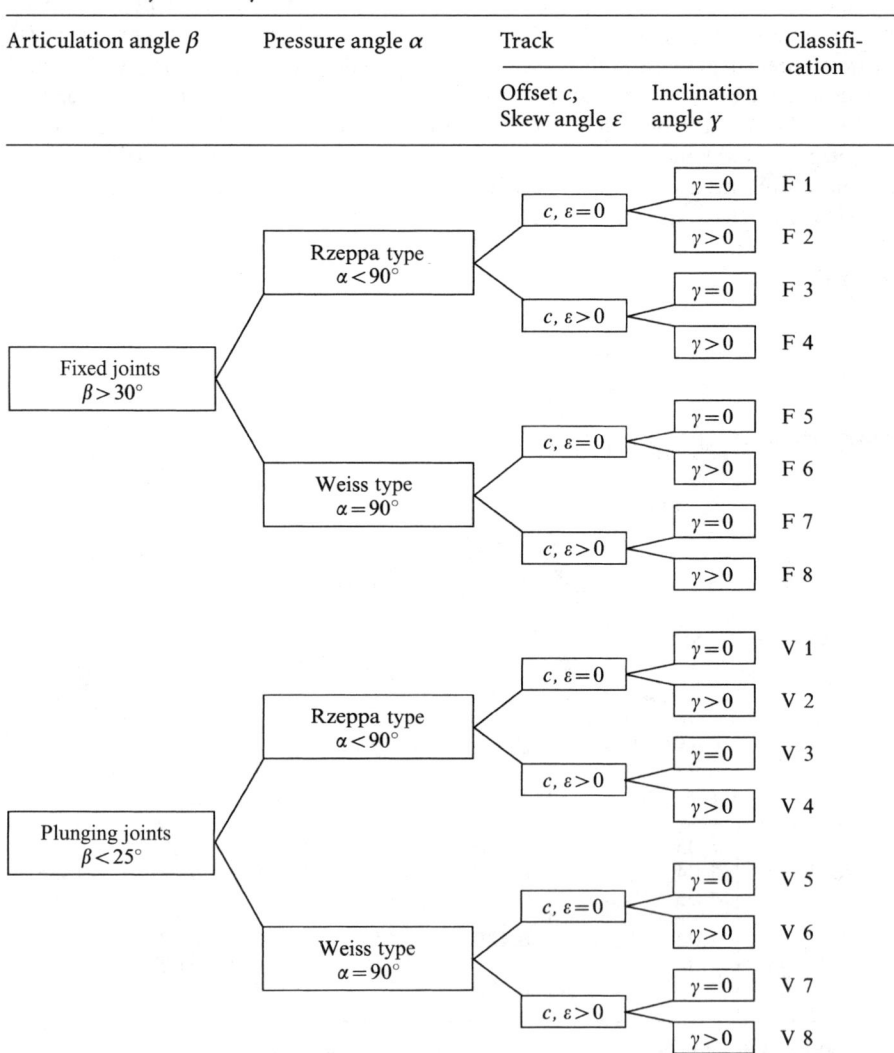

4.4.5.3 Steering the Balls

The angle of intersection χ between the symmetrical pairs of tracks is fundamental for the steering the balls. Its *size* determines the steering of the balls in the tracks: the *nature* of the ball control into the plane of symmetry π depends on the effective geometry (Sect. 2.1). Table 4.7 applies for all ball joints of the Rzeppa and Weiss types (Figs. 4.26a and b).

Intersection Angles $\chi = 50°$ *to* $60°$

These large angles steer the balls in the tracks without additional help (KR). Carl Weiss used them in 1923 for his fixed joint with circular tracks, and in 1936 for his plunging joint with straight tracks (Figs. 1.20 and 4.44).

Intersecting Angles $\chi = 18°$ *to* $40°$

These angles still steer the balls properly but need auxiliary elements (HE) to guide them. For VL-plunging joints, the tracks intersect at 32° (SR). They need a ball cage so that when the joint articulates, the balls to not fall out of the tracks. On the Devos joint with straight tracks (Fig. 1.23) and intersecting angles of about 24°, a pin guides the rotatable and axially free outer balls of its three-ball unit into the plane of symmetry π.

The ball tracks of RF fixed joints intersect at 18° to 20° due to the track offset (RO). The opening angle δ of the tracks must be larger than twice the friction angle ϱ (Fig. 1.27). The balls are steered into the plane of symmetry π by the offset (Figs. 1.28a and b). The condition for constant velocity is (2.8)

$$y'_2 = - z'_2 \cos \beta/2.$$

The offset c must not be too large (Fig. 4.55) or else the track flanks 9 and 10 become too short and are no longer able to transmit the compressive forces. Moreover, since the tracks open a cage must be used to prevent the balls from falling out.

Table 4.8. Steering the balls in ball joints

Groove intersection angle χ	Ball steering	Abbreviation
50° ... 60°	solely through intersecting tracks, i.e. without any aids for ball steering	KR
18° ... 40°	intersecting tracks plus steering pins	KR + FB
	intersecting offset tracks	KR + RO
	parallel tracks and offset cage	PR + KO
0°	parallel tracks plus pilot lever	PR + PH
	parallel tracks plus spherical surfaces	PR + KK

Ball steering: KR crossed tracks; FB steering pins; PR parallel tracks; SR inclined tracks Steering aids: RO track offset; KO cage offset; PH control (pilot) lever; KK spherical surfaces

4.4.5.4 The Motion of the Ball

consists in rolling, skiding and drilling
- A ball will roll if all the forces to through its centre; otherwise it skids
- A ball will roll if the tracks are configured parallel; intersecting tracks generate boring friction
- Contacting rolling bodies get jammed; self-locking results if the ball is forced out of its axial direction; a clearance angle $\varrho > 7°$ prevents jamming.

In order for the balls to be able to slide and roll in the tracks, the tangential angle ε at the contact point of the ball and track must be greater than the friction angle ϱ. This can be derived from the following.

The cage force K brings out the reaction forces F_N on the two points of contact of the tangents. These can be broken down into their x and y components (Fig. 4.46).

For $\Sigma P_x = 0$, the following applies for the inward pressure on the ball:

$$K - 2F_N \cdot \sin \varepsilon - 2R \cdot \cos \varepsilon = 0 \tag{a}$$

When the Force K' is reduced to zero ($K = 0$), the condition can be determined under which the ball is not pressed out, but remains stuck in the tracks.

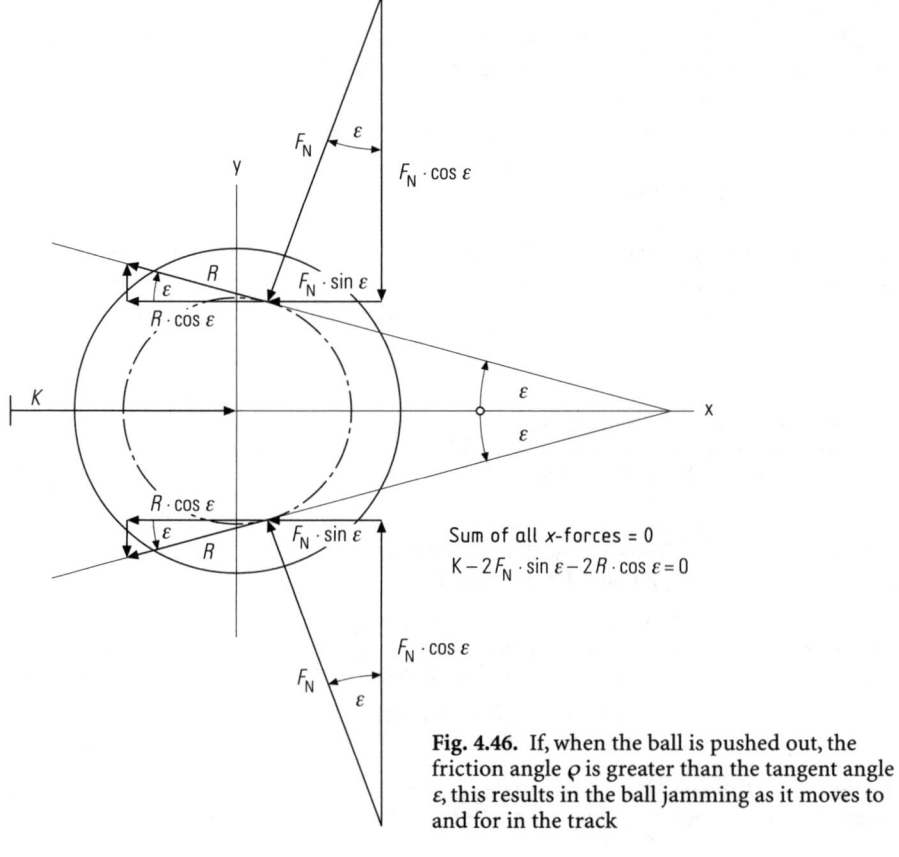

Fig. 4.46. If, when the ball is pushed out, the friction angle ϱ is greater than the tangent angle ε, this results in the ball jamming as it moves to and for in the track

4.4 Ball Joints

For outward pressure, the friction is in the opposite direction to the movement, so we get

$$0 - 2 F_N \cdot \sin \varepsilon - 2R \cdot \cos \varepsilon = 0 \tag{b}$$

For the friction force R, the following is inserted

$$R = \mu F_N = \tan \varrho \cdot F_N = F_N \cdot \frac{\sin \varrho}{\cos \varrho} \tag{c}$$

c) is inserted in b)

$$K = 2 F_N \cdot \sin \varepsilon + 2 F_N \cdot \frac{\sin \varrho}{\cos \varrho} \cdot \cos \varepsilon = 2 F_N \cdot \frac{\sin \varepsilon \cdot \cos \varrho + \cos \varepsilon \cdot \sin \varrho}{\cos \varrho}$$

Using the relationship $\sin \varepsilon \cos \varrho - \cos \varepsilon \sin \varrho = \sin (\varepsilon - \varrho)$, one gets

$$K = 2 F_N \frac{\sin (\varepsilon - \varrho)}{\cos \varrho} \tag{d}$$

The condition for the ball getting stuck or jamming is

$$\varepsilon \leq \varrho \tag{4.61}$$

Fig. 4.47 shows how the ball's tendency to roll depends on the wedge angle $\delta = 2\varepsilon$:

a) $\delta = 0\text{--}3°$, as in a grooved ball bearing. There must be a certain radial play present in order to enable the balls to roll.
b) $\delta = 7\text{--}8°$ is completely unsuitable for the ball joints. This range is used for locking mechanisms, for which secure jamming is required.
c) $>17°$, gives secure rolling and sliding, as is necessary for the offset steering on the Rzeppa principle. The tangential angle ε must be greater than the friction angle ϱ.

The oscillating movement of the ball is not true rolling but consists of a mixture of rolling, boring and sliding. Sliding in the ball tracks cannot be avoided in the articulated joint [4.22]. In 1962, on the basis of his practical experience, Erich Aucktor tried to increase the proportion of rolling of the balls in circular tracks through differing designs of the inner and outer ball tracks (German Patent 1 126 199/1962).

4.4.5.5 The cage in the ball joint

Real produced joints have a cage. In rotation mostly the smal situation passes through; here cage steering must set free the concerned ball.

The ball cage in the ball joint has to transmit high steering forces onto the sliding surfaces of the inner and outer races and the cage windows (Fig. 4.49). Stuber's 1933 offset steering system requires it to guide the balls in the meridian tracks into the

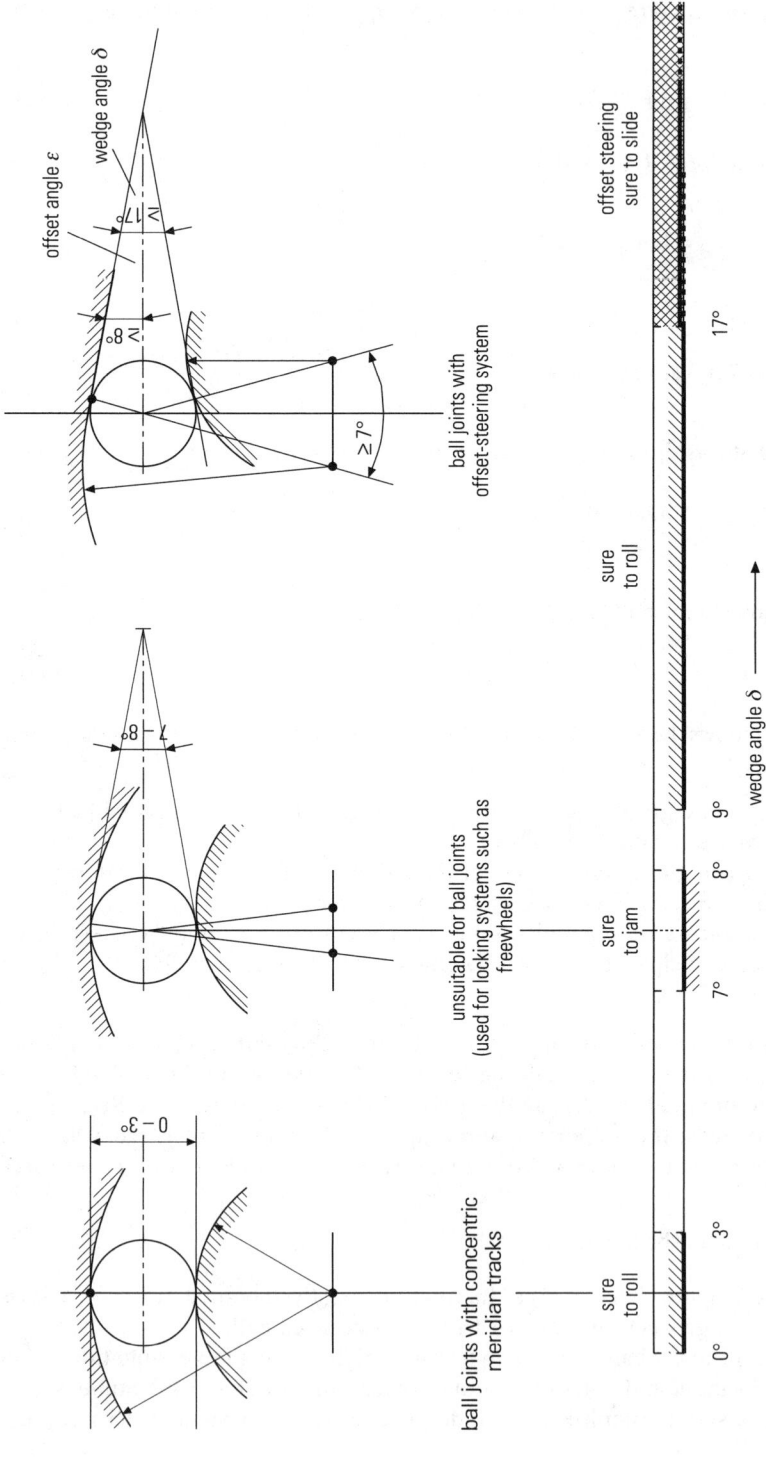

Fig. 4.47. Overview of the conditions for the wedge angle $\delta = 2\varepsilon$, for secure rolling of the ball in the direction of motion

4.4 Ball Joints

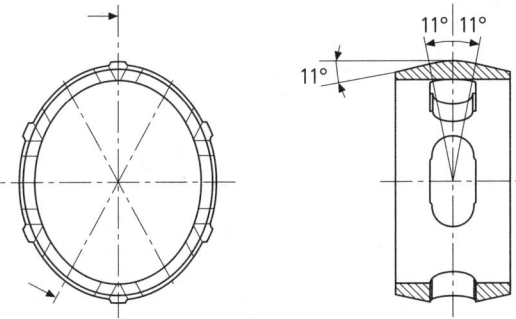

Fig. 4.48. The cage of the VL-joint

plane of symmetry, and hold them there for all articulation angles. Whereas the cage in ball bearings is only supposed to prevent direct touching of the ring-guided balls, in the rotary ball joint it *is* the medium for transmitting torque. If the joint were firmly connected at the output side, under articulation balls without cage guidance could escape out of the opening meridian tracks, without transmitting torque [2.5, p. 45; 2.11, p. 79, Note 2].

There are solid and sheet metal cages. In the ball joint one-piece solid cages predominate.

The one-piece solid cage consists of a ring with rectangular windows produced by stamping and broaching or fine stamping. Sheet metal cages, also in one-piece, are pressed stamped or drawn from flat or strip steel.

The cage forms an intermediate sliding surface in the joint (Fig. 4.48). It must take up the axial forces that come from the action of the track and the ball (Fig. 4.49a). These axial forces vary during riding a car. They originate from its different driving situations (Fig. 4.50). The amount of pushing depends on the total weight of the vehicle and its torque level.

Without the support in the cage window the ball would be squeezed out of the joint by the wedge action of the intersecting tracks (Fig. 4.49, 4.52a). The axial forces displace the cage so that the ball joint is centred via its outer and inner tracks (first centring). The radial working play S_r is removed here and between the outer surface of the cage and the inner surface of the bell, contact zone 1 results. Between the

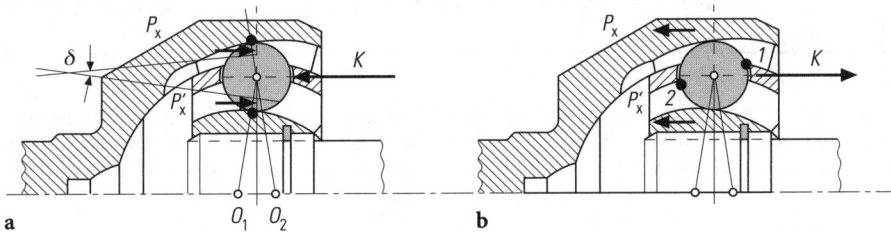

Fig. 4.49. Axial forces on the parts of the unarticulated joint during the transmission of torque. a axial forces on the ball via bell, inner race and cage. b the axial forces of the bell and the cage act externally at 1, those of the inner race on the cage internally at 2. P_x and P'_x are the x-components of the equivalent compressive force P, K is the resulting cage force.

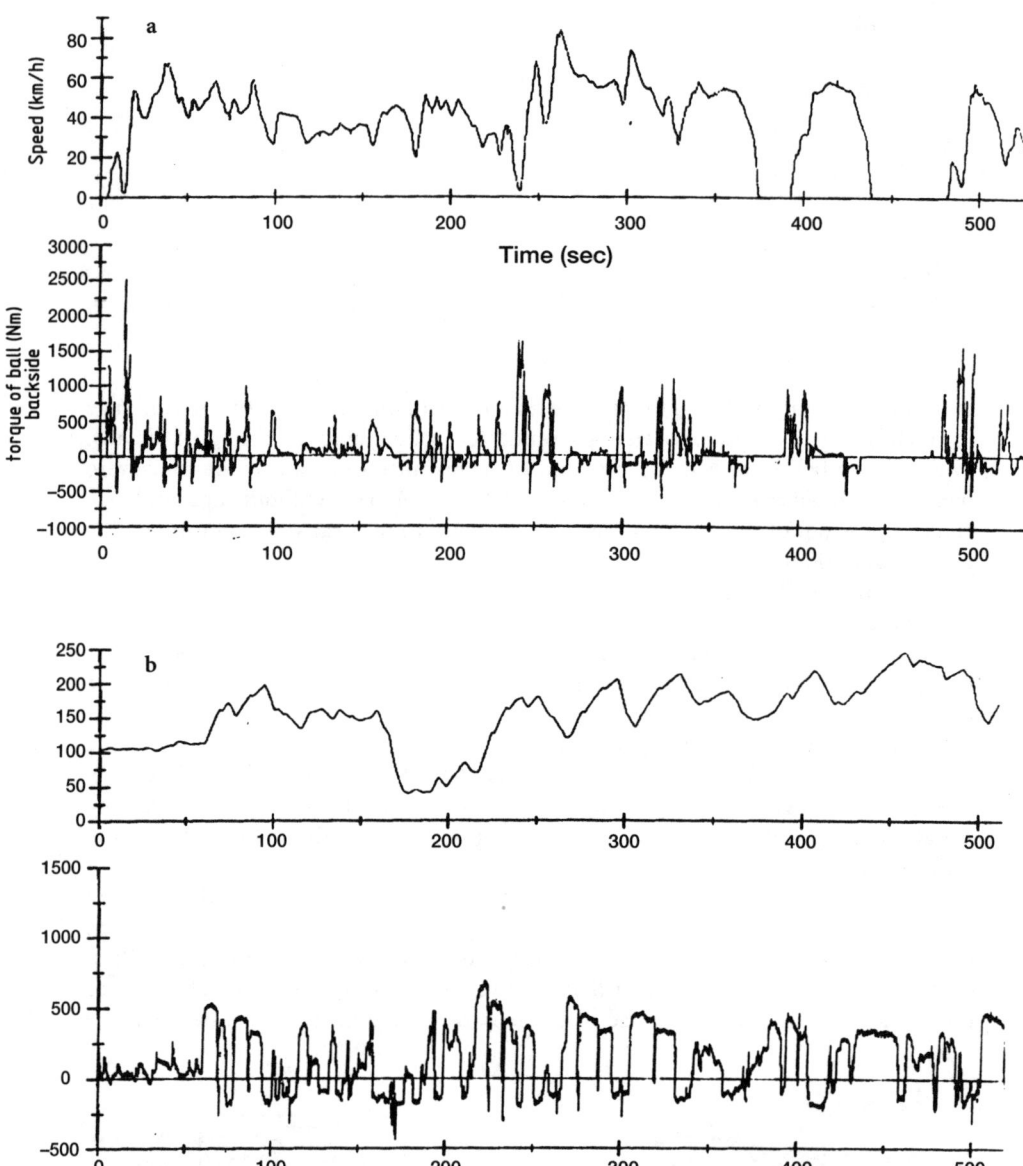

Fig. 4.50. The force variation during driving of a vehicle indifferent situations. The permanent changing of coast (−) and drive (+) produces axial forces arising through the pushing behaviour of the vehicle. **a** town traffic; **b** hilly highway. From Löbro-trials by Jacob/Paland 1993/94

spherical surface of the inner race and the inner surface of the cage, sliding zone 2 results (Fig. 4.52b), where high pressures occur under the effect of the torque. Because of this substantial friction forces arise between the cage, the bell and inner race, due to the sliding movements that occur as the joint articulates. The axial surface pressure can be reduced through a wide cage with short windows. At least one window opening must be matched exactly to the shoulders of the ball inner race so that they can be guided into the cage. The width of the window is dimensioned so that an interference of 0 to 2 μm/mm ball diameter results.

4.4.5.6 Supporting surface of the cage in ball joints

The critical operating condition occurs at the maximum articulation of the joint. With each revolution the cage rotates about half the articulation angle β, the inner race about the whole of this angle (Fig. 4.49). At the same time the direction of the axial load, that is supported on the spherical sliding surfaces, changes. The plan view in Figures 4.49 and 4.52 show the effective sliding surfaces for one of the six cage frame surfaces under torsional loading. Figure 4.52a shows the effective sliding sur-

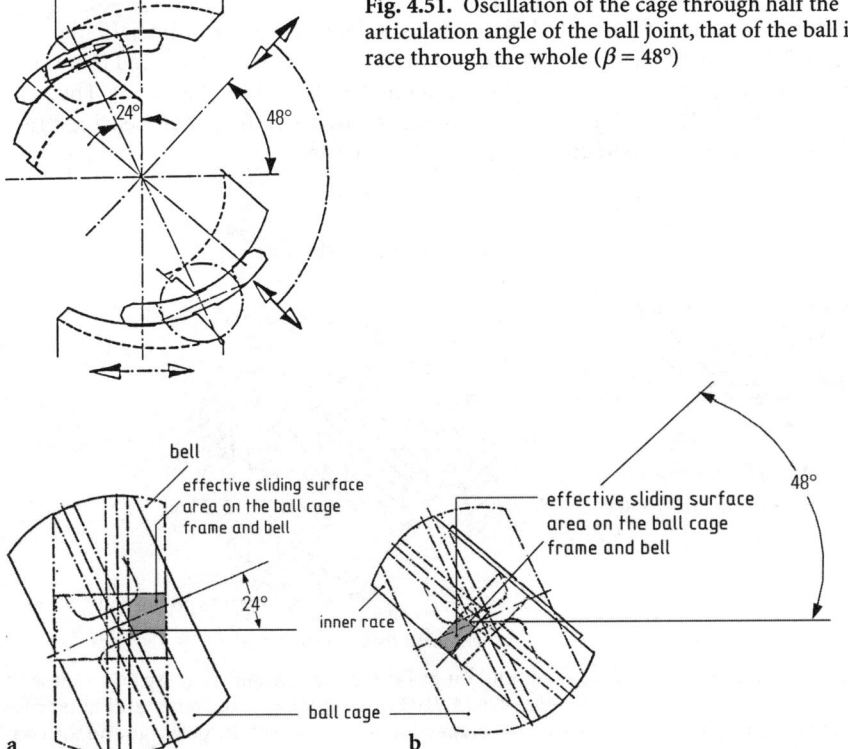

Fig. 4.51. Oscillation of the cage through half the articulation angle of the ball joint, that of the ball inner race through the whole ($\beta = 48°$)

Fig. 4.52. Plan view of the ball cage for articulation angles of 24° and 48° and the effective sliding surface for one of the six cage frame surfaces in the middle of the oscillation movement under torsional loading. **a** on the ball cage frame and of the **bell**, **b** the same on the **inner race**

face between the cage frame and the bell, Fig. 4.52b the even smaller one between cage frame and the inner race. For the largest pivoting angles the cage is the highest stressed part of the joint.

In the course of time wear occurs in the cage windows. The cage then no longer holds the 6 balls exactly in one plane, instead the balls switch during the oscillation between two close, adjacent planes. This makes itself noticeable as a periodic clicking.

4.4.5.7 The balls

The size of the balls determines the torque M_d, from which the pitch diameter can be determined. By predetermining the pitch diameter the rotation diameter can be calculated as

$$2R = D = \frac{d}{\sin\frac{180}{n}}$$

where D = pitch diameter, d = ball diameter, n = number of balls (Deutsche Auslege-Schrift (DAS) 2433 349/1974 W. Krude).

The balls experience tangential and radial accelerations as the joint articulates of the joint. They are subject to tangential and radial inertia forces that need to be lessened. With a small ball diameter d they are considerably lower (Fig. 4.55). The general load formula $M_0 = k_0 \cdot d^2 \cdot R \cdot z$ of Palmgren/Lundberg from ball bearing practice does not apply to the loading capacity of ball joints.

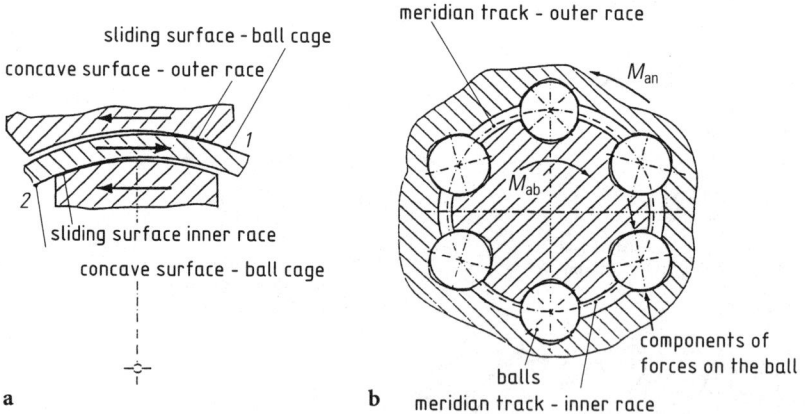

Fig. 4.53. Double centring of the outer and inner race of the constant velocity ball joint with offset-steering after Stuber (US Patent 1975 758/1933). **a** Centring of the outer and inner race by axial forces. Contact zone I between the outer spherical surface of the ball cage and the concave spherical surface of the outer race, contact zone 2 between the spherical surface of the inner race and the concave spherical surface of the ball cage. **b** Centring of the outer and inner race by the torque, after E. G. Paland 2001. Stuber's common offset has the common centre on the steering point

4.4 Ball Joints

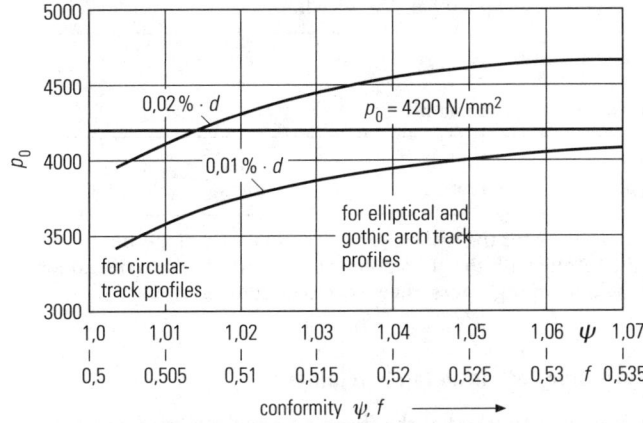

Fig. 4.54. Hertzian stress p_0 as a function of the conformity ψ or f of the ball contact and of the size of the plastic deformation (DIN ISO 76/281)

The groove radius R_K arises from the conformity ψ or f of the track, where

$$f = \frac{\varphi}{2} = \frac{\text{groove radius } R_K}{\text{ball diameter } d}$$

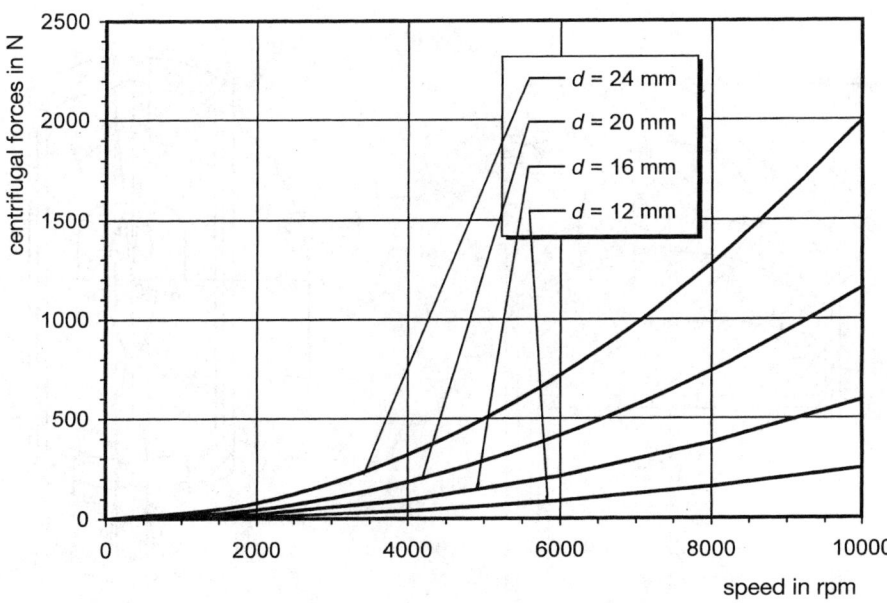

Fig. 4.55. Centrifugal forces on the balls of constant velocity ball joints with an effective joint radius $R = 32$ mm

$$F_{cf} = \frac{G v^2}{g r} \; N$$

Table 4.9. Effect of the ball size on load capacity and service life

Ball ⌀ d mm	Load capacity		Service life
	static	dynamic	
16–24	low	high	high
5–9	high	low	low

The designer must find a compromise for the ball size given the space envelope, but has to regard that the ball-⌀ raises the torque with the cube potency. The expected centrifugal forces are small in comparison to the developing forces; they have to be counted small.

4.4.5.8 Checking for perturbations of motion in ball joints

The selected effective geometry is checked at the design stage for loss of control and sliding [2.9].

If the overlap of the track flanks is too small, a cage must be provided so that the balls do not fall out if the pairs of tracks open and the torque transmission is interrupted [2.9]. It is not possible to draw up general rules about the allowable levels of track overlap because the three-dimensional movement of each joint design is different. The limits of torque transmission for spherical, straight and helical shaped tracks (in longitudinal section) were dealt with mathematically by Werner Krude in 1973. He compared the results of calculations and tests on VL joints [4.38].

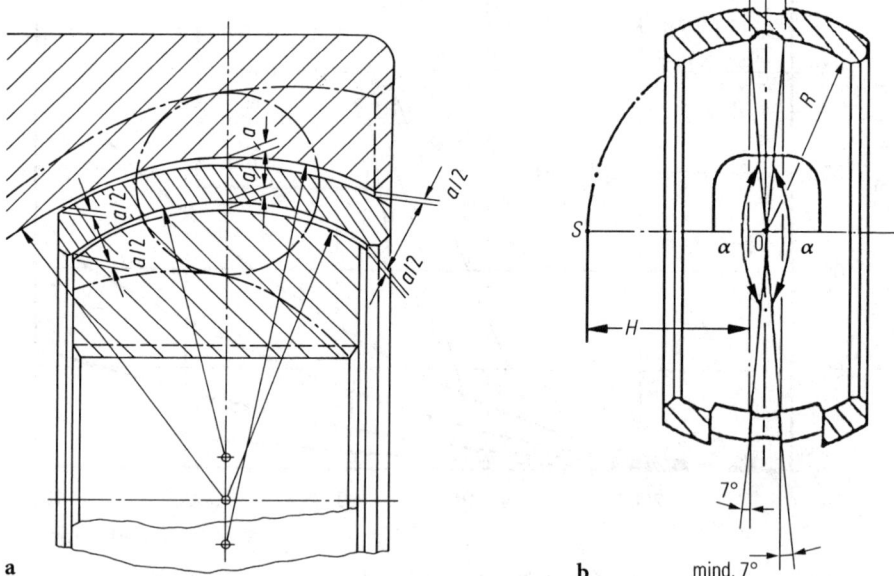

Fig. 4.56. Cage clearance in the joint: a at the centre; ~ $a/2$ at the edges. b Further solution for profile relief of the cage through relieving about 7°. $H<R$ (German Patent 37 39 868 C 2/1987 for Werner Jacob)

4.4.6 Structural shapes of ball joints

The centric fixed joints go back to Alfred H. Rzeppa 1933/34 (German Patent 624463/1933), see Section 1.3.1 and 2.2. As outer or wheel-side joints they helped William Cull's 1959 breakthrough of front wheel drive cars. Their inner centring allows simple structural shape and assembly.

It can be seen in Figure 4.3 that the highest torque capacity of a ball results if the equivalent compressive load P acts at the pressure angle $\alpha = 90°$ on the effective joint radius R, as is the case with the Weiss joint (Fig. 4.26b). This is not however possible with the Rzeppa type of joint, with its "ball and socket" construction (Fig. 4.26a), because the contact point between ball and track would fall in the gap between the inner and outer races. In spite of this, it superseded the Weiss joint in 1958, because all balls transmit the torque, the smaller intersection angle of the tracks lessens the frictional losses, the balls need a smaller preload and the joint is self-centring.

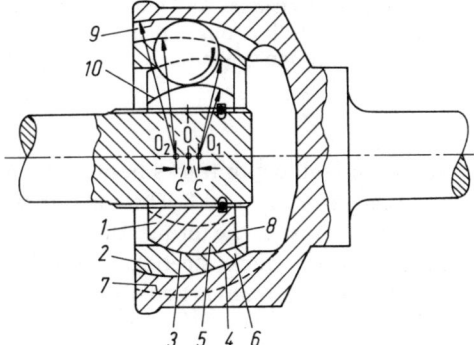

Fig. 4.56/1. Fixed joint of Honda Motors 1983 (German Patent 3 000 119). Steering both by the tracks and by the cage offset. *1* Inner track; *2* cage outer sphere; *3* cage inner sphere; *4* cage outer centring; *5* cage inner centring; *6* cage; *7* outer race; *8* inner race, *9* outer track; O_1 and O_2 offset centres

4.4.6.1 Configuration and torque capacity of Rzeppa-type fixed joints

The Rzeppa ball joint of 1927 (F1 type), with articulation angles of less than ~ 18°, did not steer the balls into the plane of symmetry π. It was improved by Bernard K. Stuber in 1933 through the track offset steering (Figure 1.26). In 1958 Alfred H. Rzeppa developed this joint design so that it was suitable for mass production; the joint operated in a trouble-free manner up to an articulation angle of 45°, which made it an ideal outboard joint for front wheel drive passenger cars (Figs. 4.34, 5.66, 5.70).

Of the four possible designs types F1 to F4 (Table 4.7), F3 has become particularly important.

These joints have track cross sections which may be circular (Fig. 4.43a) or elliptical (Fig. 4.43b). Tests have shown the static torque capacity of both joints to be about the same; however, the dynamic torque capacity is about 20% higher in the case of the circular track cross section. This is manifested in the longer life of the latter (see Table in Figs. 4.58 and 4.59).

In motor vehicles for difficult terrain, which drive at high torque even when the wheels are at full lock, the cage and balls are stressed too much by the track offset so that the (3 + 3) fixed joint can favourably be used (Fig. 4.57). Only half of the six balls

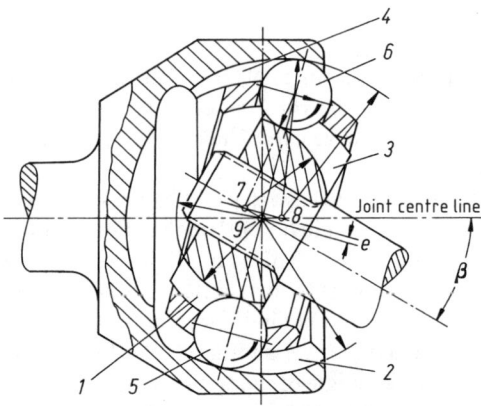

Fig. 4.57. (3 + 3) fixed ball joint with three concentric tracks and three tracks with offset generating centres. Longitudinal section of the joint articulated through β. Track 3 and 4 take over the steering into the homokinetic plane. *1* Inner, concentric track; *2* outer, concentric track, *3* inner, offset track, *4* outer, offset track; *5* "concentric" ball; *6* "offset" ball; *7* generating centre of track 3; *8* generating centre of track 4; *9* joint centre and generating centre of tracks *1* and *2*. (German Patent 3 700 868/01)

in this joint are steered by the cage, running in the full depth tracks because their generating centres 0_1 and 0_2 coincide with the centre of the joint 0. They do not exert any axial load on the cage and are well suited to transmit the torque. Up to an articulation angle of about 20° only the three offset pairs of tracks steer, then the other three intersecting pairs of tracks also steer up to the maximum articulation.

4.4.6.2 AC Fixed Joints

In terms of their "effective geometry" (see Sect. 2.2.1), AC-fixed joints (AC = angular contact) with their symmetrical offset of the track generating radii from the joint centre are "spherical meridian-joints". Their tracks run parallel to the axes of the driving and driven members, with a skew angle $\gamma = 0$; a pressure angle $\alpha = 39°$ to $42°$; and a conformity in the elliptical cross-section of the track $\kappa_Q = -0.952$ to -0.971 ($\psi = 1.03$ to 1.05).

The ratio $R/d = 1.65$ is kept constant throughout the joint series shown in Fig. 4.58/4.59. The conformity, from (4.54),

$$\kappa_L = \frac{1}{\dfrac{1.65 \cdot 2}{0.7071 \cdot 0.9877} - 1} = 0.269$$

is then also constant. The Hertzian coefficient $\cos \tau$ can be calculated from (4.57) with conformities κ_L and κ_Q

$$\cos \tau = \frac{0.269 + 0.962}{2 + 0.269 - 0.962} = 0.9418.$$

4.4 Ball Joints

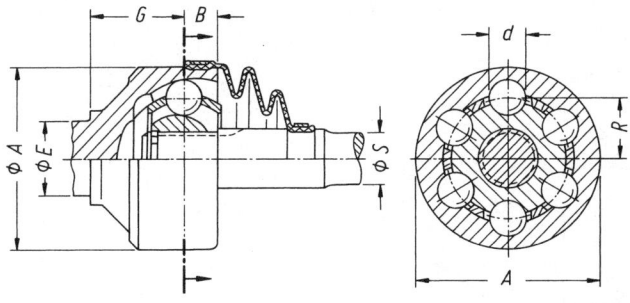

Joint AC	A	E	S	G	B	d [a]	R	M_d [b]	M_N [c] static
75	70	24	19.1	33.6	12.2	14.288	23.57	178	944
87	81	28	22.2	38.9	14.2	16.669	27.50	283	1499
95	90	30	23.8	42.2	15.3	18.000	29.70	357	1887
100	92	32	25.4	44.3	16.3	19.050	31.43	423	2237
113	103	36	28.6	49.7	18.2	21.431	35.36	602	3186
125	115	40	31.8	55.1	20.2	23.812	39.20	826	4369
150	137	48	38.1	65.9	24.4	28.575	47.15	1428	7550
175	160	56	44.5	76.7	28.4	33.338	55.00	2267	11480
200	182	64	50.8	87.4	32.5	38.100	62.87	3384	17897
225	204	72	57.2	98.2	36.5	42.862	70.72	4819	25483
250	227	80	63.5	109.0	40.6	47.625	78.58	6610	34955

[a] The decimal figures of the ball-$\varnothing$ arise through the dimensions in inches, f.i. 7/8" = 22.225 mm
[b] Permanent torque M_d with $k = 8.718$ N/mm$_2$ at $n = 100$ rpm, $\beta = 3°$ and $L_n = 1500$ hrs.
[c] Nominal torque M_N with $k = 46.23$ N/mm^2 at stretched joint.

Fig. 4.58. AC fixed joints of the Rzeppa type, according to Wm. Cull (British patent 810 289/1959) made by Hardy Spicer (GK N Automotive)

$\mu\nu = 1.59$ in Table 3.2 corresponds to this value. From (4.56) we get

$$d \Sigma\varrho = 2(2 + 0.269 - 0.962) = 2.614 .$$

The coefficient of conformity c_p can be calculated from (4.5) with these two values

$$c_p = \frac{858}{1{,}58} \sqrt[3]{2{,}62^2} = 1032 \ (\text{N/mm}^2)^{2/3}.$$

If the permissible specific load for the element transmitting the torque is known the Hertzian stress p_0 from (4.7b) can be determined. The static specific loading for the joint series in Fig. 4.58 is $k_0 = 46.23$ N/mm^2. The Hertzian stress from is then $p_0 = 1032 \sqrt[3]{46.23} = 3704$ N/mm^2.

The extreme torque M_N is obtained for $R/d = 1.65$.

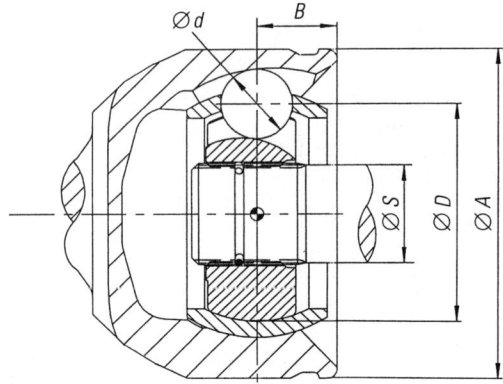

AC Joint M_N	$A \varnothing$	$\varnothing S$	B	$\varnothing d$	$\varnothing D$	splines Z
1100i	67.3	17.75	17.2	14.0	44.20	18
1300i	69.85	18.80	17.47	14.600	46.10	19
1500i	72.5	19.70	17.00	15.081	48.20	20
1700i	75.2	20.53	17.63	15.875	50.00	21
2000i	79.2	21.84	18.30	16.669	53.23	22
2300i	82.2	22.67	18.75	17.462	55.50	23
2600i	84.5	23.86	19.00	18.256	57.00	24
2900i	87.5	24.51	19.30	19.050	59.20	25
3300i	91.0	25.30	19.23	19.844	61.90	26
3700i	94.2	26.60	19.80	20.638	64.20	27
4100i	97.4	27.55	20.11	21.431	66.60	28

Fig. 4.59. AC fixed joint (improved) made by GKN Driveline 1999

Substituting in (4.53) we get:

$$M_N = 6 \cdot \underbrace{\left(\frac{3704}{1032}\right)^3}_{46.235} 1.65 \cdot 0.7071 \cdot \left(\frac{d}{10}\right)^3 \text{ Nm}.$$

The values for AC-joints in Table 4.11 are calculated from this equation and correspond to the M_N values in Fig. 4.58 [4.23, p. 18, Table 29.4].

J. W. Macielinski gave $p_0 = 3750$ N/mm^2 as the maximum value for ball joints 1970 (Table 4.10). The ISO 76/1987 permits 4200 N/mm^2 for ball bearings. However when selecting a joint it is recommended that the values in a manufacturer's catalogue be used as they reflect the state of the art.

The balls transmit the torque in the unarticulated ball joint with a constant pressure angle α and skew angle γ. With the joint at an angle β both values depend on the longitudinal shape of the tracks, arising from the "effective geometry".

4.4 Ball Joints

Table 4.10. Rated torque M_N of AC joints

Joint type	Nm	Joint type	Nm
75	944	150	7550
87	1499	175	11480
95	1887	200	17897
100	2237	225	25483
113	3186	250	34955
125	4365		

Table 4.11. Dynamic torque capacity M_d of AC joints [4.23]

Joint type	Nm	Joint type	Nm
75	178	150	1428
87	283	175	2267
95	358	200	3384
100	423	225	4819
113	602	250	6610
125	826		

Investigation of this relationship is mathematically very onerous. It has been done for straight tracks in the conical ruled uoint by Michel Orain 1976 [2.11, p. 67–75, 151]. The dynamic torque capacity of the articulated ball joint must therefore be calculated first of all without knowing the interaction of internal forces, using only the results of manufacturer's tests.

The dynamic torque capacity can be calculated if the permissible specific dynamic load k is known. In the AC joint series shown in Fig. 4.58 $k = 8.718$ N/mm². It follows from (4.53) that:

$$M_d = 6 \cdot \underbrace{\left(\frac{2124}{1032}\right)^3 1.65 \cdot 0.7071}_{8.718} \cdot \left(\frac{d}{10}\right)^3 \text{ Nm}.$$

The underlying Hertzian stress p_0 from (4.7b) here [4.25, p. 18, Table 29.4] is

$$p_0 = 1032 \sqrt[3]{8.718} = 2124 \text{ N/mm}^2.$$

The surface stress $p_0 = 2150$ N/mm², given by J. W. Macielinski applies for 1500 hours at 100 rpm and at an articulation angle of 3°.

For other values of p_0 the following applies

$$\frac{L_2}{L_1} = \left(\frac{P_1}{P_2}\right)^3 = \left(\frac{\left(\frac{p_1}{c_p}\right)^3 d^2}{\left(\frac{p_2}{c_p}\right)^3 d^2}\right)^3 = \left(\frac{p_1^3}{p_2^3}\right)^3 = \left(\frac{p_1}{p_2}\right)^9.$$

Thus small changes in the Hertzian stress lead to considerable changes in the service life [4.24].

For roller bearings these changes are, from theory:

- For an *increase* in p_0 of 25%

$$L_2 = L_1 \left(\frac{p_1}{1.25 p_1}\right)^9 = L_1 \left(\frac{1}{1.25}\right)^9 = L_1 \cdot 0{,}13.$$

The service life is *reduced* by 87%.
- For a *decrease* in p_0 of 25%

$$L_2 = L_1 \left(\frac{p_1}{p_1/1.25}\right)^9 = L_1 \cdot 1.25^9 = L_1 \cdot 7.45.$$

The life *increases* by 745%.

In (4.12) the equivalent speed n' must be inserted in the same way as in (4.23), with the difference that in Fig. 4.7 the oscillation angle of the ball joint is only $\beta/2$.

$$\frac{4\frac{\beta°}{2}}{360°} = \frac{n'}{n} \Rightarrow n' = n\frac{\beta°}{180°}.$$

With this one gets

$$\frac{L_{h2} n_2 \beta_2° \, 180°}{L_{h1} n_1 \beta_1° \, 180°} = \left(\frac{P_1}{P_2}\right)^p = \left(\frac{M_1}{mR} \frac{mR}{M_2}\right)^p,$$

$$L_{h2} = L_{h1} \left(\frac{n_1}{n_2}\right)^m \left(\frac{\beta_1}{\beta_2}\right)^n \left(\frac{M_1}{M_2}\right)^p \quad \text{[14]} \tag{4.63}$$

or written as a general function

$$L_{h2} = L_{h1} f(n)^m f(\beta)^n f(M)^p. \tag{4.64}$$

The function $f(M_1/M_2)^p$ with $p = 3$ corresponds to (4.12) from Arvid Palmgren's 1937 roller bearing theory [1.29]. Macielinski found

$$f\left(\frac{\beta_1}{\beta_2}\right)^n = \left(\frac{A_2}{A_1}\right)^3 = \left[\frac{(1 - \sin\beta_2)\cos^2\beta_2}{(1 - \sin\beta_1)\cos^2\beta_1}\right]^3. \tag{4.65}$$

The change of the pressure angle α with rotation of the articulated joint is included empirically in this.

The function $f(n)^m$ from (4.60) must be divided into two ranges:

- for speeds < 1000 rpm

$$L_{h2} \sim \frac{1}{\sqrt{n}} \sim \left(\frac{n_1}{n_2}\right)^{0.577}, \tag{a}$$

---

[14] Exponents m and n are included only for completeness, since empirical values are not available.

- for speeds > 1000 rpm

$$L_{h2} \sim \frac{1}{n} \sim \left(\frac{n_1}{n_2}\right). \tag{b}$$

Macielinski converted the life L_h at the speed range transition from 1500 hrs at 100 rpm to:

$$L_{h2} = 1500 \left(\frac{100}{1000}\right)^{0.577} = 397.275 \text{ hrs} \quad \text{at the more appropriate speed of 1000 rpm.}$$

It then follows for $\beta = 3°$

$$A = (1 - \sin 3°) \cos^2 3° = 0.945. \tag{c}$$

If the subscript 1 is omitted in (4.60) for the rig conditions, and "x" is used in place of the subscript 2, then for point contact one obtains

- for $n_x \leq 1000$ rpm

$$L_{hx} = \frac{397.275 \cdot 1000^{0.577}}{n_x^{0.577}} \left(\frac{A_x M_d}{0.945 M_x}\right)^3 = \frac{25339}{n_x^{0.577}} \left(\frac{A_x M_d}{M_x}\right)^3 \text{ hrs}, \tag{4.66}$$

- for $n_x > 1000$ rpm

$$L_{hx} = \frac{397.275 \cdot 1000^1}{n_x} \left(\frac{A_x M_d}{0.945 M_x}\right)^3 = \frac{470756}{n_x} \left(\frac{A_x M_d}{M_x}\right)^3 \text{ hrs}. \tag{4.67}$$

4.4.6.3 UF fixed yoints (undercut free)

These are the first step towards not machining the meridian tracks and further developments of non-machining manufacturing of the AC joints (Fig. 4.58). Their tracks, in the longitudinal section, are either only circular or made up of circular and straight-line parts. The basic requisite for this development was that the tracks in the bell and inner races could be made undercut free, in order to be able to use non-machining manufacturing methods.

To design without an undercut means it is necessary to place the generating centres of the tracks 0_1 and 0_2 in a line parallel to the middle axis of the joint at a distance b (Fig. 4.59). These also give a somewhat greater track depth than the AC joints. In the longitudinal direction, from the generating centre towards the front surface, the tracks are continued in a straight line and parallel to the joint axis. This allows more space for the articulated drive shaft and the articulation angle to be increased to 50°.

The skew angle γ is zero. The track cross-section is shaped like a gothic arch or elliptical (Fig. 4.43c).

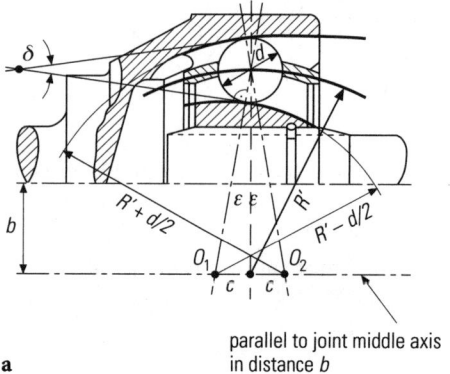

Fig. 4.60. Eccentric arrangement (III) of the generating centres 0_1, 0_2 of the tracks in a line parallel to the centre axis of the joint. Transition of the arc of a circle $R' + d/2$ into a straight line, parallel to the axis, from the centre of the joint out

Table 4.12. Data for UF-constant velocity joints made by GKN Löbro. **a** Geometrical data; **b** Hertzian stresses in N/mm² for M = 1000 Nm

		UF 95	UF 107
a Geometrical data			
Joint radius	R	29.678	32.959
Meridian track radius	R_M	29.927	33.246
Ball radius	$r = \dfrac{d}{2}$	9.000	10.000
Number of balls	z	6	6
Axial offset	V_c	± 3.95	± 4.39
Radial offset	V_b	0.00	0.00
Design pressure angle	α	45°	45°
Substitute radius	R_{ERS}	21.078	23.408
Basic steering angle	ε_0	7.581	7.587
b Hertzian stresses (N/mm²) for M = 1000 Nm			
Contact radius of the ball	R_{11} R_{12}	9.00	10.00
Track radius	outer R_{21}	− 9.405	− 10.30
	inner R_{21}	− 9.405	− 10.30
Conformity	outer f_A	0.523	0.515
	inner f_I	0.523	0.515
Running track radius	outer R_{22}	− 51.337	− 57.017
	inner R_{22}	33.337	37.017
Hertzian pressure	outer	2526.2	2081.2
	inner p_A	3070.4	2536.5
Hertzian pressure	p_0	4000	3800
Static torque in Nm	M_0	2210.9	3362.2

4.4 Ball Joints

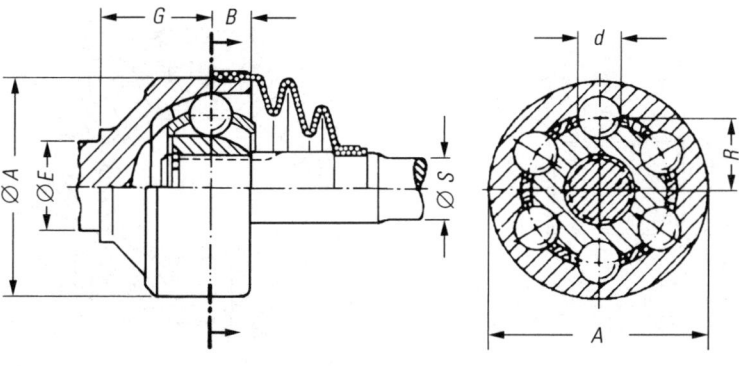

ØA mm	ØS mm	Nm$_{max}$ $\beta \leq 10°$
68		1100 i
70,1	18,5	1300 i
78	22,0	1750
88	23,0	2500
98	25,0	3400
86		2600 i
106		4600 i

Fig. 4.61. UF-constant velocity fixed joint (wheel-side) GKN Driveline. Articulation angle up to 50°

4.4.6.4 Jacob/Paland's CUF (completely undercut free) Joint for rear wheel drive > 25°

These joints are completely undercut free, as is the cage $\beta \sim 23°$. In 1972 H.-H. Welschof and E. Aucktor configured the tracks in the bell and inner race without an undercut and through eccentric generating centres (German Patent 2252827). This enabled the tracks to be manufactured without cutting and with a greater track depth. In 1975 E. Aucktor and W. Rubin also machined the concave surface of the bell, which guides the cage, without an undercut (German Patent 2522670). In 1993 W. Jacob designed the cage to be undercut free and supported the inner race on the spherical surface of a guide ring (German Patent 4317606), see Figure 4.62a.

The outer surface of the cage is separated from the bell, so that the second, redundant means of steering is avoided. The cage and the inner race are centred exclusively via the balls in the outer and inner tracks. This means there is no friction element and a joint results that can be articulated smoothly through all angles. This cage is made on a press at about 30 pieces a minute without distortion and finishing work. It, together with all other parts, can be formed efficiently, with high precision and good reproducibility. Figure 4.62b shows the substantially improved axial load absorption. The axial load is not taken via the bell and the outer contour of the cage, but is taken by a centring cup that contacts the inner race over a large spherical *surface*. Support of the centring cup via a sliding face on the joint cap that is connected to the bell largely eliminates the redundancy in the system. The cage is extended in an axial direction and is supported under a more favourable pressure angle and on a large sliding surface on the inner race of the joint.

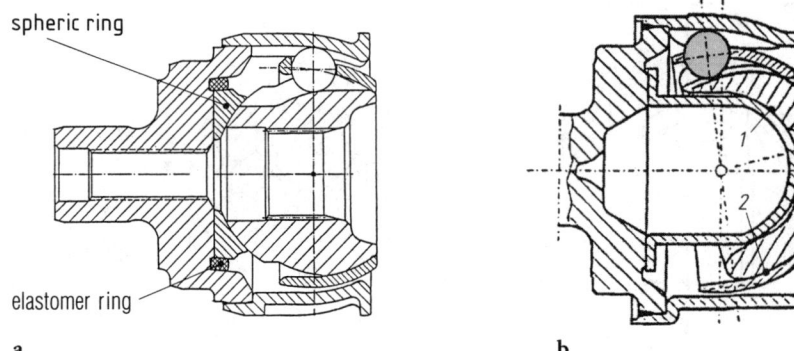

Fig. 4.62. Completely undercut free joint made up of bell, inner race and cage. **a** Sheet metal formed cage, inner race precision forged, an elastomeric ring removes the redundancy; **b** E. G. Paland's CUF 9 ball-joint with reduced surface pressure and sliding friction at 1 and 2; at 2 better axial load absorption between the sphere of the inner race and the inner surface of the cage; a cup containing lubricant does away with the radial redundancy; the cage is supported at a favourable pressure angle on a larger sliding surface area of the inner race

The inner races are deep drawn parts made of case hardening steel. Through hardened or induction or case hardening steels with sufficient depth of hardness penetration should be chosen.

The bell has a wall thickness of 3.6 mm.

The ball cage is also made out of case hardening steel and case hardened. In its windows it takes up the axial ball forces. These windows must also take up play in the circumferential direction because under articulation, they have to take up the to and for movement of the individual balls into the plane of symmetry.

A deep-drawn cup (part of the bell), apart from providing the centring, also takes the axial load from the normal forces between the ball and the bell. In addition it contains roller bearing grease for the sliding surfaces of the spherical concave profile, the inner race and on blanking cap of the joint. Sealing by means of a boot.

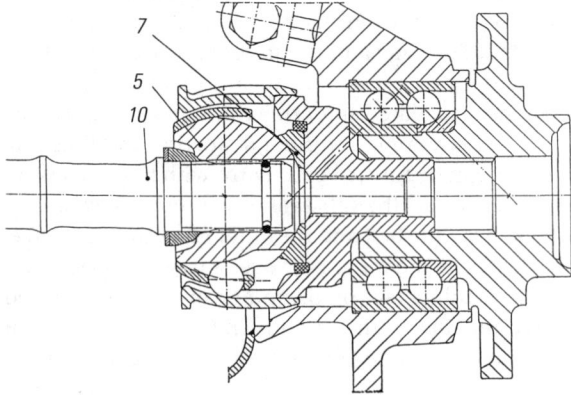

Fig. 4.63. CUF joint for a rear drive passenger car wheel bearing. 5 inner ball race with 7 outer supporting cup, 10 splined shaft

4.4.6.5 Calculation example for a CUF joint

Given:

$M_0 = 1000$ Nm	static torque
$d = 12.7$ mm	ball diameter
$R = 29.754$ mm	joint radius
$b = 6.0$ mm	radial offset
$c = 4.2$ mm	axial offset
$\alpha = 45°$	pressure angle
$\gamma = 0°$	skew angle
$\varepsilon = 6.7°$	angle of track inclination
$\varphi = 1.07°$	conformity value
$m = z = 9$	number of balls

To be calculated:

1. Hertzian compressive load Q (normal force on the balls)
2. The axes of the pressure ellipse a and b inside (inner race) and outside (bell)
3. The Hertzian stress $p_0 = p_{max}$ inside (inner race) and outside (bell)
4. The total deformation δ_0 on the inside contact surface (inner race) and outside surface (bell)
5. The static torque capacity M_0

Calculation 1: From (4.16).

$$P_x = \frac{M}{mR} = \frac{1000 \text{ Nm}}{9 \cdot 0.029754 \text{ m}} = 3734 \text{ N}$$

$$Q = \frac{P_x}{\cos(90 - \alpha)} = \frac{P_x}{\sin \alpha} = \frac{3734 \text{ N}}{0.7071} = 5281 \text{ N}$$

Solution 1: The Hertzian compressive load Q is 5281 N

Calculation 2a (inner race): From (3.45 and 3.46)

$$2a = \frac{4.72}{10^2} \mu \cdot \sqrt[3]{\frac{Q}{\Sigma\varrho}} \qquad 2b = \frac{4.72}{10^2} v \cdot \sqrt[3]{\frac{Q}{\Sigma\varrho}}$$

The values μ and v still missing here can be found through the Hertzian auxiliary value $\cos \tau$, e.g. from the Schmelz/Müller table [3.11].

For Member 1 (the ball) the given radii are equal in both directions.
Thus $r_{11} = r_{12}$.

For Member 2 (the track) the radius of the track cross-section is greater than the ball radius.

If the mid-point of the radius lies outside the member, Hertz designates this as negative. Here we have **concave** surfaces; they have a negative sign.

For the pressure ellipse in the inner race track:

$$\cos\tau = \frac{\left(\dfrac{1}{r_{11}}-\dfrac{1}{r_{12}}\right)+\left(\dfrac{1}{r_{21}}-\dfrac{1}{r_{22}}\right)}{\dfrac{1}{r_{11}}+\dfrac{1}{r_{12}}+\dfrac{1}{r_{21}}+\dfrac{1}{r_{22}}} = \frac{\left(\dfrac{1}{6.35}-\dfrac{1}{6.35}\right)+\left(\dfrac{1}{r_{21}}-\dfrac{1}{r_{22}}\right)}{\dfrac{1}{6.35}+\dfrac{1}{6.35}+\dfrac{1}{r_{21}}+\dfrac{1}{r_{22}}}$$

The radius r_{21} is shown in Figure 4.27 and is difficult to determine because in the case of ball joints the contact *does not take place at the base of the track*. For this reason no attempt is made to derive this and equation (4.52d) is used:

$$r_{21} = \frac{R+b}{\cos\alpha\cdot\cos\varepsilon} - \frac{d}{2} = \frac{29.754+6.0}{0.7071\cdot 0.9932} - \frac{12.7}{2} = 44.5605$$

The radius r_{22} is calculated with the help of the conformity value $\varphi = 1.07 \cdot r_{22} = -\varphi d/2 = -1.07 \cdot 6.35 = 6.7945$ (The negative sign of r_{22} comes from the concave surface).

$$\cos\tau = \frac{\left(\dfrac{1}{6.35}-\dfrac{1}{6.35}\right)+\left(\dfrac{1}{44.5605}-\dfrac{1}{-6.7945}\right)}{\dfrac{1}{6.35}+\dfrac{1}{6.35}+\dfrac{1}{44.5605}+\dfrac{1}{-6.7945}}$$

$$\cos\tau = \frac{(0.1575 - 0.1575)+(0.0224+0.1472)}{0.1575+0.1575+0.0224-0.1472} = \frac{0.1696}{0.1902} = -0.8917$$

The negative sign from Hertz has no significance in the present case.

With the aid of the "six-figure coefficients" the four-figure elliptical coefficients can now be found (Table 3.3) [3.11]

$\cos\tau$	μ	ν	$\mu\nu$	$2K'/\pi\mu$
0.891738	2.9908	0.4691	1.4031	0.6924

$$2a = \frac{4.72}{10^2}\,2.9908\cdot\sqrt[3]{\frac{5281}{0.1902}} = 4.2746 \qquad 2b = \frac{4.72}{10^2}\,0.4691\cdot\sqrt[3]{\frac{5281}{0.1902}} = 0.6705$$

Solution 2a: Pressure ellipse in the inner race track: $2a = 4.2746$ mm, $2b = 0.6705$ mm

$$a = 2.1373 \text{ mm}, \quad b = 0.33525 \text{ mm}$$

Calculation 2b (outer race): From (3.45 and 3.46) using the same calculationas for the inside pressure ellipse.

The half ball diameter is not subtracted here, but added.

$$r'_{21} = \frac{R+b}{\cos\alpha\cdot\cos\varepsilon} + \frac{d}{2} = \frac{29.754+6.0}{0.7071\cdot 0.9932} - \frac{12.7}{2} = 57.2615$$

$$r'_{22} = -\psi\, d/2 = -1.07\cdot 6.35 = -6.7945$$

4.4 Ball Joints

Pressure ellipse in the bell track

$$\cos \tau = \frac{\left(\dfrac{1}{6.35} - \dfrac{1}{6.35}\right) + \left(\dfrac{1}{-6.7945} - \dfrac{1}{57.2615}\right)}{\dfrac{1}{6.35} + \dfrac{1}{6.35} + \dfrac{1}{-6.7945} - \dfrac{1}{57.2615}}$$

$$\cos \tau = \frac{(0.1575 - 0.1575) + (-0.1472 + 0.0175)}{0.1575 + 0.1575 - 0.1472 + 0.0175} = \frac{-0.1297}{0.1503} = -0.8629$$

$\cos \tau$	μ	ν	$\mu\nu$	$2K'/\pi\mu$
0.862946	2.7040	0.4957	1.3405	0.7300

$$2a = \frac{4.72}{10^2} \, 2.7040 \cdot \sqrt[3]{\frac{5281}{0.1503}} = 4.1802 \qquad 2b = \frac{4.72}{10^2} \, 0.4957 \cdot \sqrt[3]{\frac{5281}{0.1503}} = 0.7663$$

Solution 2b: Pressure ellipse in the bell track: $2a = 4.1802$ mm, $2b = 0.7663$ mm
$a = 2.091$ mm, $b = 0.3832$ mm

Calculation 3: Hertzian pressure $p_0 = p_{max}$ inner and outer race from (3.54)

Inner race track

$$p_0 = p_{max} = \frac{858}{\mu\nu} \sqrt[3]{Q(\Sigma\varrho)^2}$$

$$p_0 = p_{max} = \frac{858}{1.4031} \sqrt[3]{5281 \, (0.1902)^2}$$

$$= 3521.9 \text{ N/mm}^2$$

Bell track

$$p_0 = p_{max} = \frac{858}{\mu\nu} \sqrt[3]{Q(\Sigma\varrho)^2}$$

$$p_0 = p_{max} = \frac{858}{1.3405} \sqrt[3]{5281 \, (0.1503)^2}$$

$$= 3150.9 \text{ N/mm}^2$$

Solution 3: Inner race track: $p_0 = 3521.9$ N/mm²   Bell track: $p_0 = 3150.9$ N/mm²

Calculation 4: the total deformation δ_0 at the inner contact area (hub) and the outer contact area (bell) is from (3.53)

Deformation in the inner race track

$$\delta_0 = \frac{2.78}{10^4} \cdot \frac{2K}{\pi\mu} \sqrt[3]{Q^2 \Sigma\varrho}$$

$$\delta_0 = \frac{2.78}{10^4} \cdot 0.6924 \sqrt[3]{5281^2 \cdot 0.1902}$$

$$= 0.0336 = 33.6 \text{ μm}$$

Deformation in the bell track

$$\delta_0 = \frac{2.78}{10^4} \cdot \frac{2K}{\pi\mu} \sqrt[3]{Q^2 \Sigma\varrho}$$

$$\delta_0 = \frac{2.78}{10^4} \cdot 0.7300 \sqrt[3]{5281^2 \cdot 0.1503}$$

$$= 0.0327 = 32.7 \text{ μm}$$

Solution 4: Inner race track: $\delta_0 = 33.6$ μm   Bell track: : $\delta_0 = 32.7$ μm

Calculation 5: From (4.52a and Section 4.2.3, p. 119): $M_0 = k_0 \cdot z \cdot d^2 \cdot R \cdot \sin \alpha$

First of all, from (4.7b), $k_0 = \left(\dfrac{p_0}{c_p}\right)^3$ is calculated for the point contact on the inner race

p_0 follows from (4.5) and (4.9) at:

$$c_p = \frac{858}{1.4031} \sqrt[3]{(12.7 \cdot 0.1902)^2} = \frac{858}{1.4031} \cdot 1.8003 = 1100.88$$

$$p_0 = 3300 \cdot 1100.88^{3/10} \cdot 0.0001^{1/5} = 4276 \text{ N/mm}^2$$

$$k_0 = \left(\frac{4276}{1100.88}\right)^3 = 58.60 \text{ N/mm}^2$$

$$M_0 = 58.60 \cdot 9 \cdot 12.7^2 \cdot 0.029754 \cdot 0.7071 = 1790 \text{ Nm}$$

Solution 5: $M_0 = 1790 \text{ Nm}$

4.4.7 Plunging Joints

Ball joints with straight or helical grooves are axially free. In addition to the two functions of the fixed joint (Fig. 5.1), the uniform transmission of torque and allowing the driveshaft to articulate, they have a third function, that of altering the length of the driveshaft.

Ball jointed driveshafts are recommended if the vibration requirements are stringent (Table 4.8). W or Z configurations are not needed for constant velocity jointed driveshafts. The only essential feature is that they work with the smallest possible articulation angle. They are simple to fit because no aligning is needed. Oscillating axial movements are not a problem for constant velocity plunging ball joints. The balls give low axial force even under high torque. The oscillating movement has the effect of improving the quality by polishing the tracks. Ball jointed driveshafts are generally lubricated for life and are therefore maintenance free.

4.4.7.1 DO Joints

Double-Offset joints have straight tracks parallel to the axis (Fig. 4.64). Because of this their skew angle $\gamma = 0$. The pressure angle in the elliptical cross-section of the tracks is 40° and the conformity $\varkappa_Q = -0.952$ to -0.971 ($\psi = 1.03$ to 1.05). The curvature of the tracks in the direction of movement is

$$\varrho_{21} = \frac{1}{r_{21}} = \frac{1}{\infty} = 0.$$

Thus the Hertzian coefficient $\cos \tau$ for $\varkappa_Q = -0.962$ is from (4.54)

$$\cos \tau = \frac{0 + 0.962}{2 + 0 - 0.962} = 0.9268.$$

In Table 3.3 $\mu\nu = 1.51$ corresponds to this value.
We then have for $d\,\Sigma\varrho$ from (4.57)

$$d\,\Sigma\varrho = 2(2 + 0 - 0.962) = 2.076.$$

The coefficient of conformity c_p after (4.5) is

$$c_p = \frac{858}{1.51}\sqrt[3]{2.076^2} = 925.9 \,(\text{N/mm}^2)^{2/3} \quad 2.076^2 = 4.3098; \quad \sqrt[3]{4.3} = 1.6.$$

For the permissible specific loading $k = 8.712$ N/mm², given by Macielinski [4.25, p. 21], the Hertzian stress is, using (4.7b),

$$p_0 = 925.9\sqrt[3]{8.712} = 1905 \text{ N/mm}^2.$$

From (4.53b) the dynamic torque capacity is

$$M_d = m \cdot 8.712 \cdot d^2 R \sin 40° \text{ Nm}. \tag{4.68}$$

The DO joint was further developed by Sobhy Labib Girguis in 1971 into the five-ball DOS joint (Fig. 4.65). He inverted the plunging principle kinematically and thus obtained plunge of up to 50 mm. The cage is no longer centred on a spherical inner body, but on the outer body. The plunge now takes place between the steering body of the cage and the shaft which forms, by means of integral tracks, the inner race. The

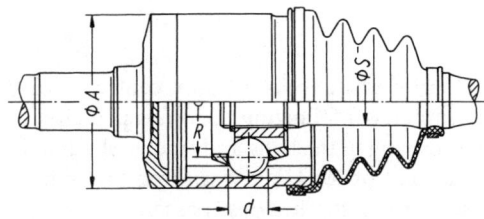

Joint St	A	∅ S	d	R	s_{max}	M_d	M_N
70	70	18.7	15.875	22.5	28	188	830
75	75	19.7	16.667	23.67	28	218	1150
79	77	20.7	17.462	24.8	26	250	1330
82	82	20.9	18.257	25.90	38	285	1370
89	87	22.7	19.844	28.18	48	367	1750

A Joint-∅　　R joint radius
S shaft-∅　　s_{max} max. plunge
d ball-∅

Fig. 4.64. Six-ball DO (Rzeppa type) plunging joints, $\beta_{max} = 22°–31°$. Hardy Spicer (GKN Driveline) design

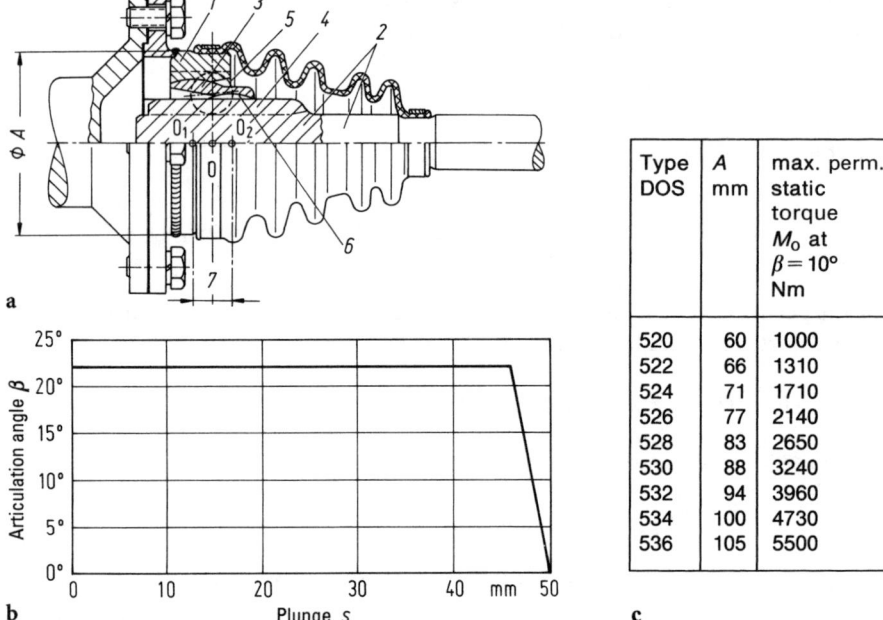

Fig. 4.65a–c. Five-ball DOS plunging joint, Girguis design 1971, produced by Löbro (GKN automotive AG). V8, according to Table 5.6 (German Patent 2114536 and US Patent 3789626). **a** Longitudinal section. *1* outer race; *2* inner race with profiled shaft; *1* offset cage; *4* tracks; *5* balls; *6* steering ring; *7* double offset; O_1 and O_2 offset joint centres; **b** plunge s in mm when the shaft is extended as a function of the articulation angle β, **c** principal dimensions in mm and torque capacity in Nm

centres O_1 and O_2 of the spheres which are offset by c from the joint centre O cause the cage to be steered into the plane of symmetry π following the offset principle.

Another plunging joint (V3) with cage steering of the balls and tracks which have parallel axes, but are inclined, is the Dana joint of 1964[15]. It was fitted in the half-shafts of the Oldsmobile Toronado passenger car, the heaviest passenger car with front wheel drive (Fig. 4.67). However, it provided only a small amount of plunge and a shallow track [5.8]. It was superseded by VL and DO joints.

4.4.7.2 VL Joints

Plunge joints like fixed joints, can be configured with various types of ball steering. Erich Aucktor's 1981 (4 + 4) joint (Fig. 4.68) with inclined and parallel tracks[16], for an articulated bus (giving a small turning circle), showed that plunging joints can achieve a high torque capacity and articulation angles up to 25°. The intersecting tracks steer the balls well in the unangled joint, but this steadily diminishes as the articulation angle increases.

---

[15] Dana-Corp., Detroit/Mich., USA (French Patent 1 318 912/1962, USA-PS 3 105 369/1961).
[16] (DE-PS 3 102 871/1981).

4.4 Ball Joints

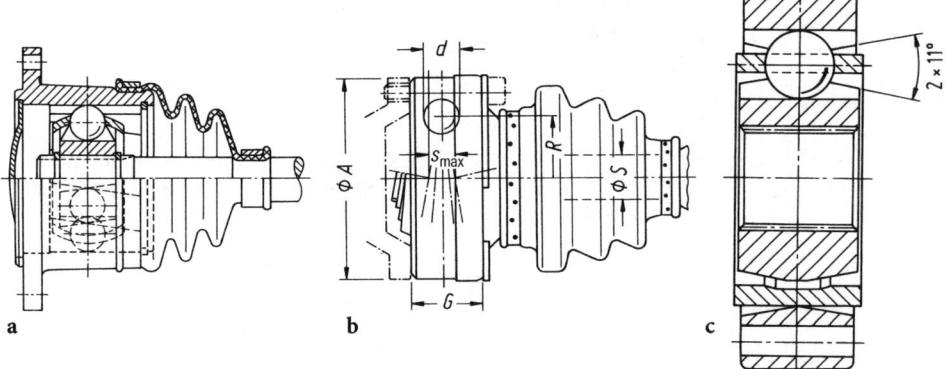

Principal dimensions in mm and torque capacity in Nm

Rear axle drive Nm	⌀ A	⌀ S	d	R
2300	86 ÷ 94	25	17,462	26,45
2900	94 ÷ 102	26		
3700i	98 ÷ 108	28		
3300i	94 ÷ 100	28	22,225	31,95
4600i	108 ÷ 116	29	23,812	35,0
front whell drive				
1500	73	20		
2000	82	22		
2300	84	25		
2600	92	26	17,62	26,45
3300i	92	28		
3700	100	28		

Fig. 4.66. VL Rzeppa type plunging joints, inclined tracks and 6 balls, Löbro design. **a** Flange version (passenger cars, differential side); **b** Disc version (machinery, passenger cars) from E. Aucktor's German Patent 1 232 411/1963; **c** VL joint with large length compensation for $\beta = 20°$. Inner and outer members are roof-shaped

Normal VL joints run very quietly under torque loading because the inner joint play is compensated for by the reciprocal axial pressure of the balls. Only at zero torque is there no compensation for play.

Cage-steered ball joints, as presently used in mass produced vehicles, produce troublesome noises under certain driving conditions because of the play between the balls and the track. Play results from the wear of the joint in operation.

In order to avoid knocking and rattling in a drivetrain at idle or under load, the play in the joints must be removed. In the VL joint there is a play of between 0.01 and 0.1 mm of the balls in the tracks and cage windows. Under the pre-tension the cage shifts the balls in its window openings axially until they are supported in the ball tracks and therefore cut out the radial and axial play in the joint.

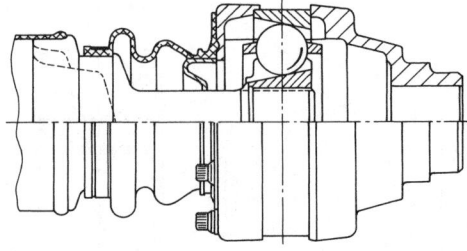

Fig. 4.67. Front wheel drive joint for the 1964 Oldsmobile Toronado, made by the Dana Corp.

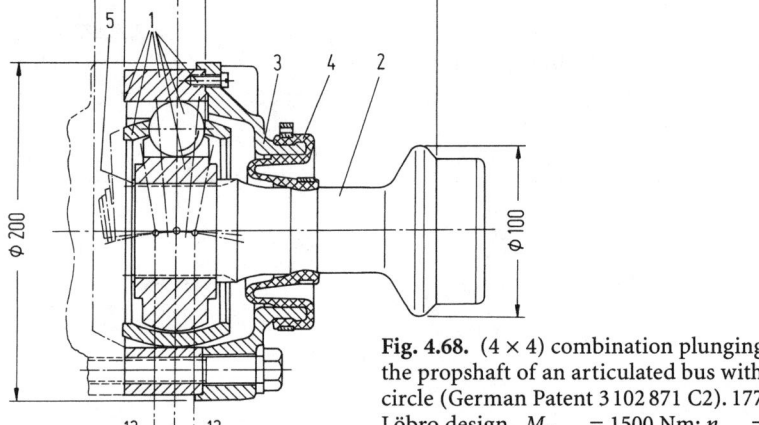

Fig. 4.68. (4 × 4) combination plunging joint VL with the propshaft of an articulated bus with a tight turning circle (German Patent 3 102 871 C2). 177 kW/2200 rpm. Löbro design. M_{Nmax} = 1500 Nm; n_{max} = 2300 rpm, β_{max} = 20°. *1* complete assembly; *2* input shaft; *3* cover plate; *4* boot; *5* Seeger ring or shall retaining ring

The parallel tracks behave in the opposite way. In the unangled joint the balls hardly steer at all, but as the articulation angle increases the steering gets increasingly better. The two track arrangements are therefore complementary.

VL joints (Löbro plunging joint) work with straight or helical tracks at 16° to the axes of the driving and driven members (Fig. 4.66). The pressure angle in the circular cross-section of the tracks is taken to be 40° and the conformity κ_Q = −0.95 to −0.97 (ψ = 1.03 to 1.05). The curvature in the direction of travel ϱ_{21} = 0. The Hertzian coefficient from (4.57) is

$$\cos\tau = \frac{0 + 0.962}{2 + 0 - 0.962} = 0.9268.$$

From Table 3.3 $\mu\nu$ = 1.51. From (4.56) we get $d\,\Sigma\varrho$ = 2(2 + 0 − 0.996) = 2.076. The coefficient of conformity is then, from (4.5),

$$c_p = \frac{858}{1.51}\sqrt[3]{2.076^2} = 925.9\ (\text{N/mm}^2)^{2/3}.$$

4.4 Ball Joints

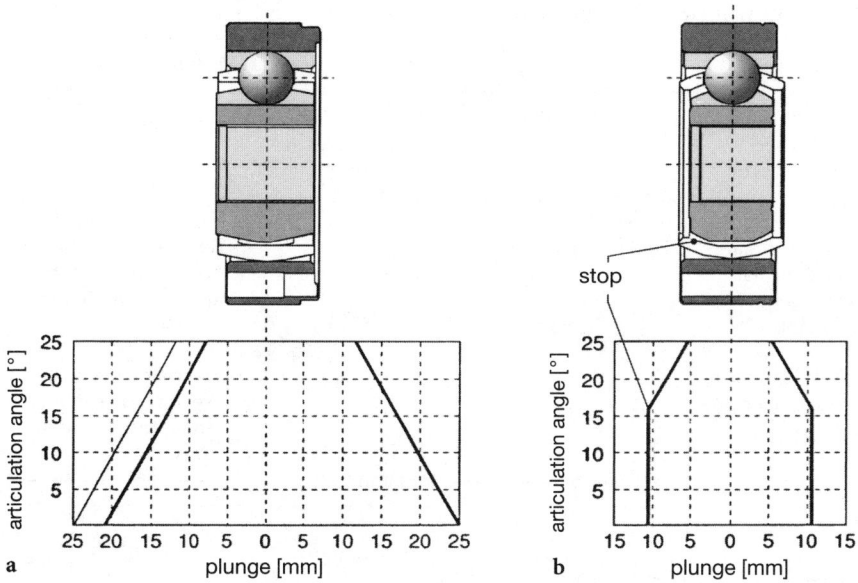

Fig. 4.69. Plunge of the VL joint, as a function of the articulation angle and how it is fitted. **a** front axle, **b** rear axle. From [5.18]

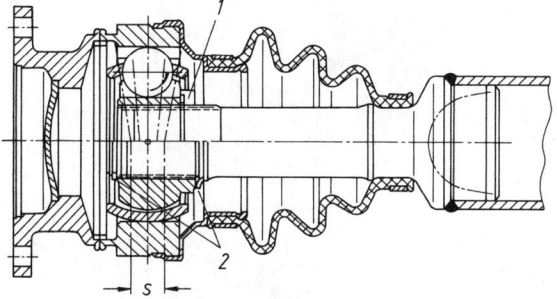

Fig. 4.70. Normal VL plunging joint with supported cage for speeds up to 3000 rpm, $\beta_{max} = 18°$. Design by Löbro (GKN Automotive AG). DIN, SAE, special flange or shaft ennection. Inner race secured with *1* snap ring (as in Fig. 5.50); *2* Seeger ring or Belleville washer

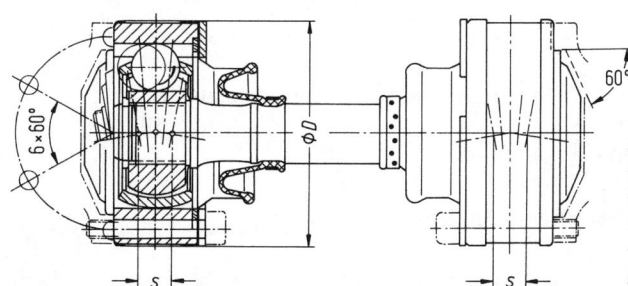

Fig. 4.71. VL high speed plunging joint for speeds up to 10000 rpm, $\beta_{max} = 10°$. Short version with two oil-tight disc joints as "floating shaft"

Table 4.13. VL plunging joint applications (Figs. 5.66 to 5.69)

Vehicle systems	Engines	Ships	Machinery
Inboard joints in front wheel drive halfshafts	Connecting shaft between prime mover and sprung attachment	Driveshafts for turning gear and ship propellers	Conveyors
Halfshafts for rear wheel drive with independent suspension	Oscillating drives		Transmission between shafts which are not parallel and do not intersect
Propshafts in trucks and passenger cars			Drives which have to work with absolutely no vibration and at constant velocity
		←———— Fan and pump drives ————→	

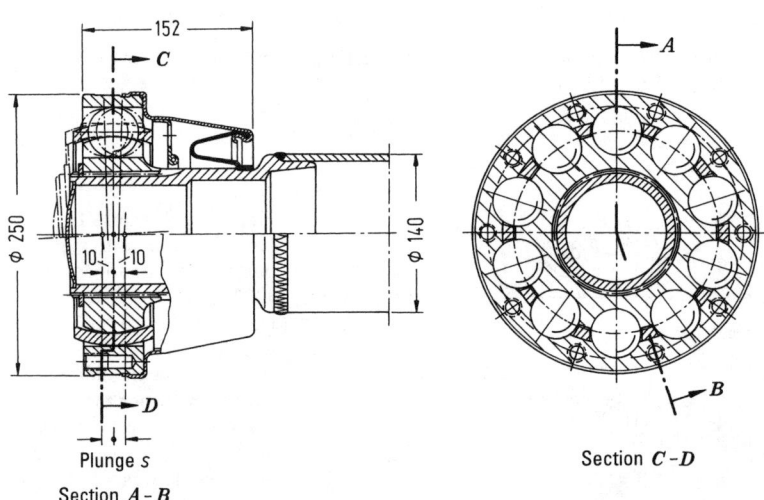

Plunge s
Section A-B

Section C-D

Fig. 4.72. Ten-ball disc joint for a turbine drive (2800 kW, n = 3000 rpm). Axial plunge 20 mm, β_{max} = 3°, ball diameter 44.45 mm, M_d = 9000 Nm, ball circle diameter 182 mm, tracks with 10° inclination angle, circular cross-section and an inverse conformity value ψ = 1.03

With the empirical value k = 9.170 N/mm² and (4.7b) the Hertzian stress is

$$p_0 = 925.9 \sqrt[3]{9.170} = 1964 \text{ N/mm}^2.$$

The dynamic torque capacity after (4.53) is then

$$M_d = m \cdot 9.170 \cdot d^2 \cdot R \sin 40° \cdot \cos 16° \text{ Nm}. \tag{4.69}$$

4.4.8 Service Life of Joints Using the Palmgren/Miner Rule

Manufacturers determine the life of a joint in hours using test rig trials under constant conditions of torque, speed and articulation angle (4.27), (4.66), (4.67), (4.82). However calculating the service life for vehicles is much more difficult. In addition to the changes in torque and speed, the different types of roads and ways of driving have to be considered.

In 1970, J. W. Macielinski put forward a method of calculating service life based on the Palmgren/Miner rule, see Sect. 1.3.3 [4.23; 1.23]. He took a simplified Central European driving programme and from this derived a procedure to work out the service life for all types of cars (Table 4.14).

For passenger cars, Macielinski used two thirds of the maximum engine torque in all gears (at the appropriate speed n_M). From the given data: maximum engine torque M_M at n_M rpm, GVW, maximum axle loads G_{ER}, laden height of centre of gravity h, static and dynamic rolling radii R_{stat} and R_{dyn}, the gear ratios i and the mean articulation angle β of the joints, it is possible to calculate:

- the shaft speed $n_x = \dfrac{n_M}{i_x}$ rpm, (4.70)

- the road speed
$$V_x = \frac{2 R_{dyn} \pi n_M}{60} \cdot \frac{3600}{1000} = 0{,}377\, R_{dyn} n_M \text{ kph}, \qquad (4.71)$$

- the torque $M_x = \dfrac{2}{3} \dfrac{\varepsilon_{E\,R}}{u} M_m i_x$ Nm, (4.72)

- the service life L_{hx}:
  for Hooke's joints (4.28)

$$L_{hx} = \frac{1{,}5 \cdot 10^7}{n_x \beta_x} \left(\frac{CR}{k_t M_x} \right)^{10/3} \text{hrs} \qquad (4.73\text{a})$$

for ball and pode joints (4.59)

$$L_{hx} = L_h \left(\frac{n}{n_x} \right)^m \left(\frac{\beta}{\beta_x} \right)^n \left(\frac{M_d}{M_x} \right)^3 \text{hrs}. \qquad (4.73\text{b})$$

For ball joints using (4.62), when

n_x(rpm)	≤ 100	> 1000
L_{hx} (hrs)	$\dfrac{25339}{n_x^{0.577}} \left(\dfrac{A_x M_d}{M_x} \right)^3$	$\dfrac{470756}{n_x} \left(\dfrac{A_x M_d}{M_x} \right)^3$

(4.73c)

Tables 4.14 and 4.15 given the utilisation of the individual gears. The following average speeds are assumed: small passenger cars: 40 to 50 kph; medium and large passenger cars: 55 to 80 kph; lorries: 30 to 40 kph; buses: 40 to 50 kph.

Table 4.14. Percentage of time in each gear on various types of road [4.25]

Type of vehicle, number of gears	Road type[a]	Gear					
		1	2	3	4	5	6
Heavy truck, 6 gears	I	0	1	2	11	24	62
	II	1	1	8	20	40	30
	III	1	1	5	22	43	28
	IV	1	1	23	16	31	28
	V	1	1	6	25	46	21
Heavy truck with trailer, 6 Gänge	I	1	1	4	12	29	53
	II	1	10	18	25	28	18
	III	1	6	30	28	24	11
	IV	1	23	18	18	28	12
	V	1	2	10	30	45	12
Medium truck, 5 gears	I	1	1	3	16	79	–
	II	1	3	23	36	37	–
	III	1	5	27	40	27	–
	IV	1	19	19	25	36	–
	V	2	4	20	42	32	–
Medium truck with trailer, 5 gears	I	1	3	12	25	59	–
	II	3	23	23	29	22	–
	III	3	30	28	30	9	–
	IV	7	28	23	21	21	–
	V	1	8	28	46	17	–
Passenger car, 4 gears	I	1	2	14	83	–	–
	II	1	7	39	53	–	–
	III	1	6	48	45	–	–
	IV[b]	–	–	–	–	–	–
	V	3	9	47	41	–	–
Passenger car, 3 gears	I	1	8	91	–	–	–
	II	2	29	69	–	–	–
	III	2	38	60	–	–	–
	IV[b]	–	–	–	–	–	–
	V	6	40	54	–	–	–

[a] Height of road difference in m: I flat 280, II hilly 400, III mountainous 500, IV high mountains 940, V town.
[b] not determined

Table 4.15. Percentage of time a_x for passenger cars

Gear	Macielinski [4.25]	Löbro		
		4 gears	5 gears	
			economy	sporty
1	1.5	1	1	1
2	6	6	5	6
3	37	18	27	18
4	55.5	75	40	30
5	–	–	27	45

4.5 Pode Joints

According to the Palmgren/Miner rule the total service life is obtained from the total fatigue damage $1/L_h$, which is the sum of the individual fatigue damages a_x/L_{hx} from (1.24)

$$\frac{1}{L_h} = \frac{a_1}{L_{h1}} + \frac{a_2}{L_{h2}} + \frac{a_3}{L_{h3}} + \ldots + \frac{a_n}{L_{hn}}.$$

From (4.67) and (1.27) the mean rotational speed is

$$\frac{V_m}{0{,}377 R_{dyn}} = \frac{1}{0{,}377 R_{dyn}} (a_1 V_1 + a_2 V_2 + a_3 V_3 + \ldots + a_n V_n).$$

From this we get

$$V_m = a_1 V_1 + a_2 V_2 + a_3 V_2 + \ldots + a_n V_n. \tag{4.74}$$

The durability in kilometers L_S is obtained as a product of the life L_h [1.25; 1.32; 4.26] with the mean driving speed V_m

$$L_s = L_h V_m \text{ km}.$$

4.5 Pode Joints

Pode joints combine elements of Hooke's and ball joints. The tangential component P_x of the compressive load P, resulting from the torque M, is transmitted by the pode trunnions of the input member via spherical members with roller or plain bearings, to the output member (Fig. 4.73). If the transmitting roller has a roller bearing, line contact occurs at point *1* between the pode trunnion and the bearing needles, and also at point *2* between the bearing needles and the roller bore. If the bearing is plain point *2* is eliminated; when assessing the trunnion stress at point *1*, a distinction must be made between the non-articulated and the articulated joint. For an articulated joint, the rollers turn, i.e. the sliding faces of the trunnion and bore are lubricated. Only low surface stresses can be permitted here ($p < 60$ N/mm^2).

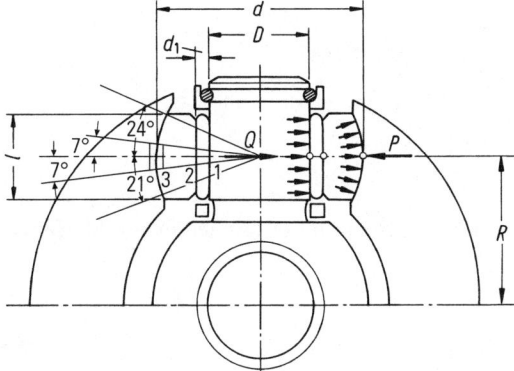

Fig. 4.73. Loads on the trunnion of a pode joint

The roller and the track are in point contact at 3; this is determined by the conformity $\varkappa_Q$ in the cross-section. The (concave) conformity $\varkappa_Q = -0.92$ to -0.98 ($\psi = 1.02$ to 1.08).

The torque capacity of pode joints is determined by the permissible stresses at points 1 to 3. The compressive load from (4.16) is distributed at points 1 and 2 to the whole needle bearing; at point 3, the equivalent compressive load P acts on the narrow contact area. Therefore

$$Q_1 = Q_2 = P_3. \tag{a}$$

For pode joints with rollers fitted with roller bearings, from (4.8) and the Stribeck load distribution (4.18) Eq. (a), gives the following for points 1 and 2

$$\frac{z}{s}\left(\frac{p_1}{c_{p1}}\right)^2 \cdot l_w \cdot d_1 = \frac{z}{s}\left(\frac{p_2}{c_{p2}}\right)^2 \cdot l_w \cdot d_1. \tag{b}$$

From this we get

$$p_1 = \frac{c_{p1}}{c_{p2}} p_2$$

or with (4.6)

$$p_1 = \frac{270\sqrt{1 + d_1/D}}{270\sqrt{1 - d_1/D}} p_2. \tag{c}$$

If the diametral ratio $d_1/D \approx 1/10$, then

$$p_1 = \frac{\sqrt{1.1}}{\sqrt{0.9}} p_2 = 1.106\, p_2.$$

Because of the less favourable conformity coefficient c_{p1}, the surface stress p_1 is about 11% higher ($p_1 > p_2$). From Eq. (b) the following applies at points 1 and 3:

$$\frac{z}{s}\left(\frac{p_1}{c_{p1}}\right)^2 l_w d_1 = \left(\frac{p_3}{c_{p3}}\right)^3 d^2,$$

$$p_1^2 = \frac{s}{z} \frac{c_{p1}^2}{c_{p3}^3} \frac{d^2}{l_w d_1} p_3^3. \tag{d}$$

The diametral ratio $d_1/d \approx 1/18$, $l_w/d \approx 3/8$ and the number of needles $z \approx 30$. Hence, from Eq. (d), we get

$$p_1^2 = \frac{5}{30} \frac{c_{p1}^2}{c_{p3}^3} \frac{d^2}{l_w d_1} p_3^3. \tag{e}$$

4.5 Pode Joints

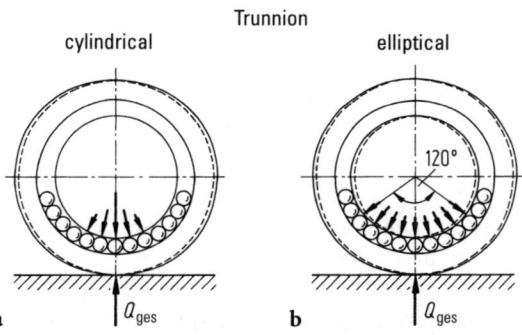

Fig. 4.74. "optimum" profiling of the trunnion of the tripode joint for uniform load distribution over 120° of the trunnion circumference. **a** Concentrated stress distribution for a normal trunnion; **b** uniform stress distribution on the trunnion with "optimum" profile, designed by Glaenzer-Spicer (GKN Automotive AG). The torque capacity is approximately doubled [2.11, p. 205]

From (4.6) the coefficients of conformity c_p are

$$c_{p1} = 270\sqrt{1 + d_1/D} = 270\sqrt{1 + \frac{1/18d}{1/2d}} = 270\sqrt{1 + 0.111}$$
$$= 270 \cdot 1.0540 = 284.6 \; (\text{N/mm}^2)^{1/2}.$$

With the conformities $\varkappa$ from (4.54), we get

$$\varkappa_L = \frac{d}{\dfrac{2(R+b)}{\cos\alpha\cos\varepsilon} - \dfrac{d}{2}} = \frac{d}{\dfrac{2(R+\infty)}{1\cdot 1} - \dfrac{d}{2}} = 0.$$

From (4.55) $\varkappa_Q = 1/-\psi = 1/-1.005 = -0.9950$, the Hertzian coefficient becomes

$$\cos\tau = \frac{\varkappa_L - \varkappa_Q}{2 + \varkappa_L + \varkappa_Q} = \frac{0 - (-0.9950)}{2 + 0 - 0.9950} = \frac{0.9950}{1.0050} = 0.9900.$$

The elliptical coefficient $\mu\nu = 2.23$ can be obtained from Table 3.3. From (4.57)

$$d\,\Sigma\varrho = 2(2 + \varkappa_L + \varkappa_Q) = 2 \cdot 1.0050 = 2.010.$$

The coefficient of conformity at point 3 therefore becomes, from (4.5),

$$c_{p3} = \frac{858}{\mu\nu}\sqrt[3]{(d\Sigma\varrho)^2} = \frac{858}{2.23}\sqrt[3]{2.010^2} = 612.8 \; (\text{N/mm}^2)^{2/3}.$$

Substituted in Eq. (e), this gives the surface stress

$$p_1^2 = \frac{5}{30} \frac{d^2}{3/8\, d \cdot 1/18\, d} \frac{284.6^2}{612.8^3} p_3^3 = 8 \cdot 3.52 \cdot 10^{-4} \cdot p_3^3$$

$$p_1 = 5.307 \cdot 10^{-2}\, p_3^{3/2}.$$

The contact stress on position 1 between needle and pode trunnion increases on position 3 between roller and tread with the $1^1/_2$ fold power of p_3.

If no special measures are taken at the pode trunnion, e.g. by improving the profile as shown in Fig. 4.76, p_1 rapidly becomes much greater than p_3, see Table 4.16. The needle bearings then determine the torque capacity of the pode joint with trunnions mounted in roller bearings, as is the case for the Hooke's joint.

For pode joints with trunnions mounted in plain bearings (Fig. 4.4), point 2 coincides with point 1. From (1.14) by Carl Bach 1891 (Sect. 1.3.2) the following applies for P_1 in Eq. (a)

$$\frac{\pi}{4} D l_w p_1 = \left(\frac{p_3}{c_p^3}\right)^3 d^2.$$

The diametral ratio $D/d \approx 1/2$ and $l_w/d \approx 4/7$, giving

$$p_1 = \frac{4}{\pi} \frac{d^2}{D l_w} \left(\frac{p_3}{c_{p3}}\right)^3 = \frac{4}{\pi} \cdot \frac{d^2}{1/2\, d \cdot 4/7\, d} \left(\frac{p_3}{c_{p3}}\right)^3. \tag{f}$$

The coefficient of conformity $c_{p3} = 612.8$ (N/mm²)$^{2/3}$ is known from (4.73b). Hence, Eq. (f) becomes

$$p_1 = \frac{4}{\pi} \cdot \frac{2 \cdot 7}{4 \cdot 612.8^3} \cdot p_3^3 = 1.94 \cdot 10^{-8} \cdot p_3^3. \tag{4.75}$$

We set the condition $\delta_b/d = 10^{-4}$ from (4.9) for a passenger car starting from rest in 1st gear. In these conditions we obtain Hertzian stresses $p_3 = 3300 \cdot 612.8^{3/10} \cdot 0.1585 = 3587$ N/mm² between the roller and the track and $p_1 = 895$ N/mm² between the roller and the trunnion from Table 4.4; these last only a short time.

Table 4.16. Surface stresses in pode joints with roller bearings

p_3 N/mm²	p_1 N/mm²
250	210
500	593
750	1090
1000	1678
1250	2345
1500	3083
1750	3885

Table 4.17. Surface stresses in pode joints with plain bearings

p_3 N/mm²	p_1 N/mm²	p_3 N/mm²	p_1 N/mm²
250	0.3	2000	155.2
500	2.4	2250	221.0
750	8.2	2500	303.1
1000	19.4	2750	403.5
1250	34.9	3000	523.8
1500	65.5	3250	666.0
1750	104.0	3587	895.0

4.5 Pode Joints

Robert Schwenke took the first step towards today's pode joints in 1902 with his inboard bipode joint (Sect. 1.3.2) [2.7]. In 1935 John W. Kittredge observed that shafts to be coupled were frequently not in true alignment, due to defective workmanship or settling of foundations. This led to undesirable strains and vibrations, especially at high speeds. His solution was a radially constrained tripode joint (US patent 2125615/1935).

Two years later the Borg Warner engineer Edmund B. Anderson reported on developments for motor vehicles, where high volumes and low cost are the order of the day. His tripode joint (US patent 2235002/1937) have the trunnions on pivots to allow oscillation in the plane of rotation of the supporting yokes. In some of his designs the ball races are formed from sheet metal, including fixed joints with arcuate races which restrain radial displacement. Other concepts have elongataed ball races which allow the joint to plunge.

4.5.1 Bipode Plunging Joints

Fixed bipode joints have not been widely used because they transmit torque just as unevenly as Hooke's joints. They were used in early, slow-running machinery and motor vehicles as plunging joints (Figs. 1.34, 1.41, 1.42) where their shortcomings did not have too great an effect. More recently, they have been replaced by ball plunging joints. The needle bearings of the rollers determine the torque capacity of these joints; their load capacities C_0 and C can be used for design purposes as for Hooke's joints.

The effective component P_x of the compressive load P in (4.16) is a result of the position of the contact faces relative to the axis of rotation of the joint. Its surface normals lie tangentially to the effective radius R and are perpendicular to the joint axis (Fig. 4.73). The pressure angle α and the skew angle γ are therefore $\alpha = 90°$ and $\gamma = 0°$ respectively. It follows from (4.16b) that

$$P_x = P \sin 90° \cos 0° = P.$$

From (4.16)

$$M = mPR.$$

The permissible static compressive load Q_{total} (see Sect. 4.2.1) for a non-articulated joint is the same as the static load capacity C_0. Thus, with (4.20) and $m = 2$, the static torque capacity becomes

$$M_0 = 2C_0 R = 2 \cdot 38 \cdot iz l_w \, dR \quad \text{Nm}. \tag{4.76a}$$

The static load capacity decreases with articulation. The output torque from (4.46) from the interaction of forces in the Hooke's joint shown in Sect. 4.3.1 is

$$M_2 = \frac{M_1}{\cos \beta} = P_2 h. \tag{4.76b}$$

P_2 is the maximum compressive load on the needle bearing of yoke 2 and should not be greater than Q_{total}. The distance between the centres of the two needle bearings $h = 2R$. With M_1 equal to the torque M_β in the articulated joint, from Eq. (4.76b) one obtains

$$M_\beta = 2Q_{total} \cdot R \cdot \cos\beta.$$

Since $Q_{total} = C_0$ it follows that

$$M_\beta = 2C_0 R \cos\beta = 2 \cdot 38 \cdot izl_w \, dR \cos\beta \, \text{Nm}. \tag{4.77}$$

Since the bipode joint has the same effective geometry as the Hooke's joint, from Poncelet's Eq. (1.1), the relevant equations in Sect. 4.2.2 to 4.2.5 apply for the bipode joint. (4.29) applies for the dynamic torque capacity.

4.5.2 Tripode Joints

Glaenzer-Spicer took up the idea of pode joints in 1953 with the tripode plunging joint. This joint provides high plunge capacity, withstands high torques and is maintenance free. It has been used successfully on the Eisenach "Wartburg" 1956, 1958 "Trabant"[1] (Fig. 4.77), the Citroën DS19, Renault Estafette and R12, and the 1965 Peugeot 204. Since then it has become the French national joint.

In 1976, Michel Orain showed that the requirement for an angle-bisecting plane in constant velocity joints is the result of a special, restrictive condition and so not all possibilities giving constant velocity are covered [2.11, Sect. 1.3.4, pp. 55–62]. Tripode joints (tripode: literally "three-footed" but more usually "three-armed", $\psi_i = 120°$) comply with the general condition giving constant velocity, that is in all practical working positions, there is an instantaneous axis parallel to the angle-bisecting plane of the joint [2.11, pp. 105–116].

A tripode joint comprises (Fig. 4.77):

- an equiangular tripode (input member *1*) on which the axes of all the trunnions lie in a plane perpendicular to the input shaft,
- a forked component (output member *2*) with three equispaced cylindrical tracks parallel to the output shaft,
- three spherical rollers, each of which makes one of the axes of the tripode intersect one of the axes of the tracks.

The effective geometry of the GE fixed joints (*Glaenzer Exterior*) from Fig. 4.77) and the GI plunging joints (*Glaenzer Interior*) from Fig. 4.79 is shown in Sect. 2.2 (Fig. 2.23). They only differ in the details of their design; on GE joints the plunge is stopped by a button *7* and a retaining clip *8* (Fig. 4.77) and the load transmission takes place from outside, rather than from inside as with GI joints. In addition, the GI joint, in contrast to the GE joint, has rollers on roller bearings.

In the tripode joint the supporting surface area is adequate if it is given a contact angle $1/d = 0.466 = \tan 25°$. The arc given by the 25° angle determines the origin for the spherical roller curvature at the contact points (Fig. 4.77).

4.5.2.1 Static Torque Capacity of the Non-articulated Tripode Joint

As with the bipode joint, the pressure angle α is 90° and the skew angle γ is 0°. Hence (4.53) applies again. The specific loading k and the Hertzian stress p_0 between the roller and track are calculated from

$$M = 3kd^2R = 3\left(\frac{p_0}{c_p}\right)^3 d^2R. \tag{4.78}$$

Because of the isostatic load transmission, the tripode spider nutates about the centre of the joint with each revolution. However, notwithstanding the oscillation, the loads on the rollers remain equal (Sect. 4.5.2). The movement has the shape of a trefoil with eccentricity

$$\lambda = (x_{11} - R)/R = 1/2(1/\cos\beta - 1), \tag{4.79}$$

from Eq. (a) in (4.81) p. 224, and a displacement

$$s = (l_1 + l_2 + l_3)/R = 3/2(1/\cos\beta + 1). \tag{4.80}$$

from Eq. (4.81), which is three times as great as the eccentricity.

4.5.2.2 Materials and Manufacture

The *tripode* is made of case hardening steel which is cold formed and hardened, the trunnions are ground.

The *tulip*, whether open or closed is made of plain carbon steel and is induction hardened. The tracks are broached only, not ground.

The *rollers* are made from case hardening steel and are hardened. The bores and their outer contours are ground and bored.

The *needle bearings* are crowded with guides.

The *end connection* for automotive use is by means of splines: for rail vehicles and machinery it is flanged.

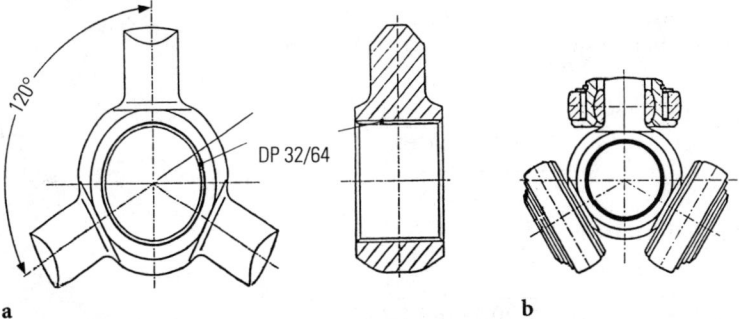

Fig. 4.75. Tripode Joint. **a** Tripode star, **b** with adjusted rollers (made by Honda)

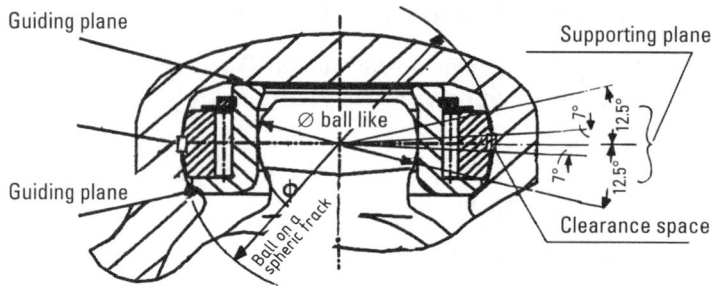

Fig. 4.76. Guided balls in the tripode joint on a spheric track and clearance space. Design by GKN Löbro

GE Fixed Tripode Joints

These are fitted as outboard joints in front and rear axle drive passenger cars (Fig. 4.77). The outboard connection is designed as part of the front wheel bearing and is thus an integral part of the axle. The GE joint allows an articulation angle 45° and is made in nine sizes. It consists of a three petal tulip *1* with the hollow cylindrical tracks for the three rollers *3* with plain bearings. The trunnions of the tripode *4* are fixed in the outer ball *5*; they transmit the torque to the stub shaft *6*. The GE joint is axially fixed by two stops on the spherical centre part of the tripode: it is lubricated for life and is sealed by a boot *9*.

Section 4.5.2 gave indications of the relationship between the GE fixed joint and the GI plunging joint because both employ the same effective geometry. They are however configured somewhat differently. Figure 4.77 shows their principal dimensions, torque capacity and application torques. The load carrying limits of the GE joint are shown in the diagram in Fig. 4.78, which was determined from the quasi-static breaking torques M_b under increasing articulation angles. Because the grease-filled boot is designed for $\beta_{max} = 45°$, a joint speed of 2000 rpm for $\beta = 4°$ should not be exceeded.

Checking the GE-fixed tripode joint with the dates:

Rated torque M_N	2850 Nm
Diameter of roller d	23.7 mm
Trunnion diameter	13.7 mm
Effective radius R	0.0202 m
Effective roller length l_w	13.0 mm

$$k = \left(\frac{p_0}{c_p}\right)^3 = \frac{M}{3d^2R} = \frac{2850}{\underbrace{3 \cdot 23.7^2 \cdot 0.0202}_{34.04}} = 83.73 \text{ N/mm}^2. \tag{a}$$

From (4.54) the conformity is

$$\varkappa_L = \frac{d}{\dfrac{2(R+b)}{\cos\alpha\cos\varepsilon} - \dfrac{d}{2}} = \frac{23.7}{\dfrac{2(20.2+\infty)}{1 \cdot 1} - \dfrac{23.7}{2}} = 0.$$

4.5 Pode Joints

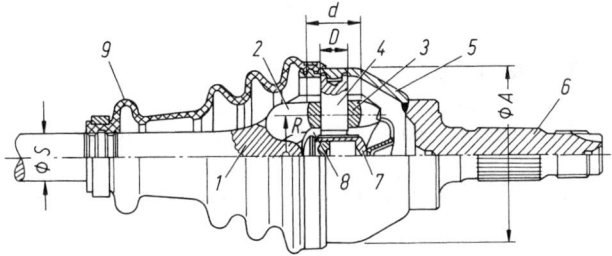

Size of joint	GE	58	66	76	82A	86	90A	92	99	113
Outer diameter	A	58	67.7	76	86	86	93	92	99	113
Shaft diameter	S	17.7	21	21.2	25.7	27	28	30	31	
Pitch circle radius	R	14.1	16.15	18.3	20.2	20.85	21.6	22.5	24.2	27.4
Roller diameter	d	16.9	19.7	21.9	23.7	24.9	25.9	26.8	28.4	31.69
Max. working angle	β	43	43	43	45	43	45	43	43	43
Max. engine capacity for passenger cars	in cm²	750	1100	1400	1800	2000	2200	2600	depends on application	
Static failure torque, (rotating joint) for $\beta = 0°$	M_b in Nm	1350	1950	2400	3000	3500	4000	4270	5250	7850
Quasi-static failure torque (rotating joint) for $\beta = 40°$	M_b in Nm	750	1050	1550	2000	2250	2500	2800	3600	5000
Max. drive torque 0.6 M_b for $\beta = 0 + 10°$	M_o in Nm	1200	1750	2100	2850	3200	3600	3900	4900	7200
Permanent torque for ≥ 500 h/ 730 min⁻¹, $\beta = 6-27°$	M_d in Nm	100	150	220	290	330	370	410	500	750
Trunnion diameter	D in mm	9.5	10.7	12.5	13.7	14	15	15	16	18.2
Effective roller length	l_w in mm	9.3	10.5	11.8	13	13.7	14	14.7	15.6	17.8

Fig. 4.77. Glaenzer-Spicer GE tripode joint with β_{max} = 43 to 45°. *1* Three-pronged tulip; *2* cylindrical roller tracks; *3* rollers with plain bearings; *4* tripode trunnion; *5* outer bell; *6* stub shaft; *7* button; *8* mushroom shaped retaining clip; *9* boot

From Eqs. (4.55 and 4.5) and Table 3.2, it follows that

$$\varkappa_Q = -0.995, \quad d\Sigma\varrho = 2.010, \quad \mu\nu = 2.23 \quad \text{and} \quad c_p = 612.8 \ (\text{N/mm}^2)^{2/3}$$

and the Hertzian stress from (4.7b) is

$$p_0 = c_p \sqrt[3]{k} = 612{,}8 \sqrt[3]{83.73} = 2681 \ \text{N/mm}^2. \tag{b}$$

From (4.53) the torque is

$$M_0 = 3 \cdot 83.73 \cdot 23.7^2 \cdot 0.0202 = 2850 \text{ Nm}.$$

The rated torque does not therefore exceed the static torque capacity M_0 from Fig. 4.78.

For the dynamic application torque $M_d = 290$ Nm, the values are considerably lower for:

- the specific loading from Eq. (a) $k = 290/34.04 = 8.52$ N/mm²
- the Hertzian stress $p_0 = 612.8 \sqrt[3]{8{,}52} = 1252$ N/mm²
- the surface stress at the tripode trunnion, from Eq. (f),

$$p_1 = 1.94 \cdot 10^{-8} \cdot 1252^3 = 38.1 \text{ N/mm}^2.$$

The static torque capacity of the non-articulated GE-82A fixed joint is determined by the breaking strength of the tripode $M_B = 3000$ Nm.

GI 72 *tripode plunging joint* from Fig. 4.79

Rated torque M_N	2600 Nm
Diameter of roller d	37.10 mm
Trunnion diameter	16.0 mm
Effective radius R	0.0233 m
Needle diameter d_1	2.0 mm

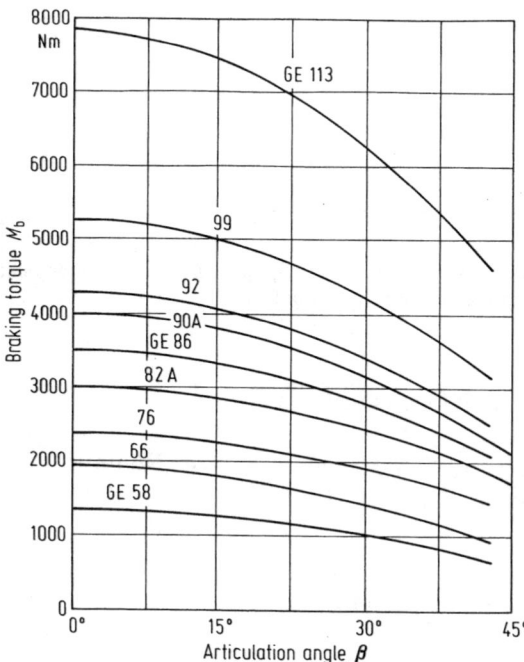

Fig. 4.78. Quasi-static breaking torques for Glaenzer Spicer GE fixed tripode joints. Determined from tests on a rotating joint subjected to increasing torque

4.5 Pode Joints

The specific loading between the roller and the track (Fig. 4.77, point 3)

$$k = \left(\frac{p_0}{c_p}\right)^2 = \frac{M}{3d^2R} = \frac{2600}{3 \cdot 37.10^2 \cdot 0.0233} = 27.02 \text{ N/mm}^2. \tag{c}$$

Since the values for calculating the Hertzian stress for this joint are the same as those for the GE-82 A joint,

$$p_0 = c_p \sqrt[3]{k} = 612.8 \sqrt[3]{27.02} = 1839 \text{ N/mm}^2. \tag{d}$$

The Hertzian stress calculated in Eq. (c) is valid because

$$\left(\frac{\delta_b}{d}\right)^{1/5} = \frac{1839}{3300 \cdot 612^{3/10}} = 0{,}0849$$

$$\frac{\delta_b}{d} = 0.0813 \cdot 10^{-4}. \tag{e}$$

The permanent deformation between the roller and the track for the rated torque $M_n = 2600$ Nm is smaller than the maximum permitted level.

The static torque capacity of the non-articulated GI-72 tripode plunging joint is limited by the permanent deformation $\delta_b/d = 10^{-4}$ at point 1 between the needle and the tripode trunnion (Fig. 4.79). Here, from (4.6) the coefficient of conformity $c_p = 270 \sqrt{1 + d_1/D} = 270 \sqrt{1 + 0.125} = 286.4$ $(\text{N/mm}^2)^{1/2}$. From (4.10) we get

$$p_1 = 2690 \cdot 286.4^{2/5} \cdot 0.1585 = 4098 \text{ N/mm}^2$$

The torque capacity from (4.8a and 4.16) is

$$M_0 = 3 \cdot \left(\frac{4098}{286.4}\right)^2 \cdot 16 \cdot 11.8 \cdot 0.0233 = 2702 \text{ Nm}. \tag{4.81}$$

If the loading capacity of the needle bearings is increased by optimising the profile of the tripode trunnions, it is safe to go up to the rated torque M_N of 2600 Nm (Fig. 4.81, p. 222).

The pair of rolling elements between the track and the roller is more highly stressed for the GE fixed joint than for the GI plunging joint because there is room for only fairly small rollers.

4.5.2.3 GI Plunging Tripode Joints

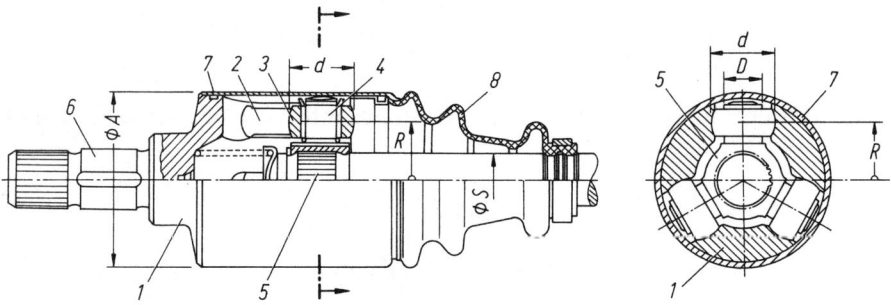

Principal dimensions in mm and Max. torque capacity in Nm

Size of joint	GI	580	62	69	720	82	87 C	100 C
Outer diameter	A	58	63	69	72	82	87	116–126
Shaft diameter	S	19.1	20.2	22.3	25.5	27.4	30.4	35.5
Pitch circle radius	R	18.64	20.84	23.2	28.15	30.5	36.8	
Roller diameter	d	29.2	31.95	29.95	37.1	33.95	33.95	33.95
Max. engine capacity for passenger cars	cm³	1100	1500	1800	2800	depends on application		
Intermediate shaft. Optimum number of splines SAE 24/48–45°		19	21	22	25	27	30	27–30
Static failure torque for $\beta = 0°$	M_b	1900	2600	3000	4500	5500	7700	9300
Max. drive torque $0.6 M_b$ for $\beta = 0 - 10°$	M_N	1040	1240	1660	2530	3150	4300	6900
Nominal torque for calculated service life $\leq$ 500 Std. at 730 min⁻¹, $\beta = 6 - 12°$; Plunge ± 10 mm	M_d	220	290	330	450	560	750	900
Trunnion diameter	D	12.5	14.9	16		18.6		26–29
Needle diameter	d_1		1.5	2		2		2
Effective needle length	l_w	7.8	10.3	11.8		13.8		13.8
No. of needles	z	29	34	28		32		32

Fig. 4.79. Glaenzer Spicer GI tripode joint, *1, 2, 4, 6* as in Fig. 4.78; *3* rollers with needle bearings; *5* hub; *7* sheet metal can; *8* boot

These are found in the front and rear wheel drivetrains of French passenger cars even more frequently than GE fixed tripode joints. They are made with an articulation angle of 25° and a plunge of 40 to 60 mm in six sizes (Fig. 4.79). Unlike the GE joint, the three rollers *3* are fitted with needle bearings. The trunnions of the tripode form an integral unit with the spider *5*, which transmits the torque by splines to the shaft. Open tulips are protected by a sheet metal can *7* and a boot *8* which retain the

4.5 Pode Joints

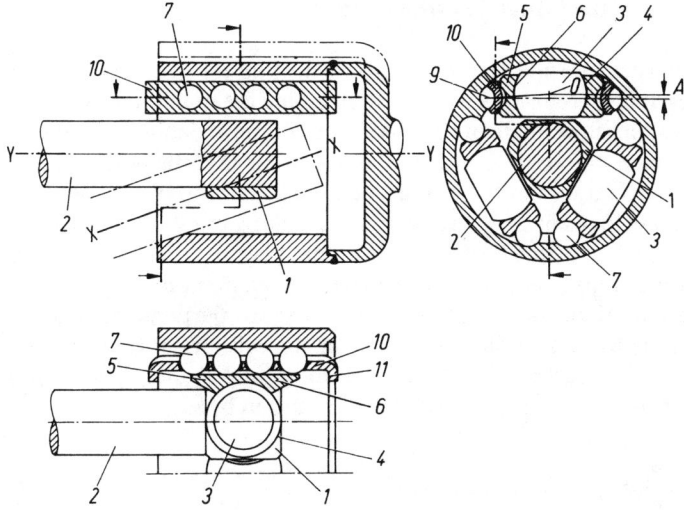

Fig. 4.80. Triplan ball joint with two axis by Michel Orain 1986 (European patent 0202968), built by Glaenzer-Spicer (GKN Automotive AG). *1* Tripode trunnion, *2* shaft, *3* radial arms, *4* spheric surfaces, *5* intermediate elements (rolling shoe), *6* spheric concaves, *7* rolling ball member (10 up to 100 microns), *8* second element, *9* cage, *10* cage with end-of-travel abutment means, *11* ball retaining means. The two elements are: first along axis X–X, the second axis along Y–Y, intersecting X–X. If the joint operates at an angle each intermediate element *5* is given an oscillatory movement around the axis of the associated tracks. Each cage *10* comprises ball retaining means *11*.

grease; for closed tulips a boot alone is sufficient. Figure 4.79 shows their principal dimensions, torque capacity and their application torques.

Principal dimensions and maximum permissible short-duration torque Nm at $\beta = 0$ to $10°$, $\beta_{max} = 25°$

Joint		diameter (mm)		Nm$_{max\,perm}$	
series	size	outside	shaft		
TG	1900	93	22.4	1700	
	2300		23.4	2100	TG = needles
	2700	100.8	24.3	2500	
	3200		24.6	3000	
TB	2300	86	23.4	1950	TB = balls
	2700	93	25.5	2530	

4.5.2.4 Torque Capacity of the Articulated Tripode Joint

During rotation at constant torque M, the three transmitting elements see periodically changing forces. It is therefore necessary to determine the maximum compressive force Q as a function of the angle of rotation φ and of the articulation angle β.

The torque M produces the equivalent compressive forces Q at the points of contact of the rollers with the tracks of the output member 2. The input member 1 can only accommodate these loads in the pode plane E_1 and along the shaft 1 (Fig. 4.81). The components A are taken by the bearing; they do not play any real part in transmitting the torque. The components F are in the tripode plane E_1 and are perpendicular to the pode trunnion. The compressive loads Q are perpendicular to the plane E_2; the plane E_2 is formed by the centre line of the cylindrical track and the respective pode trunnion axis. The angle v between the planes E_1 and E_2 is given in vectors by the angle between the normals of planes n_1 and n_2 by

$$\cos v = \frac{n_1 \cdot n_2}{n_1 \cdot n_2} = n_1^0 \cdot n_2^0 \quad [1.10, \text{S. } 177]. \tag{a}$$

The unit vectors n^0_1 and n^0_2 are normals of the planes

$$n_1^0 = \begin{pmatrix} n_{1x}^0 \\ n_{1y}^0 \\ n_{1z}^0 \end{pmatrix} = \begin{pmatrix} \sin \varphi_{1i} \\ \cos \varphi_{1i} \\ 0 \end{pmatrix}; \quad n_2^0 = \begin{pmatrix} n_{2x}^0 \\ n_{2y}^0 \\ n_{2z}^0 \end{pmatrix} = \begin{pmatrix} 0 \\ \sin \beta \\ \cos \beta \end{pmatrix}$$

$$\cos v = \frac{\begin{pmatrix} \sin \varphi_{1i} \\ \cos \varphi_{1i} \\ 0 \end{pmatrix} \cdot \begin{pmatrix} 0 \\ \sin \beta \\ \cos \beta \end{pmatrix}}{\underbrace{\sqrt{\sin^2 \varphi_{1i} + \cos^2 \varphi_{1i} + 0^2}}_{1} \underbrace{\sqrt{0^2 + \sin^2 \beta + \cos^2 \beta}}_{1}}$$

$$= 0 \cdot \sin \varphi_{1i} + \sin \beta \cdot \cos \varphi_{1i} + \cos \beta \cdot 0$$

$$\cos v = \sin \beta \cos \varphi_{1i} \;\Rightarrow\; \sin v = \sqrt{1 - \sin^2 \beta \cdot \cos \varphi_{1i}}. \tag{b}$$

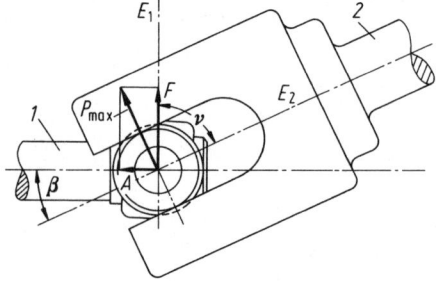

Fig. 4.81. Resolution of the equivalent compressive load $P_{\max}$ into the axial and radial components A and F on the tripode joint. 1 input member; 2 output member; E_1 tripode plane; E_2 plane formed by the centre line of the tracks and the axis of the pode trunnion

4.5 Pode Joints

It also follows from Fig. 4.80 that $\sin \nu = F/Q \Rightarrow Q = F/\sin \nu$. From Eq. (b)

$$Q = \frac{F}{\sqrt{1 - \sin^2 \beta \cos^2 2\varphi_i}}. \tag{c}$$

For $\psi = 90°$ the equivalent compressive force Q becomes a maximum

$$Q = \frac{F}{\sqrt{1 - \sin^2 \beta \cdot 1}} = \frac{F}{\sqrt{\cos^2 \beta}} = \frac{F}{\cos \beta}.$$

The lever arms l of the components F of the free tripode joint (Sect. 2.3.2) are variable because, by freeing the mid-point 0_1, relative sliding of the tripode (input member 1) can occur:

- in the x_1 direction by a,
- in the y_1 direction by b,

where a and b are positive (+) in the system $x'_1 y'_1 z'_1$ and negative (−) in the system $x_1 y_1 z_1$.

Since the forces F at the tripode which occur every 120° have to be in equilibrium, the following must apply in Fig. 4.82:

$\Sigma P_x = 0 \Rightarrow$

$$F_1 \cos \varphi_2 + F_2 \cos (\varphi_2 + 120°) + F_3 \cos (\varphi_2 + 240°) = 0 \tag{d}$$

$\Sigma P_y = 0 \Rightarrow$

$$F_1 \sin \varphi_2 + F_2 \sin (\varphi_2 + 120°) + F_3 \sin (\varphi_2 + 240°) = 0 \tag{e}$$

and, by taking moments about 0, $\Sigma M = 0$,

$$F_1 l_1 + F_2 l_2 + F_3 l_3 - M = 0. \tag{f}$$

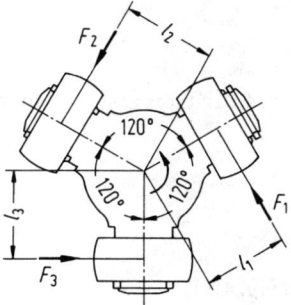

Fig. 4.82. Isostatic load transmission in the tripode joint. $F_1 = F_2 = F_3$ whatever the operating position l_1, l_2, l_3 of the rollers

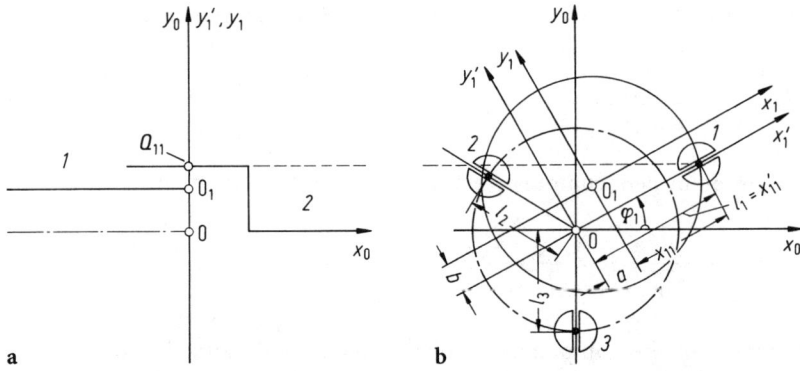

Fig. 4.83a, b. Variation of the effective lever arms for tripode joints. Relative sliding on the three pode trunnions. **a** Side view, **b** cross-section

If follows from Eq. (e) that:

$$F_1 \sin \varphi_2 + F_2 (-\tfrac{1}{2} \sin \varphi_2 + \tfrac{1}{3}\sqrt{3} \cos \varphi_2) + F_3 (-\tfrac{1}{2} \sin \varphi_2 - \tfrac{1}{3}\sqrt{3} \cos \varphi_2) = 0$$

$$F_1 \sin \varphi_2 - \tfrac{1}{2} F_2 \sin \varphi_2 + \tfrac{1}{3}\sqrt{3} F_2 \cos \varphi_2 - \tfrac{1}{2} F_3 \sin \varphi_2 - \tfrac{1}{3}\sqrt{3} F_3 \cos \varphi_2 = 0$$

$$F_1 \sin \varphi_2 - \tfrac{1}{2} \sin \varphi_2 (F_2 + F_3) + \tfrac{1}{3}\sqrt{3} \cos \varphi_2 (F_2 - F_3) = 0. \tag{g}$$

Equation (g) is only applicable for all angles of rotation φ_2 if $F_2 = F_3$, since

$$F_1 \sin \varphi_2 = \tfrac{1}{2} \sin \varphi_2 (F_2 + F_3)$$

$$F_1 = \tfrac{1}{2}(F_2 + F_3) = \tfrac{1}{2}(F_2 + F_2) = \tfrac{1}{2} \cdot 2 F_2$$

or

$$F_1 = F_2 = F_3 = F. \tag{h}$$

The three components F of the equivalent compressive loads P, lying in a plane, are always equal, this is an "isostatic transmission".

By substituting in Eq. (f) we get

$$M = Fl_1 + Fl_2 + Fl_3 = F(l_1 + l_2 + l_3). \tag{4.81}$$

The length l of the effective lever arms can be seen from Fig. 4.81. If it is assumed that $a > 0$ and $b > 0$ in the $x_1' y_1' z_1'$ system, and that $l > R$, we get:

$$l_i = R + \lambda_i R \implies \lambda_i = \frac{l_i - R}{R} \tag{a}$$

where $\lambda_i \gg 1$ is the relative sliding on the fixed x_{11}-axis (Fig. 4.83).

4.5 Pode Joints

In order to calculate the axial displacement, the global coordinates $x_0 y_0 z_0$ of the tripode are transformed into the rotating system $x_1' y_1' z_1'$ to give

$$\begin{pmatrix} \cos\varphi_1 & \sin\varphi_1 & 0 \\ -\sin\varphi_1 & \cos\varphi_1 & 0 \\ 0 & 0 & 1 \end{pmatrix} \begin{pmatrix} x_0 \\ y_0 \\ z_0 \end{pmatrix} = \begin{pmatrix} x_0 \cos\varphi_1 + y_0 \sin\varphi_1 \\ -x_0 \sin\varphi_1 + y_0 \cos\varphi_1 \\ z_0 \end{pmatrix} = \begin{pmatrix} x_1' \\ y_1' \\ z_1' \end{pmatrix}. \qquad (b)$$

Additionally, the fixed system $x_1 y_1 z_1$, offset by a and b, is transformed to give

$$x_1' - a = x_1, \quad y_1' - b = y_1, \quad z_1' = z_1. \qquad (c)$$

After these two transformations, the coordinate equations for the pode trunnion 1 with double indices [2.8, p. 176] are

$$\begin{aligned} x_{11} &= x_{11}' - a = x_{01} \cos\varphi_1 + y_{01} \sin\varphi_1 - a, \\ y_{11} &= y_{11}' - b = -x_{01} \sin\varphi_1 + y_{01} \cos\varphi_1 - b, \\ z_{11} &= z_{11}' = z_{01}. \end{aligned} \qquad (d)$$

The coordinates of the driven body 2 are transformed by articulating through the angle β, into

$$\begin{pmatrix} 1 & 0 & 0 \\ 0 & \cos\beta & -\sin\beta \\ 0 & \sin\beta & \cos\beta \end{pmatrix} \begin{pmatrix} x_0 \\ y_0 \\ z_0 \end{pmatrix} = \begin{pmatrix} x_0 \\ y_0 \cos\beta \pm z_0 \sin\beta \\ -y_0 \sin\beta + z_0 \cos\beta \end{pmatrix} = \begin{pmatrix} x_2 \\ y_2 \\ z_2 \end{pmatrix} \qquad (e)$$

and rotating through the angle φ_2 into the fixed system $x_2 y_2 z_2$ to give

$$\begin{pmatrix} \cos\varphi_2 & -\sin\varphi_2 & 0 \\ \sin\varphi_2 & \cos\varphi_2 & 0 \\ 0 & 0 & 1 \end{pmatrix} \begin{pmatrix} x_2 \\ y_2 \\ z_2 \end{pmatrix} = \begin{pmatrix} x_2 \cos\varphi_2 - y_2 \sin\varphi_2 \\ x_2 \sin\varphi_2 + y_2 \cos\varphi_2 \\ z_2 \end{pmatrix} = \begin{pmatrix} x_2' \\ y_2' \\ z_2' \end{pmatrix}. \qquad (f)$$

The coordinate equations for the reference point

$$Q_{21} \begin{pmatrix} x_{21} = R \\ y_{21} = 0 \\ z_{21} = 0 \end{pmatrix}$$

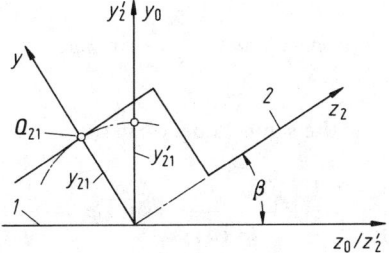

Fig. 4.84. The coordinates x_{2i} of the contact point between the mid line of the track and the pode trunnion do not alter with articulation

on the mid line of track *1* from Eq. (f) are

$$x'_{21} = R \cos \varphi_2, \quad y'_{21} = R \sin \varphi_2, \quad z'_{21} = z_{21} = 0. \tag{g}$$

From Fig. 4.85 it can be seen that

$$x_{21} = x'_{21}, \quad y_{21} = y'_{21}. \tag{h}$$

The coordinates x_{21} and y_{21} do not alter with articulation!
From Eq. (e) we get

$$\begin{aligned} x_{01} &= R \cos \varphi_2 \\ y_{01} \cos \beta - z_{01} \sin \beta &= -R \sin \varphi_2. \end{aligned} \tag{i}$$

The point of intersection between the pode trunnion *1* and the mid-line of track *1* lies on this pode trunnion; consequently it has the coordinate $z_{01} = 0$. The other coordinates are obtained from Eq. (i)

$$x_{01} = R \cos \varphi_2, \quad y_{01} = R \frac{\sin \varphi_2}{\cos \beta}, \quad z_{01} = 0. \tag{k}$$

In order to determine the sliding on the x_{11}-axis, the displacement a must be calculated. The constant velocity property of the free tripode joint was proved in Sect. 2.3.2 using (2.27) to (2.30), that is $\varphi_2 = \varphi_1 = \varphi$. With $r = R$ we now get

$$b = R \sin \varphi \cos \varphi \left(1 - \frac{1}{\cos \beta}\right) = \frac{1}{2} R \left(1 - \frac{1}{\cos \beta}\right) \sin 2\varphi. \tag{4.82}$$

Equation (4.79) is substituted in (2.28) giving

$$R \left(\sin \varphi + \frac{\cos \varphi}{\sqrt{3}}\right)\left(\frac{\sin \varphi}{\sqrt{3}} - \cos \varphi\right) - R \frac{\left(\frac{\sin \varphi}{\sqrt{3}} - \cos \varphi\right)}{\cos \beta}\left(\sin \varphi + \frac{\cos \varphi}{\sqrt{3}}\right) + \\ + \frac{2a}{\sqrt{3}} + \frac{2}{3}\frac{1}{2} R \left(1 - \frac{1}{\cos \beta}\right) \sin 2\varphi = 0.$$

Auxiliary calculation:

$$\left(\sin \varphi + \frac{\cos \varphi}{\sqrt{3}}\right)\left(\frac{\sin \varphi}{\sqrt{3}} - \cos \varphi\right) = \frac{1}{\sqrt{3}}(\sin^2 \varphi - \cos^2 \varphi) - \frac{2}{3}\frac{1}{2}\sin 2\varphi.$$

Since both the products of the round brackets have the same value, we get

$$R \left(1 - \frac{1}{\cos \beta}\right)\left\{\frac{1}{\sqrt{3}}(\sin^2 \varphi - \cos^2 \varphi) - \frac{1}{3}\sin 2\varphi\right\} + \frac{2}{3}\frac{1}{2} R \left(1 - \frac{1}{\cos \beta}\right)\sin 2\varphi = -\frac{2a}{\sqrt{3}}$$

4.5 Pode Joints

$$R\left(1 - \frac{1}{\cos\beta}\right)\left\{\frac{1}{\sqrt{3}}(\sin^2\varphi - \cos^2\varphi) - \frac{1}{3}\sin 2\varphi + \frac{1}{3}\sin 2\varphi\right\} = -\frac{2a}{\sqrt{3}}$$

$$a = \frac{1}{2}R\left(1 - \frac{1}{\cos\beta}\right)\cos 2\varphi. \tag{4.83}$$

With this equation and the coordinates from Eq. (e) the first of the three equations from (d) can be resolved to give

$$x_{11} = R\cos\varphi\cos\varphi + R\frac{\sin\varphi}{\cos\beta}\sin\varphi - \frac{1}{2}R\left(1 - \frac{1}{\cos\beta}\right)\cos 2\varphi$$

$$= R\cos^2\varphi + R\frac{\sin^2\varphi}{\cos\beta} - \frac{1}{2}R\left(1 - \frac{1}{\cos\beta}\right)\cos 2\varphi$$

$$x_{11} = R(1 - \sin^2\varphi) + R\frac{\sin^2\varphi}{\cos\beta} - \frac{1}{2}R\left(1 - \frac{1}{\cos\beta}\right)\cos 2\varphi,$$

$$x_{11} - R = R\sin^2\varphi\left(\frac{1}{\cos\beta} - 1\right) + \frac{1}{2}R\left(\frac{1}{\cos\beta} - 1\right)\cos 2\varphi$$

$$= \frac{1}{2}R\left(\frac{1}{\cos\beta} - 1\right) - \frac{1}{2}R\left(\frac{1}{\cos\beta} - 1\right)\cos 2\varphi + \frac{1}{2}R\left(\frac{1}{\cos\beta} - 1\right)\cos 2\varphi,$$

$$\frac{x_{11} - R}{R} = \lambda_1 = \frac{1}{2}\left(\frac{1}{\cos\beta} - 1\right). \tag{a}$$

This term (a) is therefore independent of the angle of rotation.

With Eq. (a) the term in (4.81) becomes

$$(l_1 + l_2 + l_3) = (R + \lambda_1 R) + (R + \lambda_2 R) + (R + \lambda_3 R) = R(3 + \lambda_1 + \lambda_2 + \lambda_3)$$

$$= R\left[3 + \frac{3}{2}\left(\frac{1}{\cos\beta} - 1\right)\right] = R\left(\frac{3}{2} + \frac{3}{2}\frac{1}{\cos\beta}\right)$$

$$= \frac{3}{2}R\left(1 + \frac{1}{\cos\beta}\right). \tag{b}$$

If this equation is substituted in (4.81) we get

$$M = F\frac{3}{2}R\left(1 + \frac{1}{\cos\beta}\right). \tag{c}$$

Using (4.77), the highest equivalent compressive load P occurs at $\varphi = 90°$ so that the static torque capacity M_β for an articulated tripode joint is:

$$M_\beta = P\cos\beta\frac{3}{2}R\left(1 + \frac{1}{\cos\beta}\right) = 3PR\underbrace{\frac{1}{2}(1 + \cos\beta)}_{M_0}$$

or with (4.78)

$$M_\beta = M_0 \frac{1}{2}(1 + \cos\beta).\tag{4.84}$$

The dynamic torque capacity M_d of the articulated tripode joint can be seen from the tables in Fig. 4.77 (GE joint) and Fig. 4.79 (GI joint). For speeds other than 730 rpm and lives other than 500 hrs, the life, derived from (4.60a) must be calculated as follows

$$L_{hx} = 500 \cdot \frac{730}{n_x} \left(\frac{\beta}{\beta_x}\right)^n \cdot \left(\frac{M_d}{M_x}\right)^3 = \frac{365\,000}{n_x} \left(\frac{M_d}{M_x}\right)^3.\tag{4.85}$$

4.5.3 The GI-C Joint

In 1991 R. A. Lloyd (GKN Technology Ltd., Wolverhampton) started to examine ways of obtaining a more friction- and vibration-free tripode joint. He let the rollers diverge obliquely at an angle of 4.6° to 7.8° to one another (Fig. 4.86). At 4.6° the axial force decreased by 39%, at 7.5° it was only 29% less. At lower speeds and smaller articulation angles the friction reduced, the vibration halved. The GI-A-joint resulted from this and the engineers from GKN then developed this into the AAR-joint.

The English ideas were developed by J. C. van Dest in his 1991 GI-C-joint, which had roller dimensions:

$$\frac{\text{width } l}{\text{outside dia. } d} = 0{,}38 = \sim \tan 21°.$$

In place of the 2 × 7° relief against self-locking in the tripode joint, Werner Jacob used a self-locking free four-joint roller (German Patent 3739 868/1987).

He changed only the outer profile in that it is relieved in the centre and has two contact points each at under 10° depending on the spherical track profile. These contact points form a radius of conformity; therefore there can be no contact in the self-locking zone.

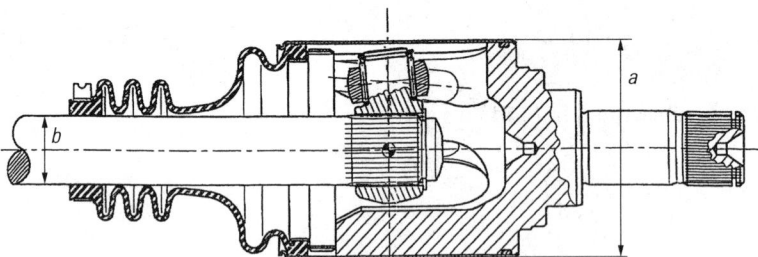

Fig. 4.85. J. C. van Dest's 1991 GI-C-joint building on the developments in England. The groove in the bell is slightly conical, the tripode star slightly angled. These changes enabled the maximum articulation angle to be enlarged to 23°

4.5 Pode Joints

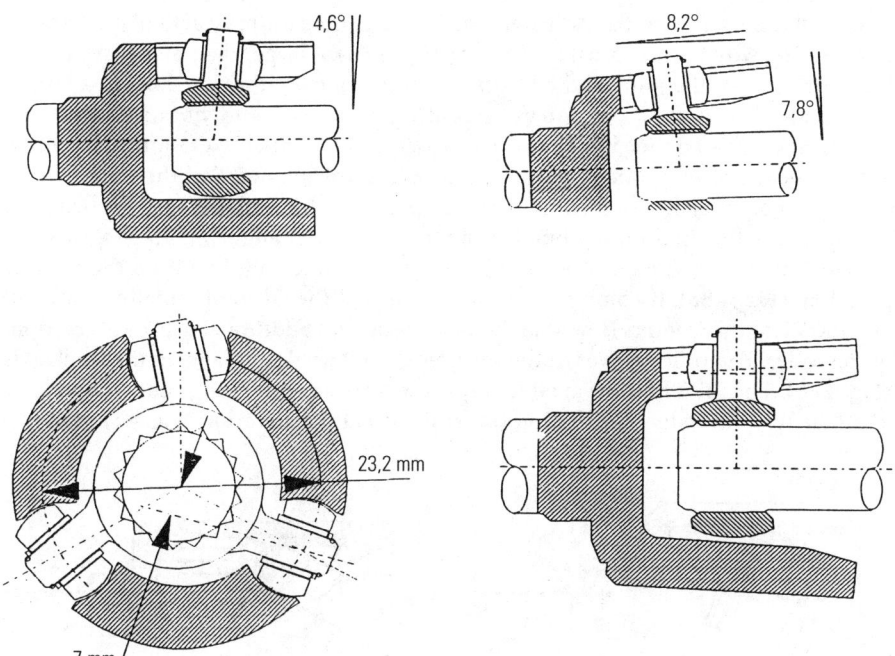

Fig. 4.86. Reduction of the axial forces in the R. A. Lloyd's 1994 GI-joint (British Patent 9322 917.7/1993) [4.41]

4.5.4 The low friction and low vibration plunging tripode joint AAR

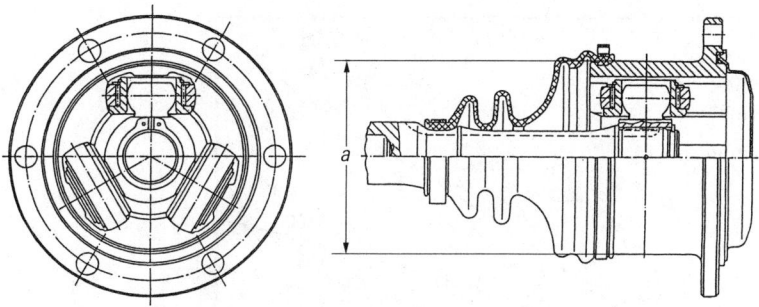

Joint AAR (Nm)	a mm	engine capacity (l)
2600i	82	
2300i	76,9	1,6–2,0
3300i	85,8	2,0–3,0
4100i	95,5	
2700	86,5	
2900	90,5	

Fig. 4.87. Plunging tripode joint, is known as Angular Adjusted Roller (AAR)

This joint, called an Angular Adjusted Roller, is designed for max. articulation angles of 23° to 26°, with perfect constant velocity (Fig.4.88). A larger diameter is chosen for the outer part of the joint, and a larger trunnion diameter, which leads to a longer service life. The joint accommodates pulsating axial loads and vibrations from the driving gear. This tripode joint has track cross-sections which are elliptical or gothic arch shaped. The two contact points are separated by about 0.6 of the roller width. 0.52% is chosen as the conformity of the ellipse or gothic arch because the origin of the curvature then lies in the contact point at the contact angle $\alpha = 25°$ (Fig. 4.89).

The ARR joint (Angular Adjusted Roller) is a double-roller joint on the tripode principle (Fig. 4.88). It combines the advantages of the GI joint (small vibrations) and the VL and DO joints (low axial loads) through an additional degree of freedom of the roller. The roller moves under all articulation angles parallel to the track axis (Fig. 4.88). The joint has low axial loads, a low resistance to plunge and a high level of efficiency, and reduced vibration transmission and generation (Fig. 4.88, 4.89).

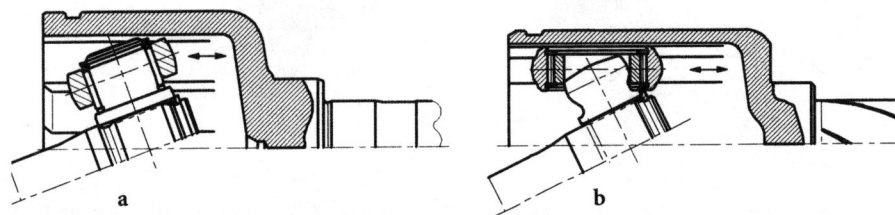

Fig. 4.88. AAR plunging joint, GKN Löbro design. **a** GI model with rigid roller and sliding friction, **b** AAR model (Angular Adjusted Roller) with pivoting roller (sliding shoe) and rolling friction (German Patent 3936 601.4/1989)

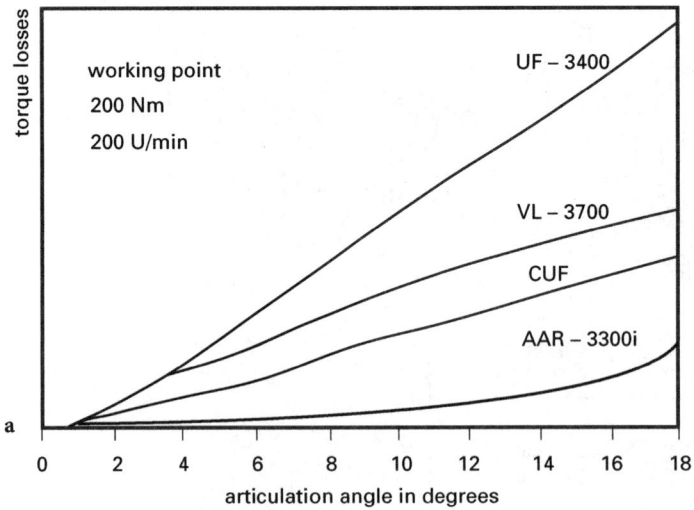

Fig. 4.89. Friction losses in different constant velocity joints. **a** 200 rpm. **b** 2000 rpm [4.35]. In % at $\beta = 6°$: UF 0.75; CUF 0,25%

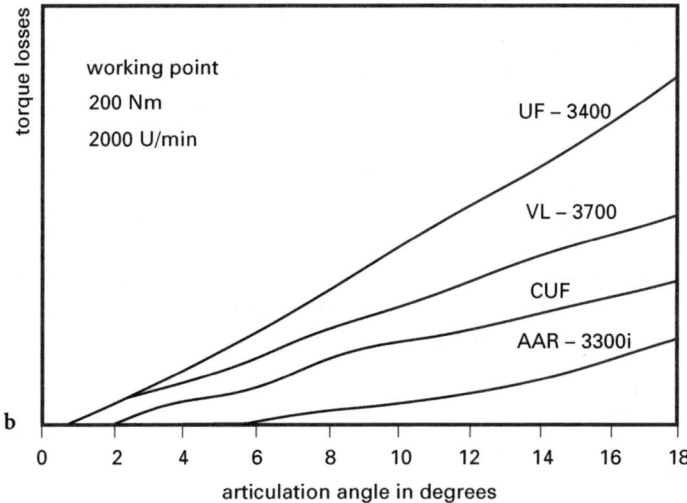

Fig. 4.89 (continued)

4.6 Materials, Heat Treatment and Manufacture

In developing ball joints the Hertzian pressure p_{max} when solid elastic bodies touch is not the only determining factor. H. Hertz [1.21] studied the size and distribution of normal stresses in the contact zone of bodies. Roller bearing technology uses

surface layer hardened materials

for its tracks and guides. With these materials the hardness and its profile into the depth z of the bodies pressed together exert a decisive influence on the transferable, static torque M_0. It is therefore necessary to know the stresses under the Hertzian pressure area in the z direction.

4.6.1 Stresses

This subject has been studied by a number of researchers since Belajef in 1917. For the German speaking world, L. Föppl [4.3] computed in 1936 that the greatest stress occurs below the contact zone at a specific depth of the bodies. He showed on p. 212, Figure 3, the curve of the stresses σ_y, σ_z, and τ along the z-axis for the contact of two rollers. The highest loading was at the depth z = 0.78a, the width of half the pressure surface area for $\tau_{max} = 0.3\,p_{max}$. On p. 215, Figure 5, he shows the relationship for the situation where a ball and a plane are in contact. The greatest stress resulted in this case at a depth z = 0.47a. In 1934, S. Timoshenko [4.4], p. 350, Fig. 178, published his findings for contact between two rollers and obtained very similar relationships for the stresses along the z-axis.

In the case of contact between ball and track, E.-G. Paland [4.29] calculated the stresses that occur σ_x, σ_y, and σ_z and τ along the z-axis for the Completely Undercut Free (CUF) joint with an computer program using T. Harris's formulae [4.40].

The greatest equivalent stress σ_v for the material on the z-axis beneath the Hertzian contact zone can be determined from Otto Mohr's 1882 shear stress hypothesis. The tri-axial stress condition is obtained by combining three orthogonal single-axis conditions. The greatest value for τ_{max} determines the equivalent stress

$$\sigma_v = 2\,\tau_{max} = \max \begin{vmatrix} \sigma_z - \sigma_y \\ \sigma_z - \sigma_x \\ \sigma_v - \sigma_y \end{vmatrix} \begin{matrix} \Rightarrow\ yz\text{-plane} \\ \Rightarrow\ xz\text{-plane} \\ \Rightarrow\ xy\text{-plane} \end{matrix}$$

From this follows the highest shear stress τ_{max}. It appears in the yz-plane: that is the narrow axis of the pressure ellipse transverse to the joint track (Fig. 4.91).

From Harris the following equations result for the stresses:

$$\begin{aligned}\sigma_x &= \lambda \cdot (\Omega_x + v \cdot \Omega_x') \\ \sigma_y &= \lambda \cdot (\Omega_y + v \cdot \Omega_y') \\ \sigma_z &= -\frac{1}{2} \cdot \lambda \cdot \left(\frac{1}{v} - v\right)\end{aligned} \qquad (4.86)$$

with the values
$\kappa = a/b$ large semi-axis of the pressure ellipse/small semi-axis of the pressure ellipse

$$\lambda = \frac{b \cdot \Sigma\varrho}{\left(\kappa - \dfrac{1}{\kappa}\right) \cdot E(\kappa) \cdot \left(\dfrac{1 - v_1^2}{E_1} + \dfrac{1 - v_2^2}{E_2}\right)} \qquad (4.87)$$

$$v = \left[\frac{1 + (z/b)^2}{\kappa^2 + (z/b)^2}\right]^{1/2} \qquad (4.88)$$

$$\begin{aligned}\Omega_x &= -\frac{1}{2} \cdot (1 - v) + (z/b) \cdot [K(\phi) - E(\phi)] \\ \Omega_x' &= 1 - \kappa^2 \cdot v + (z/b) \cdot [\kappa^2 \cdot E(\phi) - K(\phi)] \\ \Omega_y &= \frac{1}{2} \cdot \left(1 + \frac{1}{v}\right) - \kappa^2 \cdot v + (z/b) \cdot [\kappa^2 \cdot E(\phi) - K(\phi)] \\ \Omega_y' &= -1 + v + (z/b) \cdot [K(\phi) - E(\phi)]\end{aligned} \qquad (4.89)$$

The incomplete elliptical integrals are:

$$E(\phi) = \int_0^\phi \left[1 - \left(1 - \frac{1}{\kappa^2}\right) \cdot \sin^2\phi\right]^{1/2} d\phi \qquad (4.90)$$

elliptical integral of the 2nd kind

$$K(\phi) = \int_0^\phi \left[1 - \left(1 - \frac{1}{\kappa^2}\right) \cdot \sin^2\phi\right]^{-1/2} d\phi \qquad (4.91)$$

elliptical integral of the 1st kind

The stresses were recorded in the diagram Figure 4.91 with the abscissa σ/p_{max} and the ordinate z/b, as in [4.40, p. 138].

4.6 Materials, Heat Treatment and Manufacture

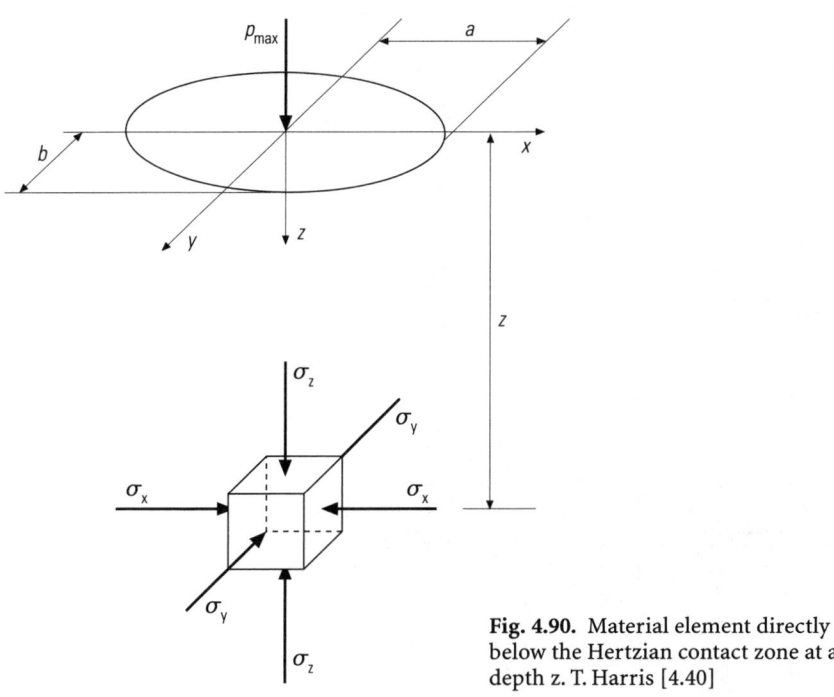

Fig. 4.90. Material element directly below the Hertzian contact zone at a depth z. T. Harris [4.40]

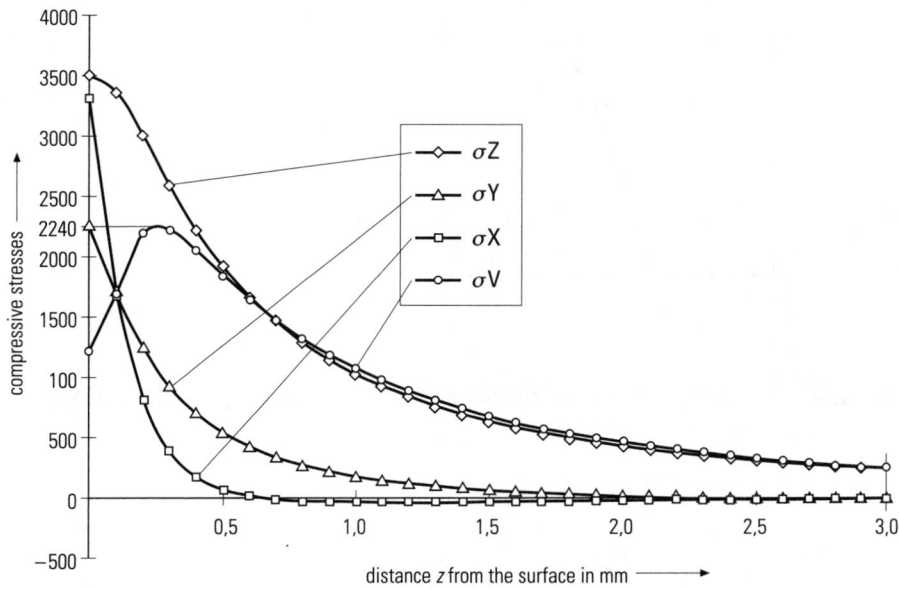

Fig. 4.91. Graph of the 3 stresses σ_x, σ_y, σ_z and the equivalent stress σ_v in Table 4.10

Table 4.18. Materials and heat treatments of inner and outer races for UF-constant velocity joints

Joint	Inner race		Outer race	
	UF 95	UF 107	UF 95	UF 107
Material	SAE 8620	SAE 8620	Cf 53	Cf 53
Hardening method	Insert	Insert	Inductive	Inductive
Case depth mm	1.1 + 0.4	1.1 + 0.4	1.1 + 1.3	1.1 + 1.3
Surface hardness HRC	58–62	58–62	60–61	60–61
Core strength N/mm^2	> 900	> 900	700–800	700–800
$R_{p0.2}$ (core) N/mm^2	≈ 630	≈ 630	≈ 510	≈ 510
Ball diameter mm	18	20	18	20

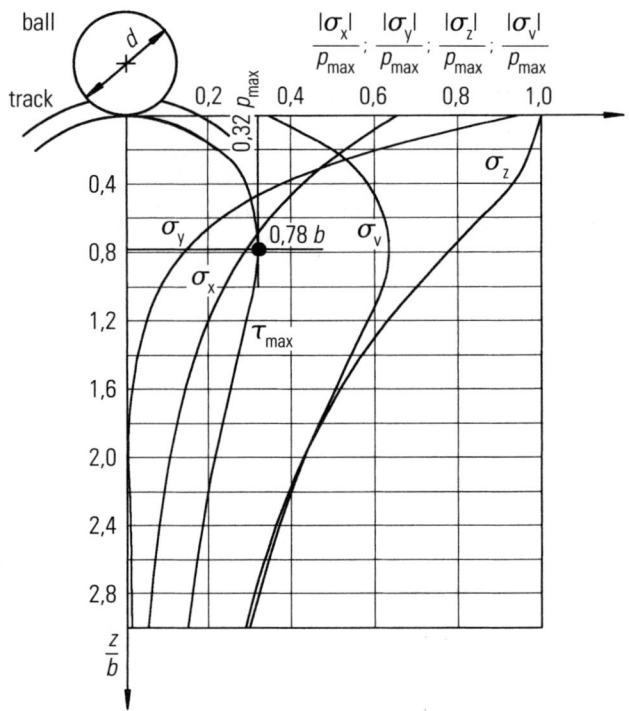

Fig. 4.92. Tri-axial stress condition of a ball in the track, calculated by E.-G. Paland in 1997

4.6 Materials, Heat Treatment and Manufacture

Table 4.19. Values for the tri-axial stress state and the required hardness of a ball in the track of the UF-joint. From a study by E.-G. Paland in 1997

Inner race

conformity	Hertzian normal load	Hertzian stress	large semi-axis	small semi-axis	elastic deformation
$f_i = 0.535$	$F_k = 5281.1$	$p_{max} = 3511.8$	$a = 2.139$	$b = 0.336$	$\delta_0 = 0.0337$

Stresses and required hardness

z	σ_z	σ_y	σ_x	τ_{max}	σ_v	HV_{erf}
0.000	−3511.8	−3321.3	−2297.6	607.1	1214.2	379.4
0.100	−3361.9	−1693.2	−1669.4	846.3	1692.6	528.9
0.200	−3003.8	−812.6	−1224.8	1095.6	2191.1	684.7
0.300	−2593.0	−379.2	−916.4	1106.9	2213.8	691.8
0.400	−2219.0	−169.4	−699.9	1024.8	2049.6	640.5
0.500	−1906.0	−65.5	−544.1	920.3	1840.5	575.2
0.600	−1650.8	−12.4	−428.8	819.2	1638.4	512.0
0.700	−1443.1	15.1	−341.7	729.1	1458.2	455.7
0.800	−1272.6	29.4	−274.5	651.0	1302.0	406.9
0.900	−1131.1	36.5	−221.9	583.8	1167.6	364.9
1.000	−1012.3	39.5	−180.3	525.9	1051.9	328.7
1.100	−911.5	40.3	−147.1	475.9	951.8	297.4
1.200	−825.0	39.8	−120.3	432.4	864.8	270.3
1.300	−750.3	38.5	−98.6	394.4	788.8	246.5
1.400	−685.1	36.9	−81.0	361.0	722.0	225.6
1.500	−627.9	35.1	−66.5	331.5	663.0	207.2
1.600	−577.4	33.2	−54.6	305.3	610.6	190.8
1.700	−532.6	31.3	−44.8	281.9	563.9	176.2
1.800	−492.6	29.5	−36.7	261.0	522.1	163.2
1.900	−456.8	27.8	−30.0	242.3	484.6	151.4
2.000	−424.6	26.2	−24.4	225.4	450.7	140.9
2.100	−395.5	24.6	−19.7	210.1	420.1	131.3
2.200	−369.2	23.2	−15.8	196.2	392.4	122.6
2.300	−345.4	21.8	−12.6	183.6	367.2	114.8
2.400	−323.6	20.6	−9.9	172.1	344.2	107.6
2.500	−303.8	19.4	−7.6	161.6	323.2	101.0
2.600	−285.7	18.4	−5.7	152.0	304.0	95.0
2.700	−269.0	17.3	−4.1	143.2	286.4	89.5
2.800	−253.8	16.4	−2.8	135.1	270.2	84.4
2.900	−239.7	15.6	−1.6	127.6	255.2	79.8
3.000	−226.7	14.8	−0.7	120.7	242.5	75.5

Values for the maximum load

| 0.254 | −2780.8 | −540.2 | −1044.5 | 1120.3 | 2240.7 | 700.2 |

These values are shown in the diagram Figure 4.92, with the abscissa σ/p_{max} and the ordinate z/b [4.1, p. 283].

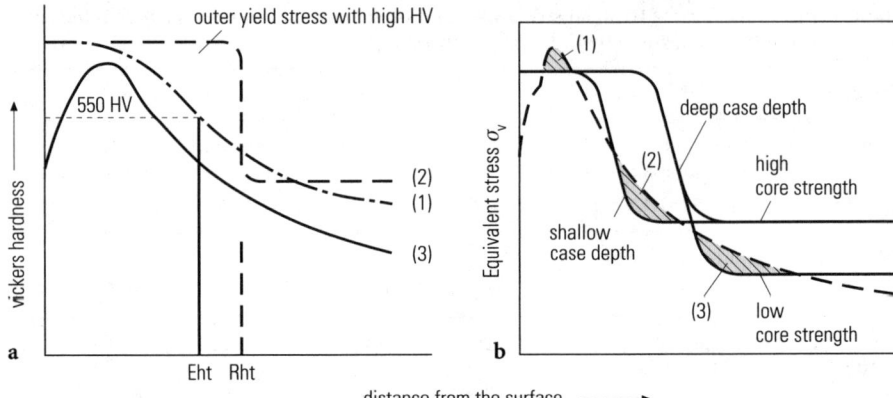

Fig. 4.93a, b. Hardness curve under the surface. **a** 1 case hardening; 2 induction hardening; 3 required hardness. **b** --- equivalent stress; —— yield stress. In the shaded area plastic deformation occurs. Eht = case depth; Rht = case depth

The highest stress occurs

at the depth $z = 0.78\, b$, for

$$\tau_{max} = 0.32\, p_{max}.$$

This is, according to Mohr's assumption the point of greatest material strain and the equivalent stress results as

$$\sigma_v = 2\, \tau_{max} = (\sigma_z - \sigma_y)_{max} = (2780.8 - 540.2) = 2240.6\ \text{N/mm}^2$$
[Dubbel, 18th edition 1.3.2]

Once the level of the equivalent stress is known, the material can be chosen.

These values are shown in the diagram Figure 4.92, with the abscissa σ/p_{max} and the ordinate z/b [4.1, p. 283].

4.6.2 Material and hardening

The material for the tracks and grooves of the rolling contact bodies must have a hard, wear-resistant surface. However, the same strength does not have to be present over the whole cross-section of the body, it can decrease with depth into the body. For this reason roller bearing technology uses

surface layer hardened materials.

Special heat treatable steels have been produced to DIN 17212 for surface hardening. After hardening they have a wear-free surface with a tough core.

For case hardening (1), the hardness curve decays slowly to the core hardness, whereas with induction hardening (2) a steeply falling curve occurs from the surfaces to the core hardness (Figure 4.93, a, b).

4.6 Materials, Heat Treatment and Manufacture

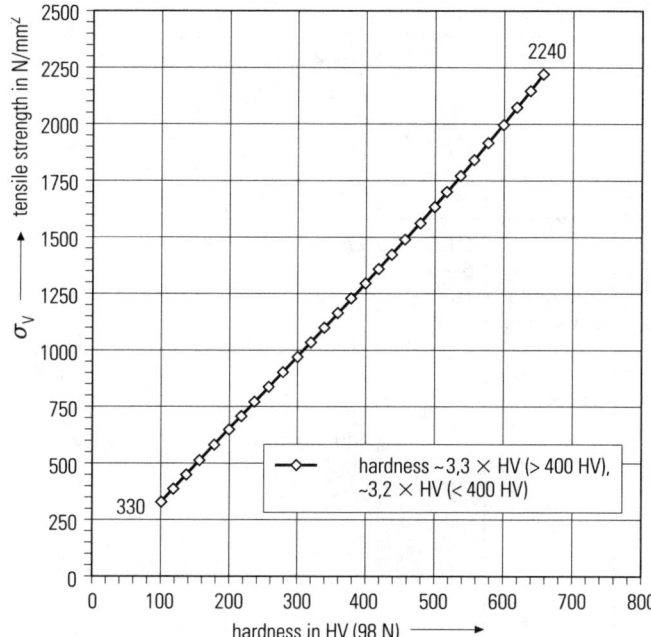

Fig. 4.94. Conversion between Vickers hardness HV and tensile strength σ_v (N/mm²) from DIN 50150

In order to avoid the plastic deformation shown in the shaded areas of Figure 4.93b, the material must be selected to have a yield point $R_{p0.2}$ (see Figure 4.95), calculated from the hardness using DIN 50150 (see Figure 4.94) higher than the equivalent stresses at all points of the curve (Figure 4.93a, b), i.e. $R_{p0.2} > \sigma_{vmax}$ and $R_{p0.2} > p_{max}$.

If this is not the case, as in Figures 4.93b and 4.94, the depth of the surface hardening must be increased. DIN 50190 gives the hardness depth as the point in the hardened layer where the hardness is 550 HV (Fig. 4.93a).

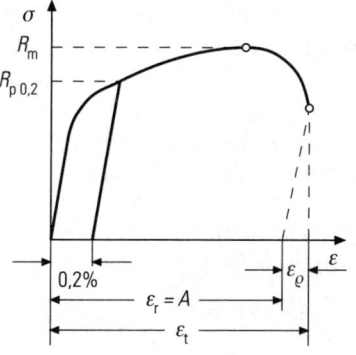

Fig. 4.95. Determining the 0.2% proof stress $R_{p0.2}$ when there is a steady transition from elastic to plastic behaviour

Table 4.20. Hardness conditions for the joint parts

Joint part	Material		Surface hardness	Hardening depths for $p_{max} = 4000$ N/mm²		Remarks
				Case hardening	Induktion hardening	
Bell and inner race	Case hardening e.g. 17 Mn Cr 5 M 16 Mn Cr 5 M (modified materials)	Induction hardening e.g. Cf 45 Cf 53	HRC 58–62 HC* ~ 700	Eht > 0.078 d	Rht > 140 $d/R_{p0.2}$	a, b
Cage	e.g. 16 Mn Cr 5			Eht > 0.078d		a
Balls (DIN 5401)	Roller bearing steel e.g. 100 Cr6 102 Cr5		63 ± 3			c
Shaft	Case hardening steel Heat treatable steel (DIN 17210/212)					

Remarks:
a Eht = case depth for case hardening: $d = 1/16"$,
  Rht = case depth for induction hardening,
  $R_{p0.2}$ = yield point of the material in the core, bell ≈ 510, inner race ≈ 630.
b Bell and inner race should have a surface hardness of *670 + 170 HV* ≈ 2144, 2688 N/mm²*.
c Surface hardening possible.
Case depth measured for a hardness of 550 HV1 as given by DIN 50190/1978, 50103.
* Conversion of the required Vickers hardness into the equivalent stress σ_v 700 · 3.2 ≈ 2240 N/mm² (DIN 50150), Figure 4.94.

4.6.3 Effect of heat treatment on the transmittable static and dynamic torque

Whilst all inner races made of case hardened material receive case depths (Eht) of 1.1–1.5 mm, the induction hardened outer races of UF joints are given a case depth (Rht) of 1.1–2.4 mm.

They are at 27% and 73% of the static torques (GKN Standard 110 001). It is sufficient to assume 50% of the static torques as the mean value.

Table 4.21. Static and dynamic torque capacities for UF constant velocity joints with outer race surface hardness depths (Rht) of 1.1 and 2.4 mm

Joint Part	Outer Race			
Joint	UF 95		UF 107	
Material	Cf 53		Cf 53	
Hardening method	Induction hardening		Induction hardening	
0.2-proof stress in the core	≈ 510 N/mm^2		≈ 510 N/mm^2	
Hardness depth min/max	1.1 mm	2.4 mm	1.1 mm	2.4 mm
p_{max} outer race N/mm^2	2130	2960	2020	2790
p_{max} inner race N/mm^2	2590	3602	2457	3402
Ball normal force N	4800	12900	6500	17250
Torque capacity Nm	600	1615	909	2413
Static torque Nm		2211		3362
%	27%	73%	26%	72%

Fig. 4.96. Tulip of the AAR joint 1998. Semi-warm, multistage formed (800–900°), then extruded. Wall thickness 3.6 to 4 mm with finished joint tracks. GKN Walterscheid Presswerk Trier design

4.6.4 Forging in manufacturing

There are advantages to using cold, semi-warm and warm forming, or a combination of these methods, in the manufacture of the constant velocity joint system: dimensional accuracy, light weight, short production times, low tolerances, good fibre orientation, material savings, some ready-to-use surfaces (or low additional machining), low scale formation and high strength.

Special unalloyed and alloyed steels are used.

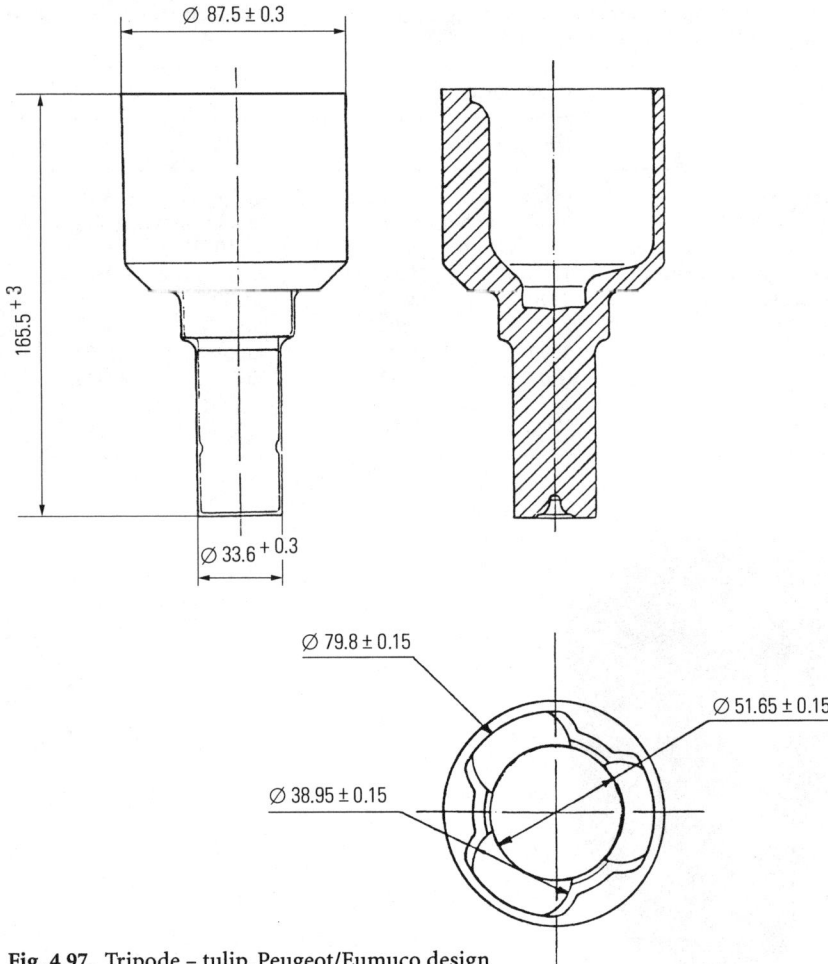

Fig. 4.97. Tripode – tulip. Peugeot/Eumuco design

The increased strength and precision requirements led to forging and pressing of components with and without undercut. High uniformity and accuracy with tolerances of 0.50 mm down to 0.15 mm were aided by forging.

In terms of joint parts, trunnions, tulips and ball races are forged. 12,500 kN multistage hot forming machines are used in the production lines.

The tulip of the AAR plunging joint with ready-to-use tracks is pressed on a multistage transfer press at 800 to 900°, then cooled down and ironed in a controlled manner. The wall thickness of the parts is 3.6 to 4 mm. The stub shafts of the fixed joint of the driveshaft are made on a 1250 t semiwarm press at 820°.

The trunnions, tulips and ball races are pressed out of round steel.

In addition, the trunnions and tulips are semi-warm pressed at about 800 to 900° using a combination process in 4–5 forming stages and are finally cold ironed. After the semi-warm pressing a controlled cooling down takes place as an interim treatment. The chamfers, centres etc are done at the same time in the pressing operation.

Ball races are cold pressed on multistage presses. During this process, depending on the type, the tracks are made ready to use or finished by hard machining.

4.6.5 Manufacturing of joint parts

Outer Joint Body

The material used for the production of large series is carbon steel Cf53; case hardening steel SAE 8620 H, which corresponds to 21 Ni Cr Mo 2, is used for medium and small series. 17 Cr Ni Mo 6 is used for large joints.

Joint component blanks are either forgings with machining allowances or precision formed parts for large scale production. Outer bodies can be made from joint parts and flanges electron beam or friction-welded together.

After processing in the soft condition the heat treatment shown in Table 4.20 is carried out. It follows the grinding of the cage windows, the tracks and the bearing points. All joint parts are surface hardened in their working areas.

CAGE

For joint sizes up to UF 107, steel tube SAE 86 A 17 is used. Cages for the large UF 140 to RF 203 joints on the other hand are manufactured out of forged rings. The window openings are punched up to size UF 107. After processing in the soft state, the cages are carburised and then ground on the inside and outside spheres and also on the window openings. The case depth on the ground surfaces, Eht $\geq 0.078\,d$, the surface hardness is Rht $> 140\ d/R_{p0.2}$ (Table 4.20). The window openings are mostly broached.

The clearance of the inner and outer spheres of the cage are $a = 0.8$ to 2.6 μm/mm cage diameter. At the edges only half the play is allowed $a/2 = 0.4$ to 1.3 μm/mm (Fig. 4.56).

The axial surface loading for high torques and large articulation angles is reduced by using a wide cage with short windows. At least one window opening must match a lobe of the inner race so that it can be inserted into the cage. The width of the window must be dimensioned to give an interference of 0 to 3 μm/mm of ball diameter.

When the joint is running in, the ball presses lightly into the window face and thus creates the necessary working clearances. Sharp edges on the window openings are carefully radiussed, possibly by barrelling, so that they do not strip off the lubricant.

Securing the Inner Race to the Connecting Shaft

For the closed ball joint the most secure connection, which absorbs a high axial load and can easily be undone, has proved to be the square ring made of drawn spring steel (Fig. 4.100a). As a connection which is easy to undo the round circlip DIN 5417 is also used, but this is not as secure against axial load as the square ring (Figs. 4.100b–d).

Fig. 4.98. Outer race of a joint as a deep drawn part made out of case hardened material with 9 axial, arch-shaped ball tracks. GKN Löbro

Outer Bodies of the VL-Joint
were manufactured from forged rings in carbon steel Cf 53 until the size VL 125, further sizes from case hardening steel SAE 8620 H. All forms of connections are practicabel according to Fig. 5.67.

The hardness values of the races for carbon steel Cf 53 and for case hardening steel SAE 8620 H are shown in Table 4.21 and Fig. 4.94.

The Balls
are made from roller bearing steel 100 Cr 6.

Connection Shafts
are manufactured out of layer hardened tubes 38 B 3.

The Joint's Inside Body
has not to be grinded, getting the outside as a pointed arch shape with the same arcs as the inside of the cage (Fig. 4.98). For high plunging joints the outside of the ball hub is roof like (Fig. 4.48).

Cages of the VL-Joints
for standard plunge with inner reinforcement are shaped like a shell ball. For greater plunge they are roof like. The windows were finished through broaching for 0 to 3 µm/mm of the ball-$\varnothing$. For joints until $\beta = 18°$ the windows are grinded after the worm process. The grinded window size has to allow a stress of 0 to 2 µm/mm ball-ϕ. For the hardness conditions look on Table 4.20 and Fig. 4.93.

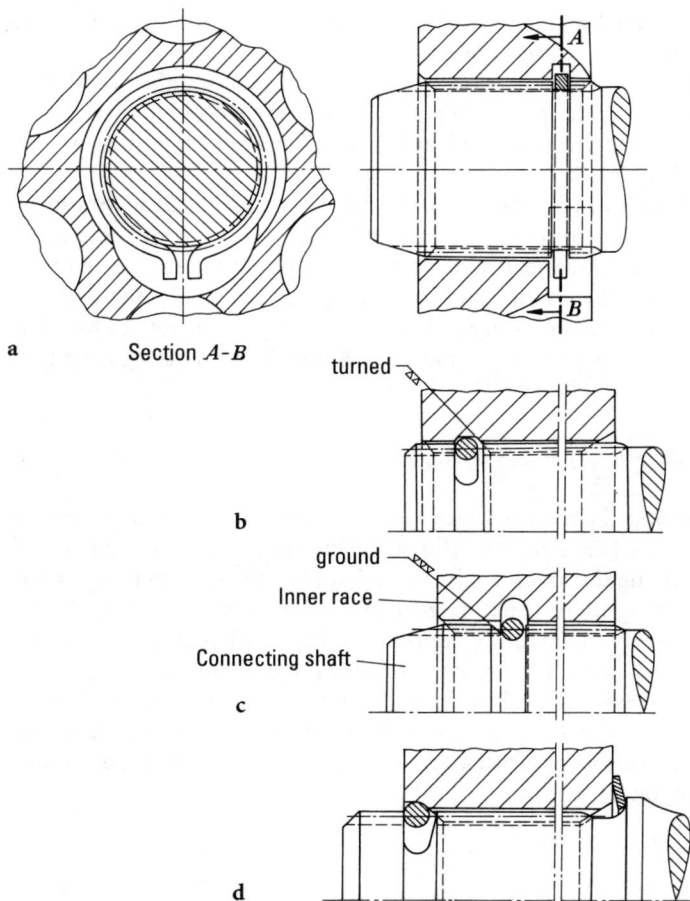

Fig. 4.99a–d. Various types of detachable, axial connections between the inner race and the connecting shaft. **a** Square retaining ring, guided by the inner race. Very secure union; undone with expanding tool. Used in Figs. 5.67, 5.68 and 5.69; **b** round retaining ring, guided by the shaft, springs into the inner race annular groove; **c** round retaining ring, guided by the inner race, when the shaft is insertred it springs into the ground annular groove. Low scatter of the uncoupling force compared with **b** due to good surface of the slope of the groove. Connections **b** and **c** have axial play; **d** round circlip, similar to **b**. Locking groove at front end, shaft collar preloaded by Belleville washer to prevent axial play

4.7 Basic Procedure for the Applications Engineering of Driveshafts

The loading of driveshafts depends both on the prime mover and the machinery which is connected. This is given by the shock or operating factor f_{ST} (Table 4.2).

The objective is a balanced relationship between the maximum transmitted torque M_0 (permanent deformation $\delta_b/d < 1 \cdot 10^{-4}$) and the required durability L_{hx}. If the life L_{hx} is too short, the speed is increased by altering the gear ratios to reduce the torque M_0. The effect of the change in the speed on the life L_{hx} is linear; the effect of changing the torque is to the third power.

The basic procedure for the applications engineering of a constantly loaded drive shaft is shown in Table 4.22. The decision on whether to use a Hooke's joint or a constant velocity joint, i.e. the permissible non-uniformity, is made with the help of (1.2) in Sect. 1.2.1 (Fig. 1.7).

For variable loading, as in a motor vehicle, the life is calculated with a mean torque M_m obtained by transforming (1.22) using $P = M/R$ (Sect. 1.5.3). The individual joints are examined using a mean articulation angle β_m. The life is derived from the total fatigue damage $1/L_{hx}$, made up of the damage for each individual gear, see (1.20) and the calculation examples in Sects. 4.4.5 and 4.5.2.3. The total, equivalent torque M_m is made up from a series of constant torques M_i each acting for a time a_1 (%), where i = 1,2, ... n as follows

$$P_m = \sqrt[3]{a_1 M_1^3 + a_2 M_2^3 + \ldots + a_n M_n^3}.$$

For non-uniform loading, the required parameters can be measured and analysed to give load spectrum [4.23; 4.25; 4.26]. Traditional analysis has resulted in a block programme (Fig. 5.77) giving the loads on the driveshafts for the different gear ratios. Nowadays it is becoming more usual to generate more sophisticated block programmes, based on the numbers of revolutions at torque and angle combinations, from basic data gathered on comprehensive service or proving ground measurement exercises. Computer analysis is used to break down the complex data to allow the construction of meaningful test schedules or the prediction of service life for the driveline components. This work is often carried out in partnership by the automobile makers and component suppliers.

4.7 Basic Procedure for the Applications Engineering of Driveshafts

Table 4.22. Applications Engineering procedure for a driveshaft with uniform loading

Stage	Parameter	Equation	Equation No.	Section	Figure Table
All types of joint					
1	Design torque	$M_B = k_a M_{erf}$ Nm	(3.54, 3.57)	4.6, 3.10.2	Table 4.2
2	Choice of joint size	$M_B \leqq M_N \geqq M_0$ Nm	–	4.6	Figs 4.10, 4.24 to 4.27
3	Speed articulation angle	$n\beta \leqq (n\beta)_{Tab}$ °/min	(4.36)	4.2.8	Tables 4.4
4	Critical bending speed	$n_{max} = 1{,}22 \cdot 10^8 \cdot d/l^2$ rpm	(4.39c)	4.2.9	Figs 5.8, 5.9
	For a plunging tubular shaft	$n_{I\,crit} = (0.7 \text{ bis } 0.9) \cdot 1{,}22 \cdot 10^8 \dfrac{\sqrt{D^2+d^2}}{l^2}$ rpm	(4.39b)	4.2.7	Figs 4.15, 4.19
Hooke's joints					
5	Static torque capacity	$M_0 = \dfrac{2}{S_0}(1-\dfrac{d}{D_m}) \cdot i \cdot z \cdot d \cdot l_w \cdot R$ Nm	(4.21a, b)	4.2.1	Figs 4.10, 5.11
6	Load capacity	$2CR = M_X k_1 \sqrt[10/3]{\dfrac{L_{hx}\, n_x\, \beta}{1.5 \cdot 10^6}} \leqq 2CR_{req}$	(4.29)	4.2.5	Figs 4.10, 5.10 Table 4.2
Ball and tripode joints					
5	Specific loading	$k = \dfrac{M_N}{md^2 R \sin\alpha \cos\gamma}$ N/mm^2	(4.53)	4.4.1	Table 4.4
6	Conformities	$\varkappa_L = \dfrac{d}{\dfrac{2(R+b)}{\cos\alpha \cos\varepsilon} \pm \dfrac{d}{2}}$; $\varkappa_Q = 1/-\psi$	(4.53)/(4.54)	4.4.1	Figure 4.27
7	Hertzian coefficient	$\cos\tau = \dfrac{\varkappa_L - \varkappa_Q}{2 + \varkappa_L + \varkappa_Q}$	(4.57)	4.4.1	Figure 4.27
8	Product $d\Sigma\varrho$	$d\Sigma\varrho = 2(2 + \varkappa_L + \varkappa_Q)$	(4.56)	4.4.1	–
9	Identifying the elliptical coefficients	μ, ν	(3.47–3.50)	3.8	Table 3.2
10	Coefficient of conformity	$c_p = \dfrac{858}{\mu\nu} \sqrt[3]{(d\Sigma\varrho)^2}$ (N/mm^2)$^{2/3}$	(4.5) (3.54)	4.1.1	Table 3.2
11	Hertzian stress	$p_0 = c_p \sqrt[3]{k}$ N/mm^2	(4.7b) (3.57)	4.1.1	Table 4.5
12	Function of the articulation angle (°)	$A = \cos^2\beta\,(1 - \sin\beta)$	(4.65c)	4.4.6.2	–
13	Life $n \leqq 1000$ rpm	$L_h = \dfrac{25339}{n^{0{,}577}} \left(\dfrac{AM_d}{M_{req}}\right)^3$ hrs	(4.73c)	4.4.8	–
	$n > 1000$ rpm	$L_h = \dfrac{470756}{n} \left(\dfrac{AM_d}{M_{req}}\right)^3$ hrs	(4.73d)	4.4.8	–
	Tripode joint	$L_h = \dfrac{365000}{n} \left(\dfrac{M}{M_{req}}\right)^3$ hrs	(4.85)	4.4.2	Figs 4.81–4.82

4.8 Literature to Chapter 4

4.1 Karas, F.: Der Ort größter Beanspruchung in Wälzverbindungen mit verschiedenen Druckfiguren. (The location of the greatest stress in rolling connections with various pressure distributions). Forsch. Ingenieurwes. 12 (1941) p. 237–43
4.2 Beljajef, N. M.: Memoirs on Theory of Structures. St. Petersburg 1924
4.3 Foeppl, L.: Der Spannungszustand und die Anstrengung des Werkstoffes bei der Berührung zweier Körper. (The stress state and strain of the material when two bodies touch) Forsch. Ingenieurwes. 7 (1936) p. 209–221
4.4 Timoshenko, St.: Theory of elasticity. New York: McGraw-Hill 1934, 339–352
4.5 Mohr, O.: Abhandlungen a.d. Gebiete d. techn. Mechanik, 3rd edition (V). Berlin: Wilh. Ernst & S. 1928; 1st edition 1905, 2nd edition 1914 (Proceedings on Techn. Mechanics)
4.6 Gough, H. J.: The fatigue of metals, London: Benn 1926
4.7 Moore, H. F.; Kommers, J. B.: The fatigue of metals, New York: McGraw-Hill 1927
4.8 Niemann, G.: Maschinenelemente (Machine Elements). Vol. 1. Berlin: Springer 1963 p. 210
4.9 Loroesch, H. K.: Influence of Load on the Magnitude of the Life Exponent for Rolling Bearings. American Society for Testing and Materials (ASTM), Special Technical Publication 771, p. 275–292. Philadelphia 1982
4.10 ISO 76, 1987, Sect. 6.1, p. 5
4.11 Fischer, W.: Die Berechnung der dynamischen Tragfähigkeit von Wälzlagern (Calculating the dynamic load capacity of roller bearings). 2nd Edition. Schweinfurt: SKF Kugellagerfabriken GmbH 1958, p. 19, 21
4.12 DIN ISO 281, 1979, Part 1, sect. 6.1, p. 9
4.13 Jante, A.: Kraftfahrt-Mechanik, 3 Der Kraftschluß m. d. Fahrbahn (Motor vehicle mechanics, 3. Friction at the road surface). In: Bussien, R. (Publisher): Automobiltechnisches Handbuch (Handbook on Automotive Engineering). 14th Edition: M. Krayn, Techn. Verlag 1941, p. 9–15
4.14 Reuthe, W.: Die Bewegungsverhältnisse bei Kreuzgelenkantrieben (Relationships of movement for Hooke's jointed drives). Konstruktion 2 (1950) 305–312
4.15 Kleinschmidt, H. J.: Gelenkwellen. In: Anwendungen der Antriebstechnik (Driveshafts. In: Applications of driveline technology). Vol. II Kupplungen (Couplings). Mainz: Krausskopf 1974, p. 45/46
4.16 Dunkerley, S.: The Whirling and Vibration of Shafts. Phil Soc. Trans. London 185 (1894), p. 270
4.17 Grammel, R.: Neuere Untersuchungen über kritische Zustände rasch umlaufender Wellen (Recent investigations into the critical states of rapidly rotating shafts). Ergeb. de. Exakten Naturwiss. 1 (1922) p. 92–119
4.18 Biezeno, C. B.; Grammel, R.: Technische Dynamik (Engineering Dynamics) Berlin: Springer 1939
4.19 Foeppl, A.: Technische Mechanik. (Engineering mechanics) Vol. IV Dynamik. 2nd Edition. Munich: Oldenbourg 1901, p. 248–252
4.20 Stodola, A.: Die Dampfturbinen (Steam turbines), 3rd Edition Berlin: Springer 1905, p. 193–196
4.21 Dietz, H.: Die Übertragung von Momenten in Kreuzgelenken (Torque transmission in Hooke's joints). VDI Z. 82 (1938) 825–828 and 83 (1939) p. 508
4.22 Lehr, E.: Die umlaufenden Massen als Schwingungserreger (Rotating masses as vibration sources) Masch. Bau/Gestaltung 1 (1982) p. 629–634
4.23 Richtlinie VDI (VDI Directive) 2060: Beurteilungsmaßstäbe für den Auswuchtzustand rotierender starrer Koerper (Criteria for assessing the state of balance of rotating rigid bodies). Düsseldorf: VDI-Verlag 1966
4.24 Phillips, J. R.; Winter, H.: Über die Frage des Gleitens in Kugel-Gleichganggelenken (On the question of sliding in constant velocity ball joints). VDI Z. 110 (1968) p. 228–233

4.8 Literature to Chapter 4

4.25 Macielinski, J. W.: Propeller shafts and universal joints – Characteristics and methods of selection. Conf. Drive Line Engineering 1970. Inst. Mechanical Engineers, Automotive Division, Session 3, Paper 29
4.26 Stellrecht, H.: Die Belastbarkeit der Wälzlager (The load carrying capacity of roller bearings). Berlin: Springer 1928
4.27 Spehr, Eugen: Der Kardanfehler bei einem System mit zwei Gelenken. VDI-Zs. 103 (1961) 247–250
4.28 Reister, D.: Lebensdauerberechnung für Klemmrollen-Freiläufe. Industrie-Anzeiger 92 (1971), Nr. 71, S. 1805–11 (Ausgabe „Antrieb-Getriebe-Steuerung", 8. Heft)
4.29 Paland, E. G.: INA-Techn. Taschenbuch, 5. Aufl. Herzogenaurach: INA-Wälzlager Schaeffler KG, 1998, S. 169/170
4.30 Stölzle, K.; Hart, S.: Freilaufkupplungen, Konstruktionsbücher 19. Berlin: Springer-Verlag 1961, S. 69–73
4.31 Karde, Kl.: Die Grundlagen der Berechnung und Bemessung des Klemmrollen-Freilaufs. Automobiltechn. Zs. 51, (1949), Nr. 3, S. 49–58 u. 52 (1950), H. 3, S. 85
4.32 Thüngen, H. Frhr. v.: Der Freilauf. Automobiltechn. Zs. 59 (1957), Nr. 1, S. 1–7; ders., in: Bussien, R.: Automobiltechn. Hdb., 18. Aufl., 1965, S. 280–282
4.33 Wieland, G.: Kardanfehler bei Kreuzgelenk und Gelenkwelle, und Mittel zu ihrer Beeinflussung. Glasers Annalen 80 (1956), H. 1, S. 19–24
4.34 Duditza, Fl.: Doppelkardangelenk-Getriebe. Antriebstechnik 11 (1972), Nr. 3, S. 91–97
4.35 Jacob, W.; Paland, E. G.: Das neue Gleichlauf-Kugelgelenk NT-01 und sein Einsatz in Gelenkwellen. Rodgau: Eichler Electronic 2001
4.36 Miller, F. F.: Constant Velocity Universal Ball Joints – Their Application in Wheel Drives. SAE-Paper 958 A (1965)
4.37 Enke, K.: Konstruktive Überlegungen zu einem neuen Gleichlaufschiebegelenk. Automobil. Ind. 15 (1970) 33–38
4.38 Krude, W.: Theoretische Grundlagen der Kugel-Gleichlaufgelenke unter Beugung. IV. Eppan-Tagung Uni-Cardan AG 1973 (MS)
4.39 Kutzbach, K.: Quer- u. winkelbewegliche Wellenkupplungen. Kraftfahrtech. Forschungsarb. 6. Berlin: VDI-Verlag 1937
4.40 Harris, T. A.: Rolling Analysis. New York: John Wiley & Sons, 3. edition, 1990
4.41 Taniyama, K.; Kubo, S.; Taniguchi, T.: Effect of Dimensional Factors on the Life of Rzeppa Universal Joint. SAE Technical Paper Series, 850 355 Passenger Car Transmissions, SP 619, 1983/85
4.42 Lloyd, R. A.; Bartlett, S. C.: Improvements to the noise-vibration-hardness performance of tripode CV joints through mathematical modelling. Inst. Mechan. Engin. C 487/029, 1994

5 Joint and Driveshaft Configurations

Driveshafts are systems made up of a number of elements arranged and linked together. A system is characterised in that it can be delineated from its surroundings, with the links to the surroundings at the boundary of the system [5.1; 5.2]. A variety of factors affects the selection of the sub-systems.

The elements of the driveshaft systems – the individual parts – are shown in Fig. 5.1 together with their functions. These are the connecting shaft *1*, the joints *2* and the intermediate shaft *3*.

The type of joint gives the driveshaft its name, leading to three families:

- Hooke's jointed,
- ball jointed,
- pode jointed.

The function of driveshafts is to transmit torque and rotation from the driving unit to the driven unit as uniformly as possible, with at least one unit being able to change

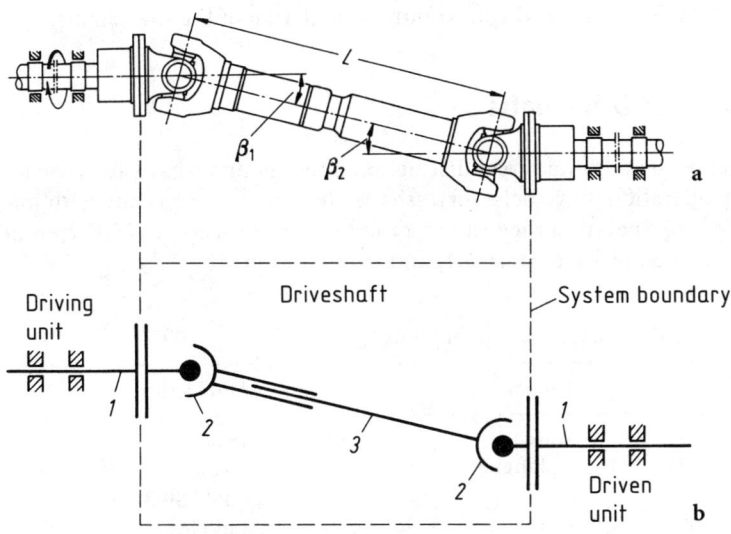

Fig. 5.1 a, b. Driveshaft system. **a** Hooke's jointed shaft in a Z-configuration, **b** diagrammatic representation of the system elements *1* to *3*

its spatial position within set limits. This overall function can be divided into three individual functions:

- 1 the uniform transmission of torque and rotation,
- 2 the ability to alter the distance between the input and output,
- 3 the ability to alter the angle between the input and output.

In general, the aim is to make one element perform several functions. In the three families of driveshafts, the following elements play a part in functions 1 to 3:

System element	Driveshaft family		
	Hooke's joint	ball joint	pode joint
Connecting elements	1	1	1
Joints	1, 3	1, 2, 3	1, 2, 3
Intermediate shaft	1, 2	1	1

Hooke's jointed driveshafts need a torsionally stiff plunging element in the intermediate shaft; in driveshafts having ball or pode joints this is an inherent characteristic. Furthermore, in Hooke's jointed driveshafts, the uniform transmission of rotation and torque is only partially fulfilled even at moderate articulation angles, see Chaps. 1 and 2.

In the case of driveshafts, the underlying function is the transfer of a compressive force between the curved surfaces of bodies at a distance R from an axis of rotation. The equations needed for this are derived in Chap. 3; the universal torque equation (4.16) is given in Chap. 4. The points of action and the position of the active areas are shown in Fig. 4.3. Table 5.1 shows the maximum articulation of the shaft joints.

5.1 Hooke's Jointed Driveshafts

Manufacturers and users draw a broad distinction between driveshafts for vehicles or for stationary operation. In vehicles, driveshafts run at high speeds and with low torques. In stationary operation they run at a range of speeds (up to 12 000 rpm in test rigs and balancing machines) [5.3, 5.4] and at high torque.

Table 5.1. Maximum articulation angle β_{max} of joints

Hooke's joints		Ball joints		Pode joints	
single	20 ... 35°	Weiss, plunging	20 ... 25°	tripode, fixed	40 ... 45°
double	42 ... 50°			plunging	20 ... 25°
agriculture, double	70 ... 80°	Löbro, plunging	20 ... 25°	bipode	20 ... 25°
		Rzeppa, fixed	45 ... 50°		

5.1.1 End Connections

The only function of the end connections is to transmit torque. They should centre the driveshafts well and link, by friction or positive connection, or both, the input to the output via the driveshaft.

There are five practical possibilities for connecting driveshafts (Fig. 5.25a–g).

1. *Flange*. The driveshaft ends in a flange with yoke lugs to DIN (Fig. 5.2a), or SAE standard (Fig. 5.2b), or in a flange with cross serrations (Fig. 5.3c).
2. *Slip yoke*. The driveshaft ends in a hub which is either closed or split with lugs (Figs. 5.28, 5.30). It is standard practice in agricultural engineering to have a splined quick-disconnect hub (Fig. 5.55c). Splines are used as plunging elements on the output shafts of passenger car transmissions (DIN 5461–63).
3. *Stubshaft*. Driveshafts with a single or double joint end in two stubshafts with keys or splines (Fig. 5.24), e.g. in steer drive axles of commercial vehicles.
4. *Wing bearing*. The driveshaft ends in two unit packs with small wing bearings which are replaceable as an assembly (Figs. 5.28–5.30), e.g. for American construction vehicles, and which are centred on the positively driven end yoke (Fig. 5.5d) and bolted in position.
5. *End yoke*. The driveshaft ends in a cross which is mounted in round bushes. It is secured with U-bolts to the hub of the driven unit and is centred by the bushes. Used in the USA.

Comparison of the Possible Methods of Connections

By adjusting the axial play in the unit pack and by exact, play-free centring, flanged driveshafts can achieve a balancing quality of Q 16 and therefore give very smooth running. If replacement is necessary shafts having the same balancing quality will restore the driveline to its original state.

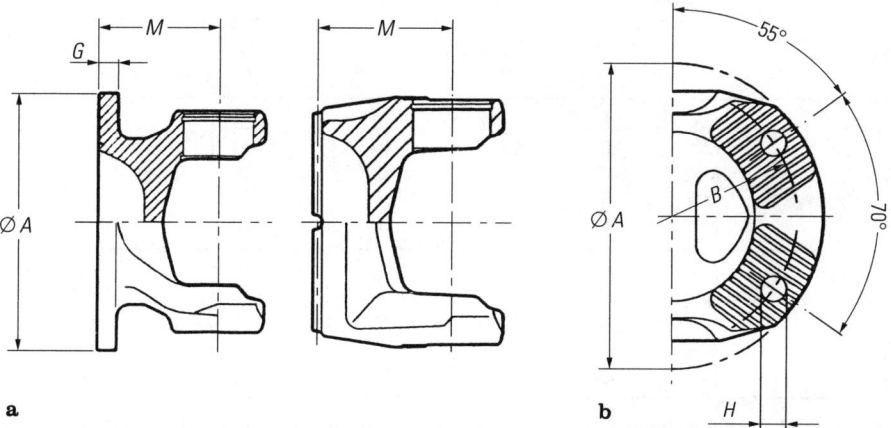

Fig. 5.2. Driveshaft flange connections. **a** DIN 15450–53; **b** positive drive with face splines for 120, 155, 180 mm ∅ (ISO 12667)

DIN 15450-53/ISO 7646   SAE/ISO 7646   ISO 12667

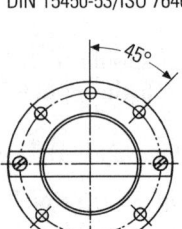 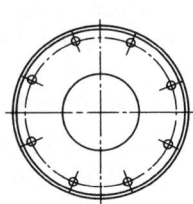 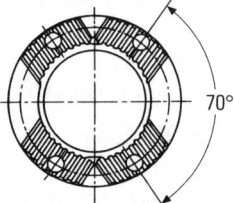

a internally centred   **b** externally centred   **c** Hirth-Klingelnberg-spur cut

d Cross geared flanges
Kreuzverzahnte Flansche

Flange ⌀ A [mm]	Thread ⌀ d [mm]	Length l [mm]	Flange thickness G_1 [mm]	G_2 [mm]	Conn ect. length $G_1 + G_2$ [mm]	Nut high h [mm]	Tightening torque M_a [Nm]	
120	M10	40	14	10	24	9	62	
150	M12	45	16	12	28	11	105	±20%
180	M14	50	18	14	32	12	170	

DIN-Flange   SAE-Flange

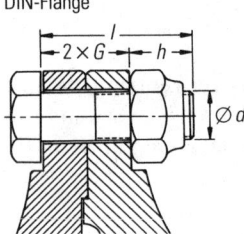   Hex bolt short
similar to DIN 931/10.9
self-saving DIN 980/10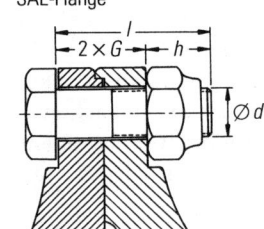

Driveshaft Connection	Thread ⌀ d [mm]	Length l [mm]	Grip 2 × G [mm]	Nut highth h [mm]	Tightening moment M_a, T_a [Nm]	
DIN	M8	25	12/14	8	35	
	M10	30	16	10	70	
	M12	40	20	12	120	±7%
	M16	45	24	16	295	
	M16	50	30	16	295	
SAE	M8	30	16	8	35	
	M10	30	16	10	70	
	M10	35	19/21	10	70	±7%
	M12	35	16	12	120	
	M12	40	21	12	120	

Fig. 5.3a–d. Driveshaft flange connections. **a** Innercentering, cone key (DIN 15450-53), transverse taper (DIN ISO 7646; **b** Outercentering (ISO/SAE 7647); **c** Spur cut Hirth/Klingelnberg (ISO 12667); **d** Cross geared flanges

5.1 Hooke's Jointed Driveshafts

Flange connection ⌀	GWB Series	Swing diameter in mm	β_{max}	Rated torque M_N in Nm	Main use
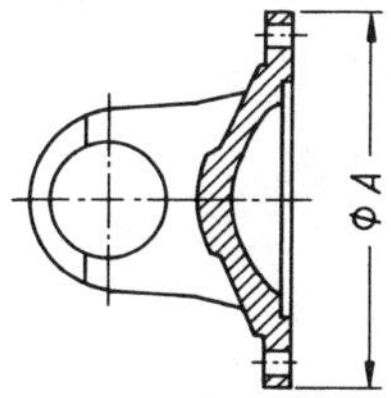	687	90 – 206	25° – 44°	2400 – 35000	Road and rail vehicles over 2,5 t GVW. ships drives, general machine tools
	587	215 – 265	20°	43000 – 57000	Rail vehicles, ships calender, rolling mill and crane drives. (DIN 15450)
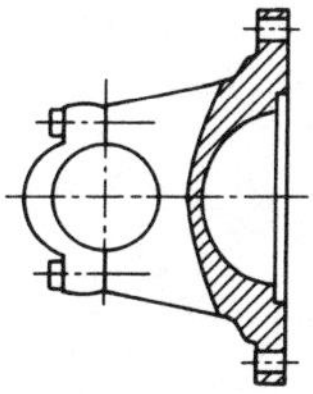	390	240 – 370	15°	60000 – 255000	Rail vehicles, ships calender, rolling mill and crane drives. (DIN 15450)
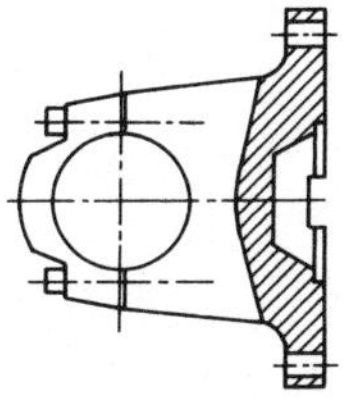	392	225 – 550	10° – 15°	70000 – 115000	Calender and rolling mill drives with high torques and low speed
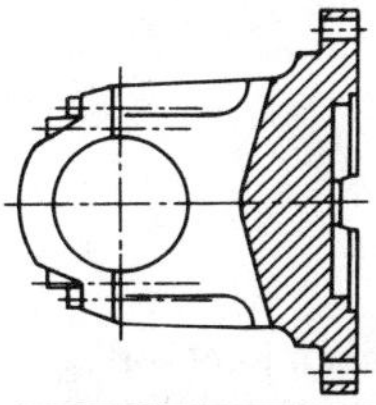	498	600 – 1200	5° – 15°	1880000 – 15000000	Slow running main drives for rolling mills, general heavy machinery

Fig. 5.4. Lug designs for cross trunnion needle or roller bearings. Flange diameter A = swing diameter

The slip yoke connection centres on the input and output shafts which have a lower centring quality due to the tolerances. The non-correctable residual unbalance of this driveshaft therefore drops to the level $Q \approx 40$.

The situation is similar for wing bearing and U-bold connections. The axial play of the cross trunnion is not adjustable, hence a high level of balance cannot be achieved.

5.1.2 Cross Trunnions

The most important element of the Hooke's joint is the cross trunnion bearing. The replacement of the plain bearing by the roller bearing in 1928–30 (Figs. 5.7a to c)

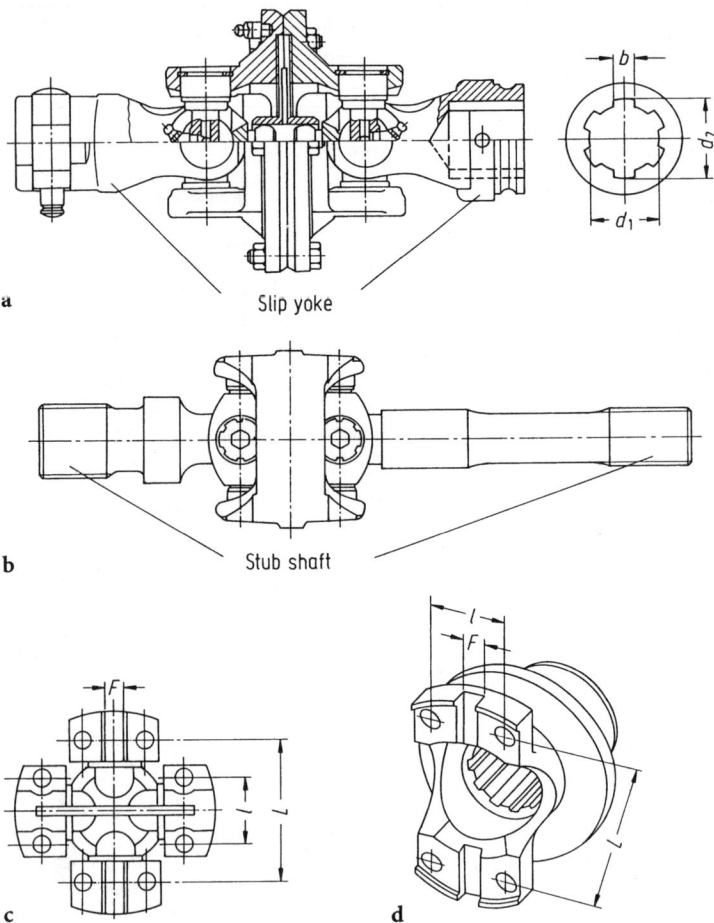

Fig. 5.5a–d. Ways of connecting driveshafts. **a** Homokinetic double Hooke's joint for agricultural machines with slip yoke connection at both ends, disconnected by a sliding pin; **b** double Hooke's joint for steer drive axles with stub shafts at both ends (see also Fig. 5.15); **c** Hooke's joint with wing bearings, Mechanics design (Rockford Powertrain Inc, USA, and Stieber Div., Ketsch); **d** end yoke for wing bearings – Mechanics type (Type C)

greatly advanced the service life of the joint. As a result the life for commercial vehicles installations has increased to between 500 000 and 1 000 000 km and the efficiency to 98–99%.

Improved lubrication and sealing (Figs. 5.7b and c) are other advances. The lubricant is conveyed by centrifugal force from a central reservoir *1* through bores *2* and spiral grooves in the thrust washer *3* to the ends of the trunnion and the bearings *6*. The sealing must keep the lubricant in the trunnion bearing and prevent contaminants, including water, from getting into the bearing. Contacting single and double lip seals are generally used for this purpose (Fig. 5.7c). With this tried and

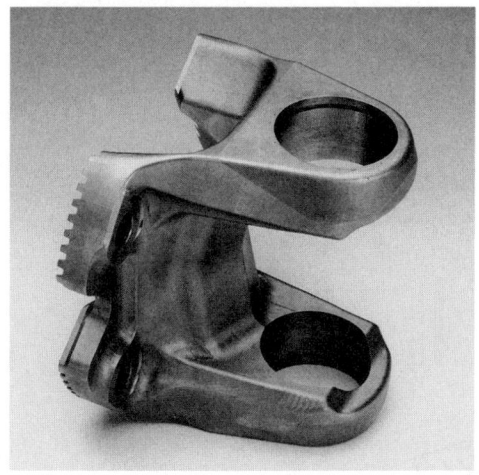

Fig. 5.6. Flange driving dog from forged steel to reduce weight of the driveshaft by 1 kg. GWB design. For a Hooke's driveshaft of 19 000 Nm the weight is reduced by 10 %, that means from 43 to 39 kg. Example: Flange driving dog by 2 kg (see foto), cross trunnions by 2 · 0.3 kg and the trunnion flange driving dog by 0.8 kg

tested method, if pressure-resistant greases are used, a supporting lubricant film also forms at the contact points of the rolling bodies and in their highly stressed rolling areas (Fig. 5.8).

The bearing assemblies are held securely in the yoke by:

- plastic injection (e.g. Opel),
- circlips, positioned externally (GWB, USA and agricultural engineering) or internally (Mercedes-Benz),
- bolted caps or hoop straps (Glaenzer-Spicer and GWB),
- U-bolts or wing bearings, with external centring (USA),
- staking (INA).

In 1970 the torque capacity was increased up to 10 000 000 Nm by the introduction of single and multi-row roller bearings for the trunnions instead of needle bearings, Figs. 5.7a and b (suggested as early as 1953 by Heinrich Roessler, German Patent 936 141, as "crowned bearing needles" against edge pressure). The roller bearings are made to rotate by axial preloading, reducing the harmful sliding of the rollers.

In the driveshaft is to run smoothly, its overall centre of mass and its principal axis of inertia must lie on the axis of rotation [4.20]. It must also be centred exactly. This is achieved by reducing as far as possible:

- the radial play of the trunnion, e.g. by the use of spacers or circlips of different thicknesses,
- the unbalance of the rotating driveshaft in two given planes by means of balancing machines. If the operating speed is less than 300 rpm, balancing is not usually carried out.

A high balancing quality Q from Fig. 5.8 allows a closer approach to the critical bending speed.

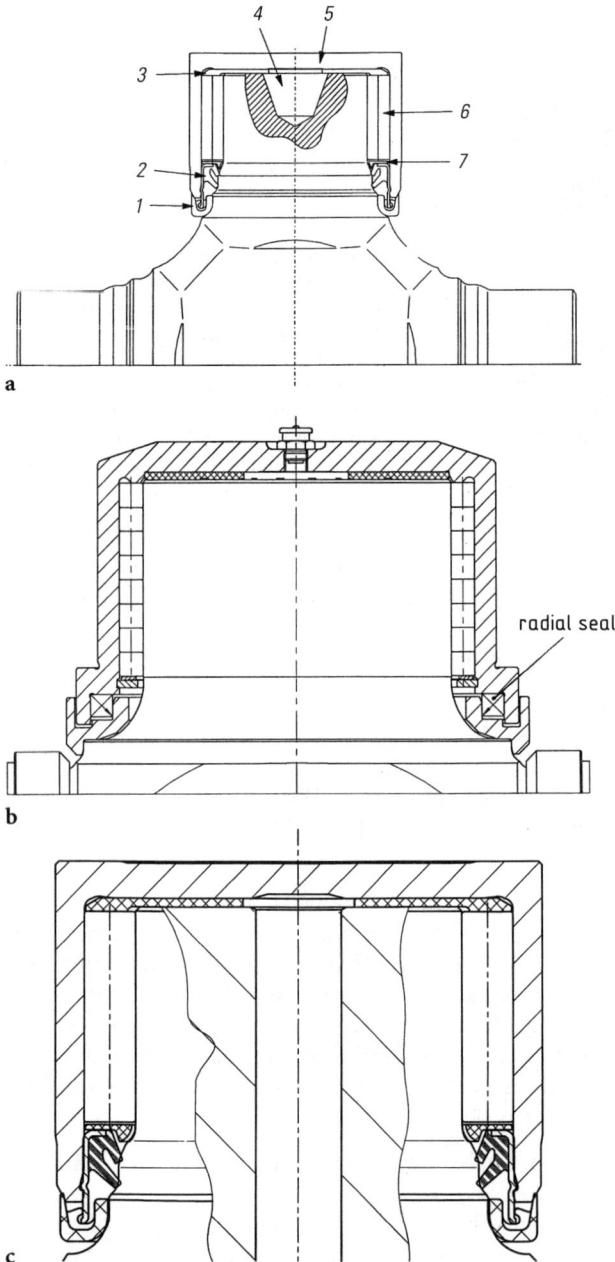

Fig. 5.7a–c. Maintenance-free cross trunnion pack for commercial vehicles with grease pocket. *1* front seal, *2* main lip seal, *3* thrust washer with spiral lubrication grooves, *4* grease pocket, *5* bearing cup, *6* needle or roller, *7* support washer as axial bearing against the floor of the cup. **b** joint cross pack with eight-row, guided roller bearings and front end thrust washer. **c** joint cross pack of the heavy series of industrial shafts with medium-walled solid bush, GWB design

5.1 Hooke's Jointed Driveshafts

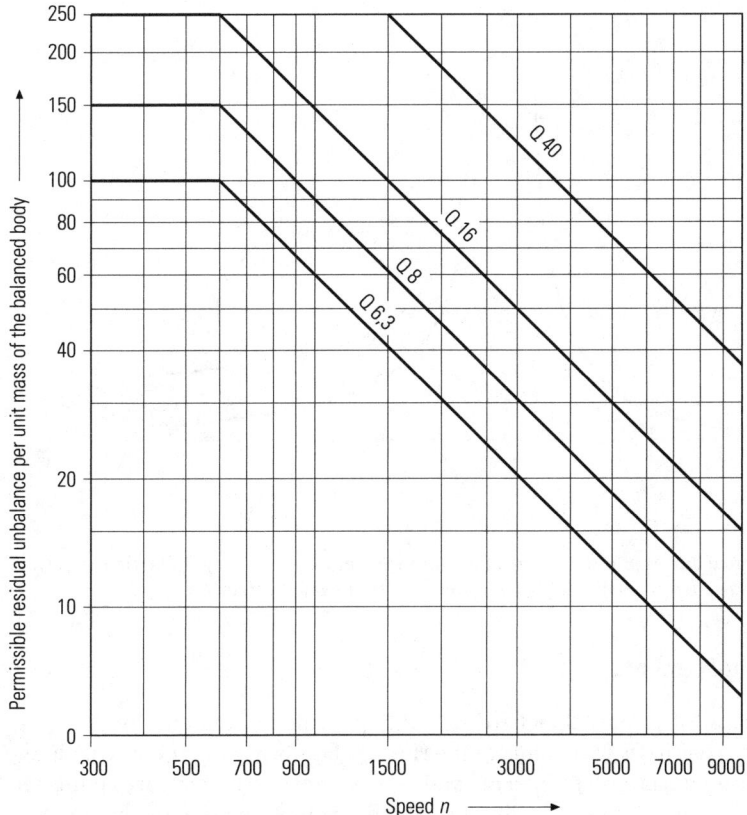

Fig. 5.8. Levels of balancing quality Q for driveshafts according to VDI Directive 2060

The following quality levels Q are possible with the different designs of the trunnion bearing [4.21]:

Curve in Fig. 5.8	Design	Q
1	Roller bearing without adjustment of axial play	≈ 40
2	Roller bearing with adjustment of axial play, up to 0.12 mm	16
3	Ditto, up to 0.05 mm	6.3

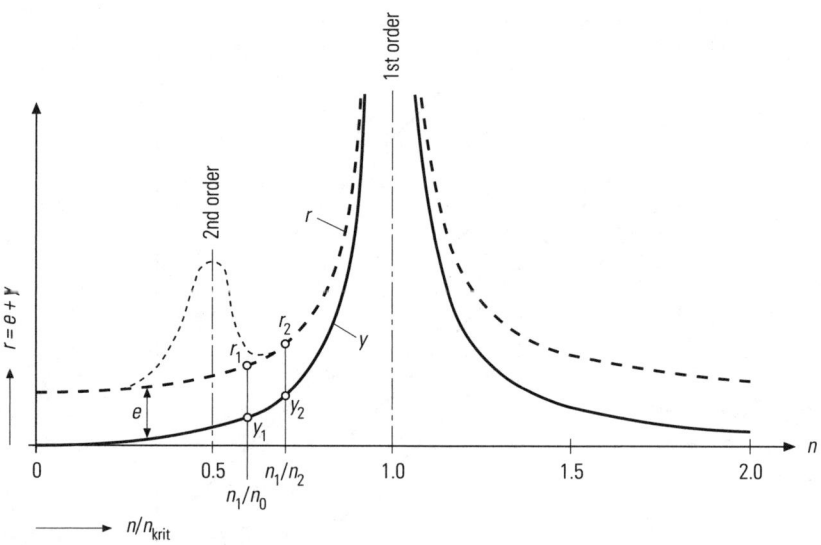

Fig. 5.9. Resonance curve of a driveshaft, first natural frequency. —— y = elastic deflection, ----- r = amplitude e from elastic deflection y and centre of mass offset e

5.1.3 Plunging Elements

The pivoting joint of Robert Bouchard can only provide limited end motion (Fig. 1.16). All other driveshafts with joints of the Hooke's family require their own plunging element. This consists of an internal and an external body which are connected so that they can move longitudinally and which are torsionally stiff due to their geometry. The mating parts can have involute splines to DIN 5480, lemon-shaped or star-shaped profiles (Fig. 5.11).

5.1.4 Friction in the driveline – longitudinal plunges

Longitudinal or linear plunge must be provided if the driveshaft has to adjust to fairly long working lengths, due to the position of the driven units. The shaft must be able to plunge in a sliding or rolling manner, and must have as little friction as possible (similar to a linear ball bearing DIN 636).

Spline shaft profiles to DIN 5463/64 allow sliding longitudinal plunge on the driveshafts as shown in Figure 5.11.

Sliding

– grooved spherical ball jointed driveshafts;
– European designed driveshafts with rilsan coated splines for vehicle and machinery,
– American designs with SAE 16-tooth splines;
– driveshafts for agricultural machines: special driveshafts with longitudinally guided slideways.

The plunge resistance here is 90 N, particularly with a fairly long travel s_w.

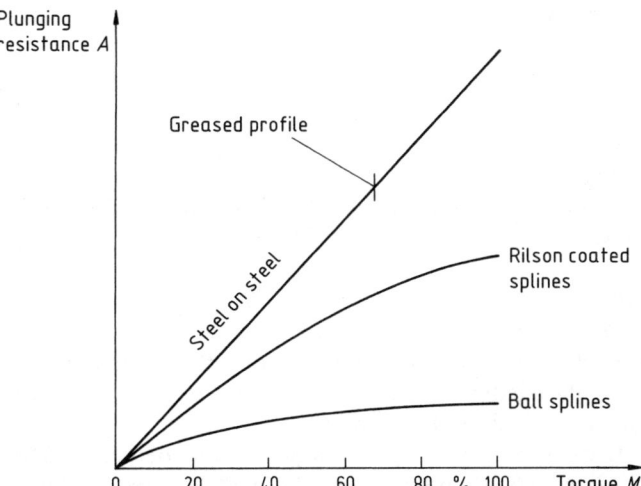

Fig. 5.10. Plunge resistance A for sliding and rolling elements as a function of the torque and surface conditions

Rolling

Guided longitudinal plunge using roller bodies appeared as early as 1888 in the USA, 1899 at DWF and 1900 at Fichtel & Sachs. In more recent time the advantages of rolling friction have been recognised, because of the halved grease consumption. It is also said to absorb high-frequency axial vibrations (Table 5.2). Here the spline shaft forms a unit with the inner part of the constant velocity fixed joint. A combined boot is fitted because its axial plunge resistance is lower than of the bellows-type boot. A plunge resistance less than 20 N is achieved, even at zero speed. Consideration must be made for two distinct cases of halfshaft plunge:

– transverse engines with automatic transmission
– longitudinal engines.

Splines allow the greatest length compensation for driveshaft. However, while they move easily at zero torque, under high torques splined shaft with straight sides can jam because:

– the surface pressure on the sides of the spline is high on account of the small lever arm,
– the axial distribution of the surface pressure along the sides of the splineis too uneven because of helical deformation.

An increase inengine performance and engine torque, lighter but stronger structures and greater demands from end users have resulted in further developments of freely plunging driveshafts. The plunge compensates for the varying vibrations in the driveline.

The axial resistance to the longitudinal plunge

$$A = M \frac{\mu \cdot N}{R \cdot \cos \alpha} \qquad (5.1)$$

where $\alpha = 30°$ is the gearing pressure angle.

It acts on the cross trunnion and results in radial loading of the needle or roller bearing. This has to be taken into account in the design process using

$$Q_{res} = \sqrt{Q^2 + A^2} \text{ N}. \qquad (5.2)$$

In (5.1) and (5.2), M is the driveshaft torque in Nm, R is the effective radius of the plunging element in m, μ is the coefficient of friction (which depends on the conditions within the element), and Q is the radial load in N.

Since the plunge resistance A from (5.1) depends largely on the coefficient of friction μ of the faces which slide on one another, attempts are made to reduce this by

- good lubrication and sealing,
- applying molybdenum disulphide MoS_2,

Table 5.2. Examples of longitudinal plunge via balls in drive shafts

Year	Patent, Figure	Inventor, Company	Design
1908	US Patent 1022909 Figure 2.10	William Whitney	Longitudinal plunge through balls without a cage
1914	British Patent 14129	Victor Lee Emerson	Roller with ball race
1960	Figure 2.11	BTB based on Whitney	Halfshaft with ball race, length ratio of the ball guide only 0.44
1960	US Patent 1854873	Mechanics/Borg Warner	Driveshaft with spring-guided balls
1961	French Patent 1341275	William Cull/Birfield	Plunging via a ball race
1963	Figure 5.9	Edmond B. Anderson, (Borg Warner), Kurt Enke (Daimler-Benz)	Roller rotation with 2 to 4 roller cartridges
1980	Figure 5.29d	Zahnradfabrik Schwäb. Gmünd (ZF)	Telescopic steering wheel shaft, free from play. Plunge resistance < 20 N
1981	German Patent 3103172	Honda/Audi	Longitudinally guided tripode rollers
1982	Figure 5.15	GKN/Löbro	cage-guided rollers (Audi 80 and 100)

For shorter travels s_w the tripode joint provides adequate plunge.

5.1 Hooke's Jointed Driveshafts

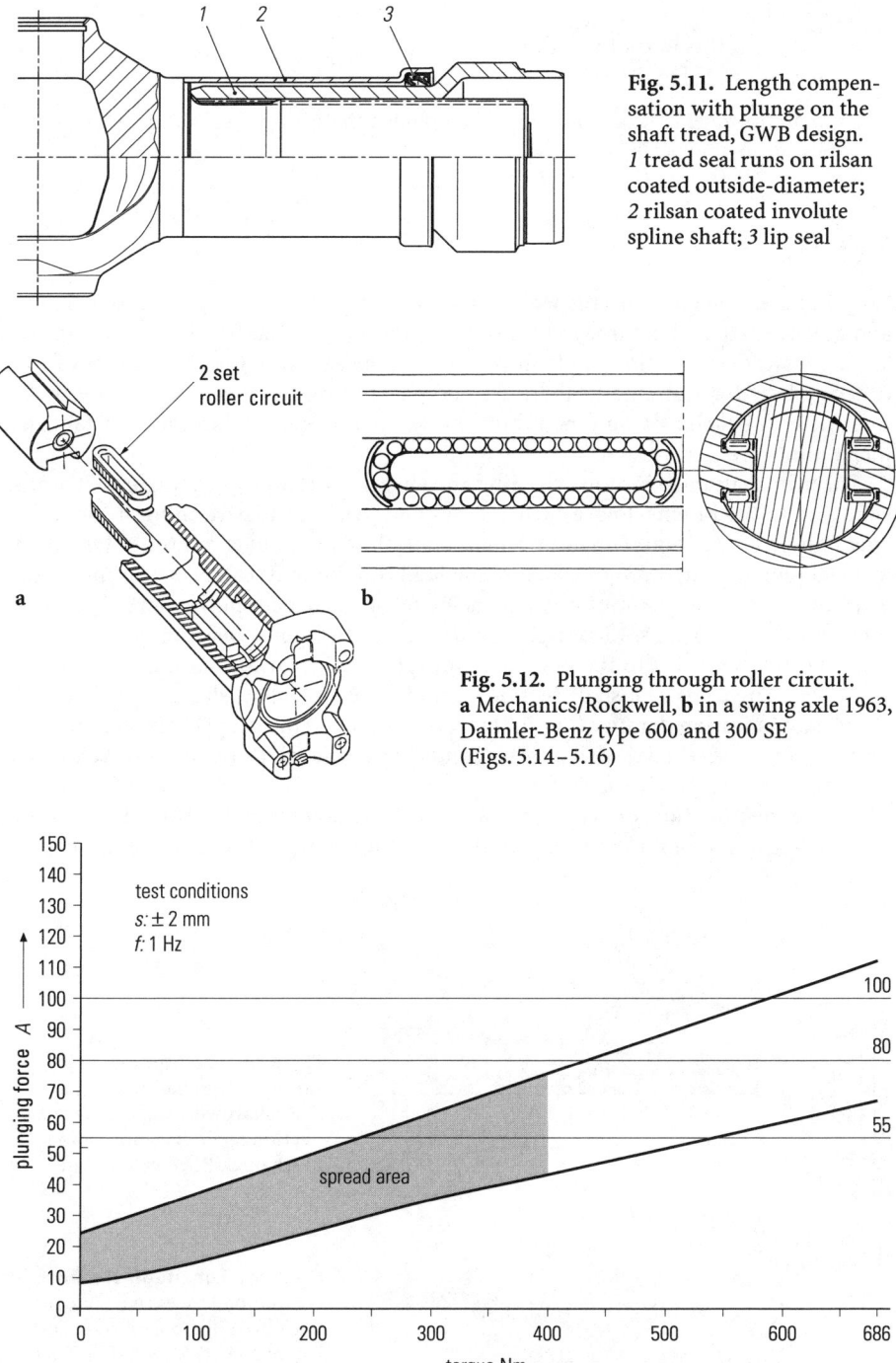

Fig. 5.11. Length compensation with plunge on the shaft tread, GWB design. *1* tread seal runs on rilsan coated outside-diameter; *2* rilsan coated involute spline shaft; *3* lip seal

Fig. 5.12. Plunging through roller circuit. **a** Mechanics/Rockwell, **b** in a swing axle 1963, Daimler-Benz type 600 and 300 SE (Figs. 5.14–5.16)

Fig. 5.13. Planging unit with ball. Plunging force *A* sloping from torque Nm

- phosphating,
- coating with Rilsan (polyamide),
- the use of balls (Fig. 5.12).

The most common is Rilsan coating. This allows the plunge resistance A to be reduced by almost 50% (Fig. 5.11).

5.1.5 The propshaft

Propshafts are needed for rear wheel drive to transmit the torque from the engine and gearbox. They turn three to four times faster than halfshafts (up to 10 000 rpm). Their greatest articulation angle in continuous operation is $\beta = 7°$. Because of the differing length requirements there are one-part to three-part shafts, the latter with intermediate bearings. Combined shafts made up of Hooke's joints and constant velocity joints are also known.

The propshaft must be quiet running and absorb vibrations. The single part propshaft (up to 2.6 m wheel base) achieves a good level of quiet running if there are small-angled fixed joints (up to 23°) on its ends; the front joint is combined here with an axial plunge unit, through which radial loads on the shaft can be accommodated. This allows different vibration frequencies to be absorbed, particularly high frequency axial vibrations. In *two-part-shafts* an elastic centre bearing positions the propshaft with respect to the bodywork and reduces the transmission of vibration and noise. The *front* propshaft section embody a flexible disc, elastically centred at $\beta = 0°$, as a vibration damper (Fig. 5.71). The rear part of the propshaft from the centre bearing to the rear axle drive corresponds to the one-part shaft VL joints can also be used for the ends of propshafts.

The connecting tube of every driveshaft is a resonant body that transmits noises. For this reason the tubes are lined with foam, and rubber stoppers or star-shaped

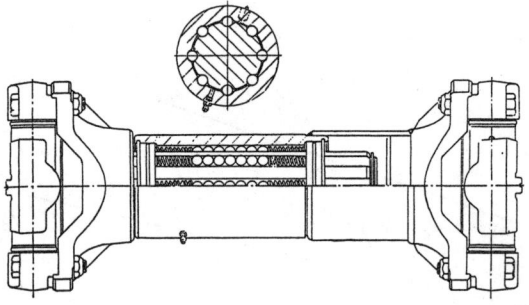

Fig. 5.14. Longitudinal plunge through eight ball races in a driveshaft with small yoke connection facilitates lubrication, Mechanics/Rockwell design

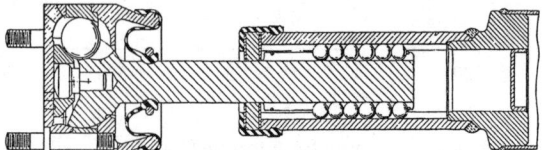

Fig. 5.15. Longitudinal plunge via four ball races in a CV driveshaft by William Cull (French Patent 1341 275/1961–62). There is no cage which leads to friction losses between the balls

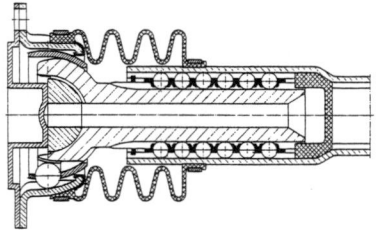

Fig. 5.16. CUF-Sideshaft joint with linear plunge via cage-supported ball race.
Static breaking strength M_B = 4000 Nm, rigidity 220 to 240 Nm/°, torsional play: depending on build tolerances 4–14', as a maximum 10–12'.

sections are pressed in, and steel rings are pressed on to the tube. This reduces the noise emission by 3 to 6 dB(A). The extruded aluminium tubular shaft has good properties as regards noise, crash behaviour, temperature resistance (130–145°) and is also inexpensive. The wall thickness and yoke have to be reinforced in places. Through friction welding aluminium tubes can be connected to steel stub shafts.

5.1.6 Driveshaft tubes made out of composite fibre materials

Composite fibre materials are well-known in aerospace construction and machinery. In the sixties they were also used in motor vehicle construction because of their lightweight properties (Fig. 5.17). Since 1985 Glaenzer Spicer, Uni-Cardan/GKN Lohmar and Nordiska Kardan AB (Sweden) have developed propshaft tubes made from resin, carbon and glass fibres with a torque capacity up to 25,000 Nm, a maximum torque of 32,500 Nm, with lengths from 0.5 to 2.6 m and a diameter of 190 mm. They have been tested at speeds from 2500 to 8000 rpm, temperatures of –50 to +140°C. Other designs with steel and aluminium reinforcement have been produced. If the carbon glass fibre ratio goes up to 50 : 50, the weight saving is 50%. A one-part "composite" shaft can replace two-part steel shafts (Fig. 5.17) and suitable flanges and joints have been designed (Figures 5.18 to 5.21). A composite driveline can withstand heavy loads, is rotationally elastic, and absorbs noise and vibrations. Along with the airbag it increases accident safety. Composite shafts can also be produced economically in small quantities [5.13, 5.14, 5.20, 5.21].

GKN Technology in England developed a constant velocity flexible disk joint, hot pressed out of glass fibre fabric reinforced polyamide resin, for the ends of compos-

Table 5.3. Data for composite propshafts for motor vehicles

	Engine speed (rpm)	Failure torque (Nm)	Length (mm)	Fibres	Percentage (%)
Sports cars 4 × 4	8000	1800	2100	carbon	55
Touring cars 4 × 4	6200	1600	1650	glass & carbon	50
Transporters/lightweight lorries	5000	4000	2250	carbon	55
Medium-sized lorries	2600	8500	2550	glass & carbon	60

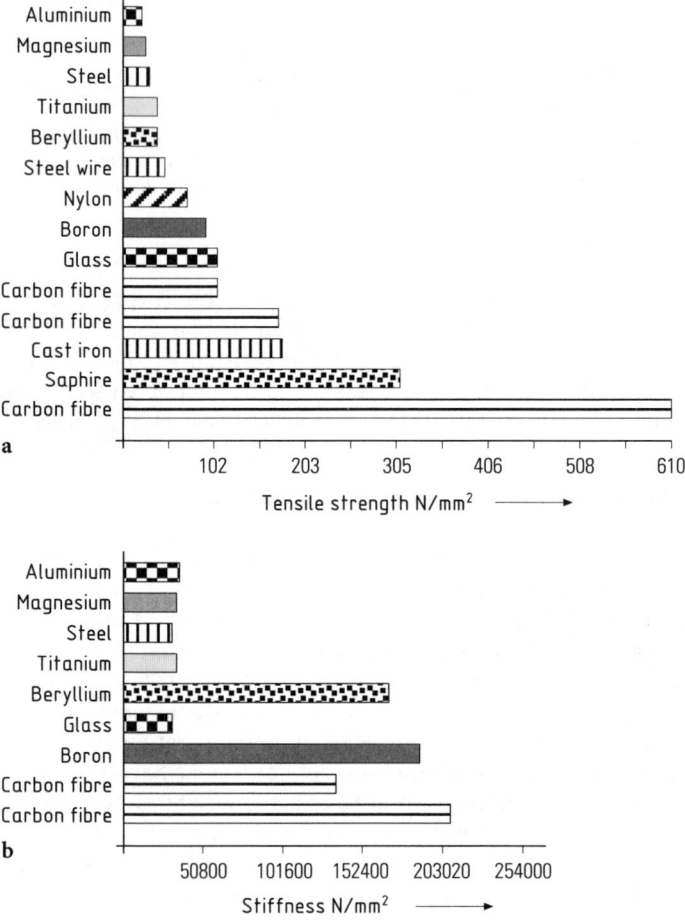

Fig. 5.17. The properties of filaments and fibres compared with metals. a specific tensile strength; b specific stiffness

ite propshafts and halfshafts. A disc of this kind of 140 mm dia., 5 mm thick, weighs 50 g and works at an articulation angle of 2° to 5°. In combination with the composite propshaft, it provides a good lightweight driving mechanism. Imbalance can be reduced by absolutely straight tubes, small welding offsets, improved flange centring and minimum concentricity errors of the centring seats.

To summarise the following can be said:

- composite shafts do not rust, are not sensitive to fuels and lubricants, are corrosion-resistant and free from maintenance,
- they remain dimensionally stable when processed,
- they have good loading capacity and possess limp home properties,
- they dampen vibrations,
- they allow a high level of driving comfort.

5.1 Hooke's Jointed Driveshafts

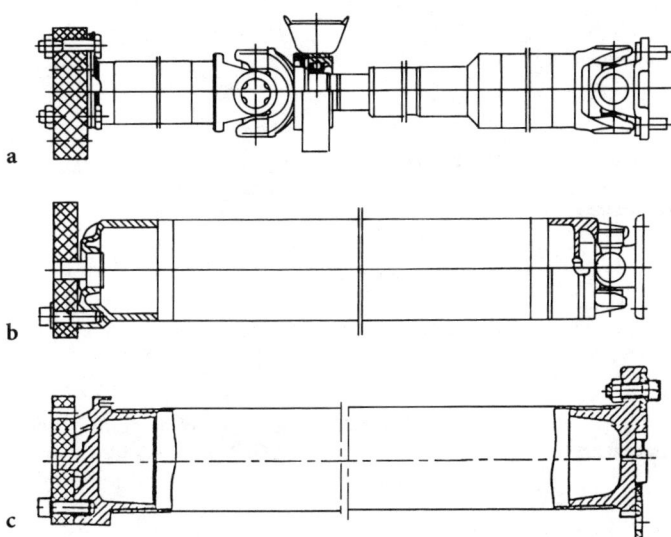

Fig. 5.18. Comparison between steel and composite propshaft. **a** Conventional two-part steel shaft (10 kg), **b** single-part mixed metal composite shaft (5 kg), **c** shaft made solely of composite material (2.5 kg)

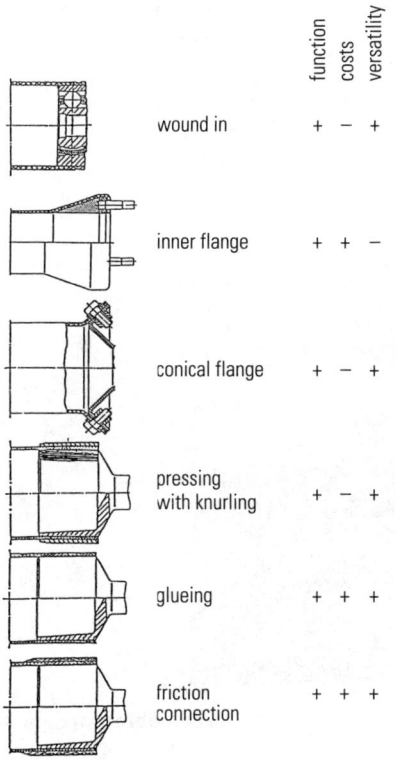

	function	costs	versatility
wound in	+	−	+
inner flange	+	+	−
conical flange	+	−	+
pressing with knurling	+	−	+
glueing	+	+	+
friction connection	+	+	+

Fig. 5.19. Possible connections for composite fibre (glass fibre reinforced plastic) drive shafts. GKN Glaenzer Spicer design

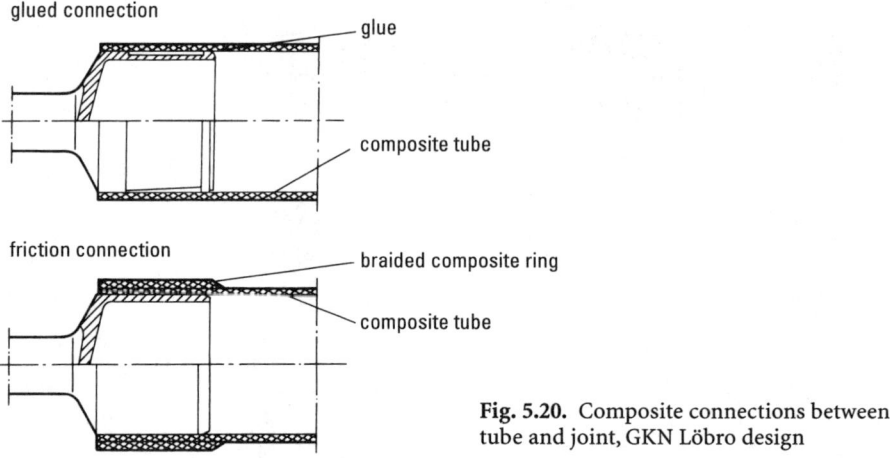

Fig. 5.20. Composite connections between tube and joint, GKN Löbro design

Fig. 5.21. Propshaft wound from 12,000 km of very thin glass and carbon fibres, (weight = 2.5 : 5 kg). GKN Automotive Ltd design

Table 5.4. Comparison of steel and glass fibre reinforced plastic propshafts for a high capacity passenger car, GKN design

Propshaft	
Glued in connections	
Length, joint centre to joint centre	1585 mm
Weight	5 kg
Weight of a comparable steel shaft	10 kg
Composite tube:	
Matrix: epoxy resin	
Fibres: HT type carbon fibres	75 volume %
E type glass fibres	25 volume %
Outside diameter	85 mm
Structural wall thickness	2.2 mm
Length	1529 mm
Weight	1.5 kg
Application temperature	-40 to $+140°C$

5.1.7 Designs of Driveshaft

In order to cover as large a range of applications as possible, series of driveshafts are made. A series includes driveshafts of the same design, but in several sizes governed by the end-users' needs. The aim of a driveshaft series is to achieve a high level of design commonality for the different sizes.

5.1.7.1 Driveshafts for Machinery and Motor Vehicles

The most frequently used driveshafts are those with two single joints on a tubular shaft with a plunging element (Fig. 5.26). They are manufactured in light, medium, heavy, super-heavy and ultra-heavy series. Table 5.5 gives the torque capacity ranges. The light and medium series is common in motor vehicle construction (Figs. 5.26 and 5.31), the heavy to ultra-heavy range is used in slow running, stationary drives. The largest driveshafts operate in rolling mills in slabbing mill trains (Fig. 5.37).

The design of shafts is governed by the purpose for which they are used and by their loading. The width of the roller bearings increases in the cross trunnion bearings; the number of rows of roller bearings increases from one to five (Fig. 5.7b); yoke lugs are split and bolted to enable the roller bearings to be changed quickly (Fig. 5.7c); and their flanges are fitted with keys or 70° cross serrations to ISO draft standard (Figs. 5.3c).

In order to extend the range of standard series with two single joints (Fig. 5.27), shafts with one single and one centred double joint or two centred double joints have been designed (Figs. 5.22a–c).

Double Hooke's joints with self-centring can be used for driveshafts in special cases where constant velocity is required in spite of different input and output angles. The double Hooke's joint with self-centring (Fig. 5.22c) consists of two single joints, the outputs of which are pivoted together by a centring bearing and are able to move towards one another. The torque is transmitted by the connecting ring be-

Table 5.5. Torque capacity of Hooke's jointed driveshafts, GWB design

Series	Rated torque M_N in Nm	Plunge in mm	Flange diameter in mm
light 473, 287	150 ... 2400	28 ... 100	58 ... 120
medium 587	3000 ... 23000	35 ... 180	120 ... 250
587 double joint	6000 ... 12000	110	165 ... 180
heavy 190	25000 ... 160000	40 ... 190	225 ... 435
super heavy 292	44500 ... 720000	40 ... 355	225 ... 550
ultra heavy 398	1060000 ... 10000000	250 ... 300	600 ... 1270

tween the two joints. If the double joint is angled, the two single joints adjust so that each is at about half the total angle of articulation. The rotary movement is thus transmitted with near constant velocity up to the maximum total articulation angle of 42°. The centring bearing is sealed by means of a sleeve and filled with grease.

It is possible, for instance, to use the combination of a centred double joint with a single Hooke's joint and a Rilsan-coated plunging element (Figs. 5.22a and 5.29) as the propshaft in articulated vehicles (Fig. 5.23a–d). The advantages are: a large angle about the articulation axis; eliminate the intermediate shaft and the second support bearing (Fig. 5.23b); no need to align the joints with the articulation axis; vibration-free operation because of the quasi-homokinetic double joint and the low angle of the single joint. One disadvantage is the axial force which is passed on by the plunging element to the bearings of the shaft in the case of high steering lock under load. This disadvantage can be avoided if a constant velocity plunging joint is fitted instead of the Hooke's joint with a plunging element (Figs. 5.14, 5.15).

If torques and speeds are to be transmitted uniformly at different articulation angles, single joints are unsatisfactory because the conditions for constant velocity (the Z- or W-configuration) can no longer be met. Driveshafts with two double joints must be employed (Figs. 5.22b and c). They permit any desired changes in separation and angle, and transmit torque and rotary movement almost homokinetically up to 42°.

5.1.8 Driveshafts for Steer Drive Axles

Special designs of double Hooke's joints with external centring are used in steer drive axles, e.g. on lorries, cranes, agricultural tractors and construction vehicles. They take up little space and transmit high torques from the differential to the driven wheels, at articulation angles up to 40° (see Fig. 4.41) or 50° (see Fig. 5.22).

If the vehicle is driven solely by the steer axle, the service durability must be calculated: the size of the double jointed driveshaft is selected using the load spectrum. If several axles are driven, the effect on the torque distribution of any differential locks must be taken into account. Although the torque capacity of a double joint decreases with increasing articulation angle, this restriction is unimportant for lorries because, in practice, the maximum torque cannot be transmitted at full wheel lock. The loading of the drivetrain by high axial forces, due to the change in length of the driveshaft when large wheel deflections occur, is a disadvantage. A fixed constant velocity joint can be fitted instead of the externally centred double Hooke's joint.

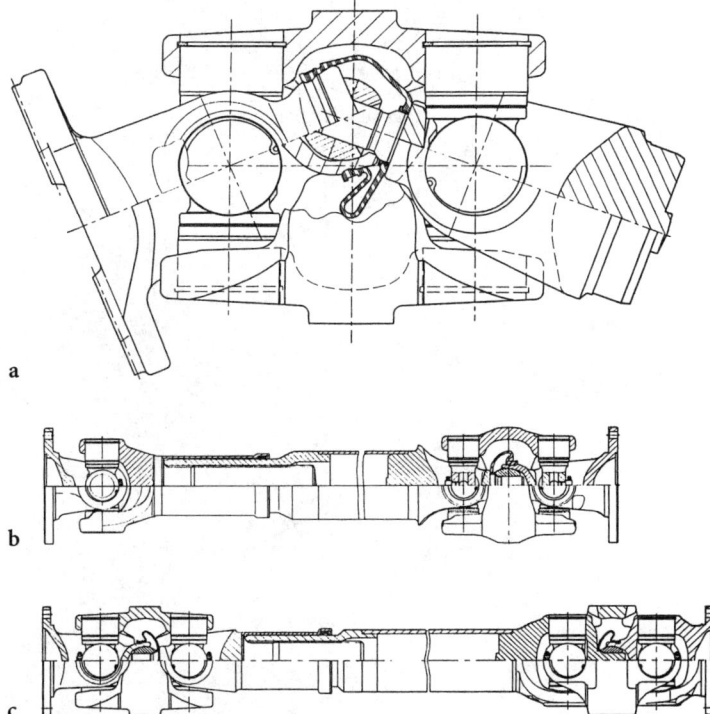

Fig. 5.22a–c. Double Hooke's jointed driveshafts. GWB design. **a** With a single joint and a centred double joint, $\beta = 20/42/50°$; **b** with two centred double joints, $\beta = 42°$; **c** = double joint with self-centring

5.2 The Cardan Compact 2000 series of 1989

Motor vehicle manufacturers require the drivelines of their vehicles to have the following properties: lightweight, easy and safe to assemble, a small number of individual parts and smooth in their running [5.10].

The Cardan Compact 2000 shaft is designed to meet these requirements. This is a series of steel and aluminium universal joint shafts with a lightweight construction, achieved by new production methods using forging and pressing, small wall thickness tubes, end face flange serrations and smaller diameters. The vibration behaviour also improved. Instead of a two-piece steel shaft, a one-piece aluminium shaft with a larger diameter and corresponding wall thickness is possible. With its lower weight, the balancing quality improves, and the rotational excitation and the dynamic load on the bearings decrease.

The requirements were:

- higher quality with a better price-performance ratio
- longer life with longer service intervals
- fewer but more versatile models.

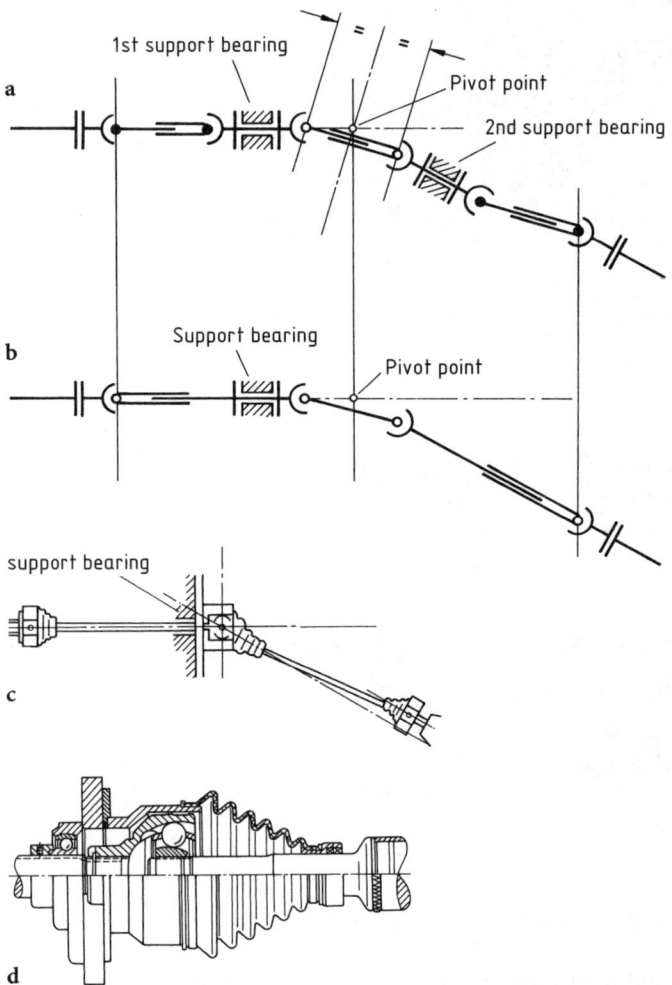

Fig. 5.23 a–d. Driveline in vehicle steered by articulation. **a** Hooke's joints on two shafts in Z- and one shaft in W-configuration with two support bearings; **b** two Hooke's jointed driveshafts with one support bearing; **c** three constant velocity joints on one two-piece shaft with two plunging joints, a wide-angle fixed joint and only one support bearing; **d** support bearing with constant velocity RF fixed joint and stationary boot

The aims set by the designers who developed the Compact 2000 were:

- greater torque transmission with a service life of 0.5 million km, lower weight and less maintenance,
- a standardised series of shafts and flanges, with fewer individual parts and easier assembly.

The 13 different sizes are capable of handling torques from 2,400 to 35,000 Nm, which makes the range applicable to large passenger cars, all sizes of heavy goods vehicles and stationary machinery.

5.2 The Cardan Compact 2000 series of 1989

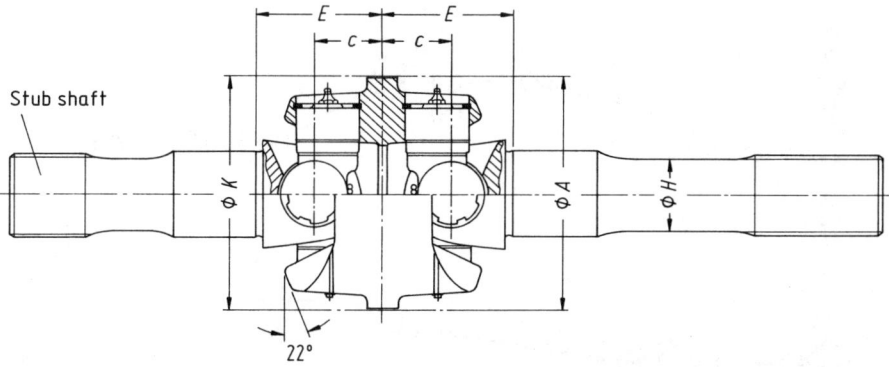

Specifications in mm and torque capacity in Nm. Articulation angle = 50°

Joint size		469.10	469.20	469.340	469.37	469.40
M_N	Nm	4000	6000	8000	13000	16000
A		112	128	138	158	168
E		68	75	81	92	97
H		35	40	44	51	55
K		115	131,5	142	163	173
2c		72	82	90	102	108
G	kg	3,1	5,3	6,4	9,7	12,0
x		7,42	8,44	9,21	10,29	11,25
y		1,76	2,0	2,18	2,49	2,67

Fig. 5.24. Non-centred double Hooke's jointed driveshaft with stubshaft for steer drive axles, GWB design. H Equivalent shaft diameter for steel heat-treated to 1200 to 1350 N/mm²; K swing diameter for max. articulation angle; G weight of joint over 2E; c distance between joints; M_N max. torque without plastic deformation of the joint parts for a straight joint ($\beta = 0°$). x and y, computed with equ. 4.33 and 4.40, multiplied with the safety factor $f_S = 1{,}25 \cdot \varkappa$ out of Fig. 4.21, y out of Fig. 4.22

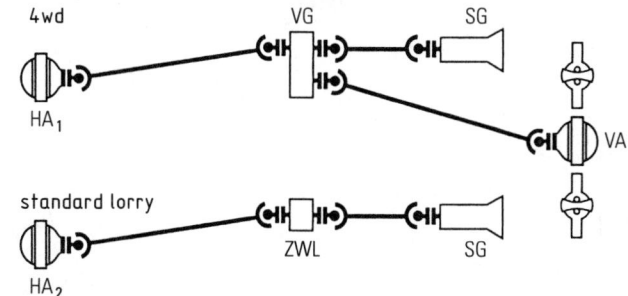

Fig. 5.25a–g. Hooke's driveshafts in commercial vehicles. VA = front axle, HA_{1-2} back axle, SG = manual gearbox, VG = transfer box, ZWL = intermediate bearing. **a** 4wd 2axled car with double Hooke's jointed driveshaft to the front axle

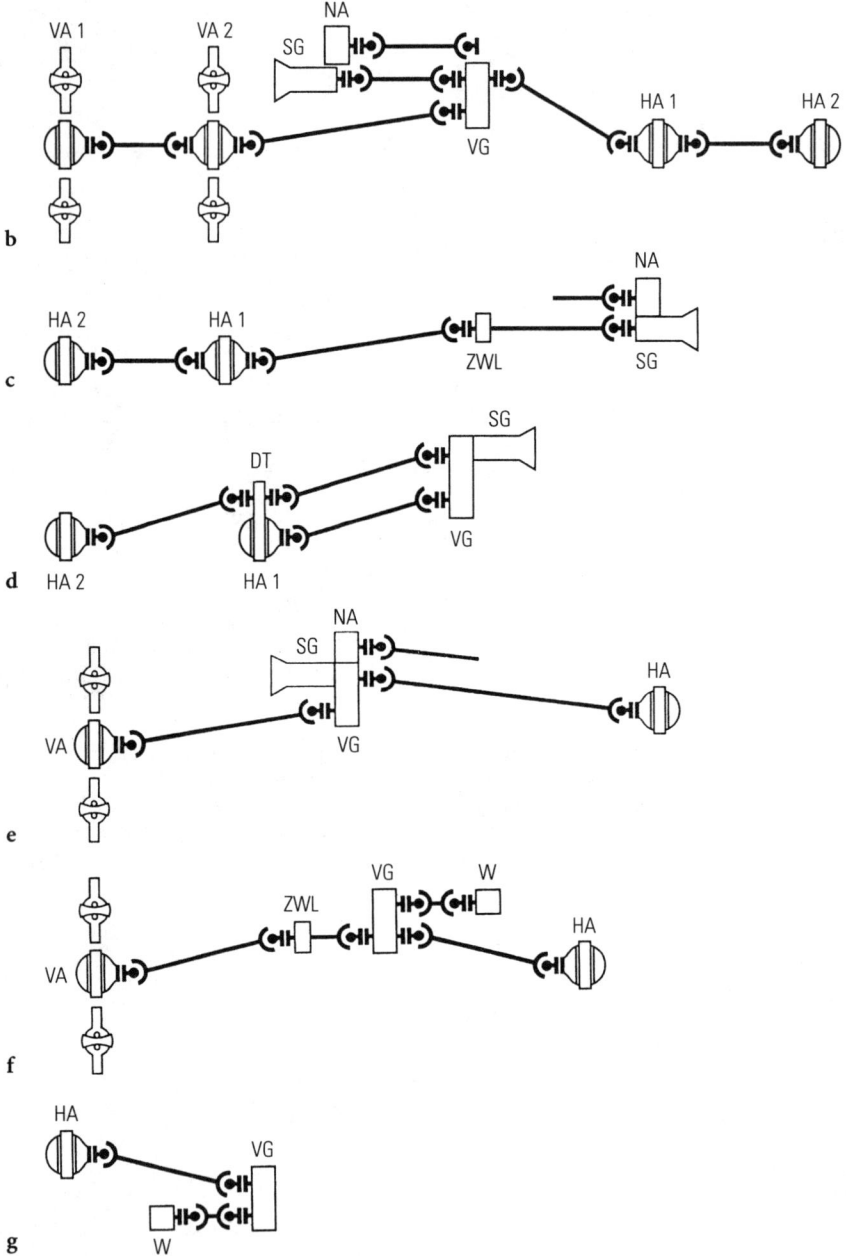

Fig. 5.25 b–g (continued). **b** Car with two double axles with transfer box NA and double Hooke's jointed driveshaft to the front axles, **c** 3axled dust-cart with drive-through axle DT to the 2nd back axle and CVJ-driveshaft to the dust concentrator or hydraulic pumps, **d** idem, **e** fire truck with CVJ-driveshafts to the aggregates, **f** buckle vehicle with centric double joint and wing bearing driveshafts, **g** soil compactor with wing bearing driveshafts

5.2 The Cardan Compact 2000 series of 1989

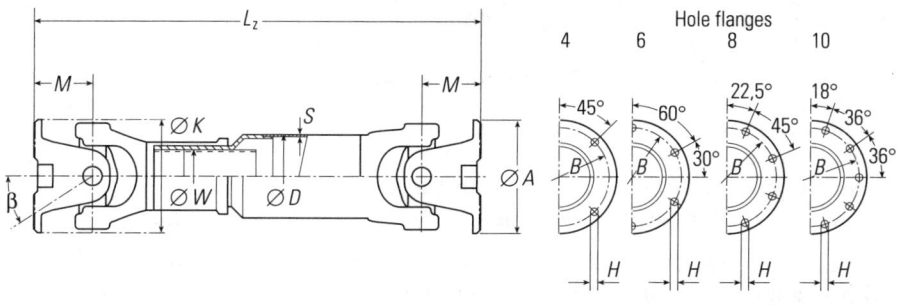

Joint size	C	2015	2020	2040	2035	2045	2055	2060
M_N	Nm	2400	3500	14000	10000	17000	21000	30000
Flange-⌀	A	100	120	155	120	180	180	180
Hole circle-⌀	B	84	101.5	155.5	155.5	155.5	155.5	155.5
Flange bolt KV	H	4 × M 10		4 × M 12	4 × M 10		4 × M 14	
Specifications (mm)	K	90	98	158	142	172	176	194
	M	48	54	82	75	87	92	100
	D · s	63.5 × 2.4	76.2 × 2.4	120 × 3	100 × 3	120 × 4	120 × 6	130 × 6
	$L_{z\,min}$	346	379	546	542	579	616	635
	L_a	60	70	110	110	110	110	110
Physical data	G_W kg	8	11.2	33.3	23.7	43	51.7	60.5
	G_{tube} kg	3.62	4.4	8.65	7.17	11.44	16.82	18.34
	J_R kgm²	0.0034	0.0059	0.0296	0.0169	0.0385	0.055	0.0932
	J_1 kgm²	0.0024	0.0047	0.0346	0.023	0.0525	0.0722	0.1208
	C_{tube} Nm/rad 1×10^5	0.34	0.60	3.02	1.72	3.93	5.6	7.2
	C_{joint} Nm/rad 1×10^5	0.37	0.57	3.21	1.74	4.58	6.49	7.68
cross trunnion data	6 R Co Nm	1587	3984	13596	9210	18105	24045	28428
	2 C R Nm	928	1281	4832	3308	6348	8136	9475
	D mm	17.83	20.24	31.75	27.16	34.929	38.1	40
	d mm	2.55	2.55	5	4.5	5.5	6	6
	l mm	11.8	14.6	17.8	16	19.8	24	24
	z	25	28	23	22	23	23	24
	d_B mm	27	30.2	47.64	42	52	57	59
	R mm	28.1	30.3	53.55	46.9	58.25	58.25	66

Fig. 5.26. Specifications and physical data of Hooke's joints for commercial vehicles with length compensation. GWB design. $\beta = 25/35/44°$. 2 CR = dynamic transmission parameter, G_W = weight of a 1 m-shaft, K = rotation-⌀, D · s = size of intermediate shaft, $L_{z\,min}$ = collapsed length, L_a = length compensation through slip, J_r = mass moment of inertia for L_z, J_i = idem with length compensation, C = torsional stiffness, d = needle/roller-⌀, 1 = idem, length, z = number of rolling bodies per row, 6 RC = woring torque, K = working joint-⌀, d_B = bush-⌀° of the cross trunnion, i = number of rolling rows

5.2.1 Multi-part shafts and intermediate bearings

Drivelines with multi-part drive shafts in cars and lorries are needed when:

- the wheelbase is over 2.5 m,
- auxiliary units are attached,
- single-piece drivelines can provide an adequate critical speed,
- bending and torsional vibration is adversely affecting the drive behaviour.

A multi-part shaft comprises a main shaft with plunge and an intermediate shaft with an appropriate intermediate bearing. It is also possible to have two intermediate shafts combined with the main shaft.

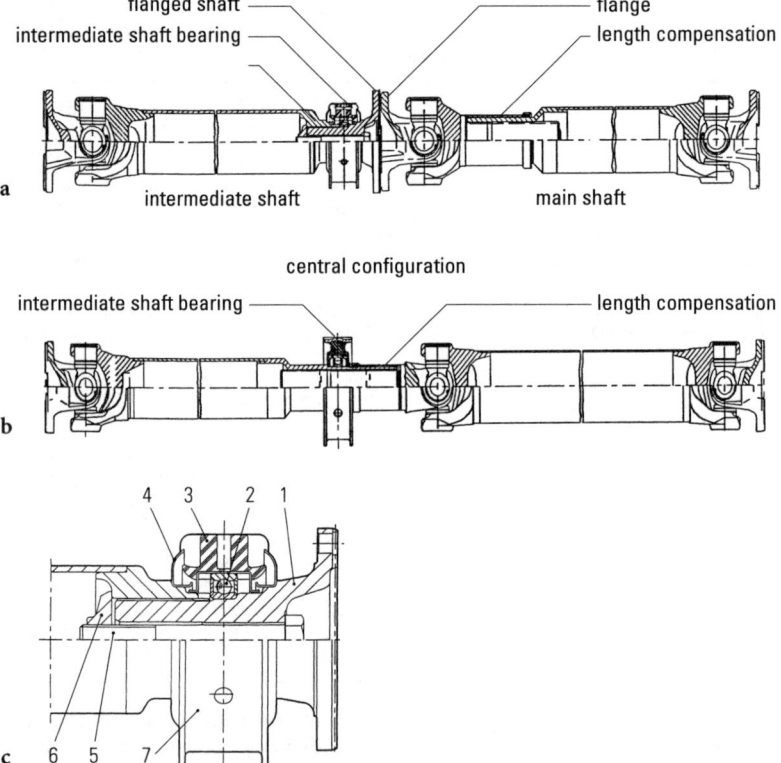

Fig. 5.27. Intermediate bearing in the driveline. Versions: **a** standard, **b** central configuration; both on the intermediate shaft. GWB design, **c** intermediate bearing in detail. GWB design. *1* flanged shaft with bearing seat and flange connection on the same component, *2* permanently lubricated deep groove ball bearing with grease-filled labyrinths protecting against dirt and moisture, *3* elastomeric and isolating vibration and noise, absorbing axial and angular movements and accommodating misalignment, *4* plates to shield against spray, *5* central securing bolt, *6* weld nut securing the assembly, *7* strap holding the elastomeric unit and providing mounting to the vehicle frame

5.2 The Cardan Compact 2000 series of 1989

The intermediate shaft bearing in the driveline has two tasks (Fig. 5.27):

- to limit the degrees of freedom; a driveline with three shafts would otherwise not function,
- it should influence the vibration behaviour.

A multi-part driveshaft goes through at least one resonance in almost all applications. With a "soft" bearing the resonance occurs at low speed; in the operating speed range it isolates the vibrations. It works because of its progressive spring characteristics. A "hard" bearing on the other hand shifts the resonance into a higher speed range.

5.2.2 American Style Driveshafts

At the beginning of the 20th century Hooke's jointed driveshafts were introduced in motor cars to drive the rear axle, replacing chain drives. While car manufacturers in Europe made their driveshafts themselves, in the USA Wilfred Spicer started in 1904 to mass-produce not only his joint with cross trunnions and with plain bearings but also complete driveshafts [5.8]. They were easy to standardise and simple to fit. In

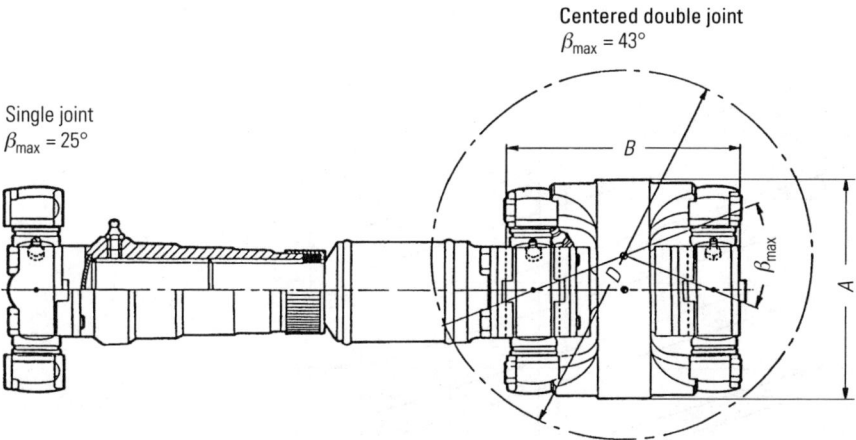

Specifications in mm

Standard (Nm)	Serie	M_{max}	$\beta°_{max}$	A	B	D
6 C	140 M 60	3400	43°	155	162	255
7 C	148 M 70	5700	43°	163	204	300
8.5 C	165 M 85	14000	43°	190	221.6	330

Fig. 5.28. Centered double-joint for high and variable articulation angles by Edmund B. Anderson (US-PS 2698 527/1951), manufactured by Italcardano. Constant velocity quality and smoothed vibrations. Basically two standard-single joints are connected in phase by a coupling yoke. A centering device divvides the angle between the shafts into two equal angles; the result is equal or constant velocity between the driving and the driven shaft

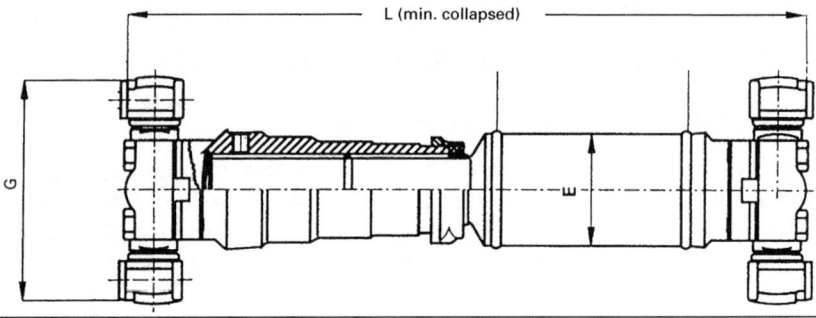

Size of joint		Rotary-Ø	Conne-xion-Ø	Articu-lation angle	Length	Max. static torque M_N	Tube dimensions		M_d
							Diameter	Thickness	
		A	G	β_{max}	L_{min}	Nm	E	S	Nm
2	C	87	79.4	20°	165–280	800	50	3	480
4	C	116	108.0	15°–23°	178–307	1500	50	3	1020
5	C	123	115.1	7°–30°	155–405	2650	58	4	1820
6	C	150	140.5	10°–25°	165–395	3400	76	4	2300
7	C	158	148.4	17°–25°	225–550	5700	88	4	3350
8	C	216	206.4	30°	271–500	8500	100	6	5000
8,5	C	175	165.1	13°–25°	275–525	14000	100	6	5000
9	C	223	209.6	25°	325–700	18600	115	7.5	7600
10	C	225	212.7	15°–25°	440–800	26000	140	5	11000
12,5	C	295	280	16°	538–850	43000	140	10	
14,5	C	326	310	16°	505–1000	62000	160	12.5	

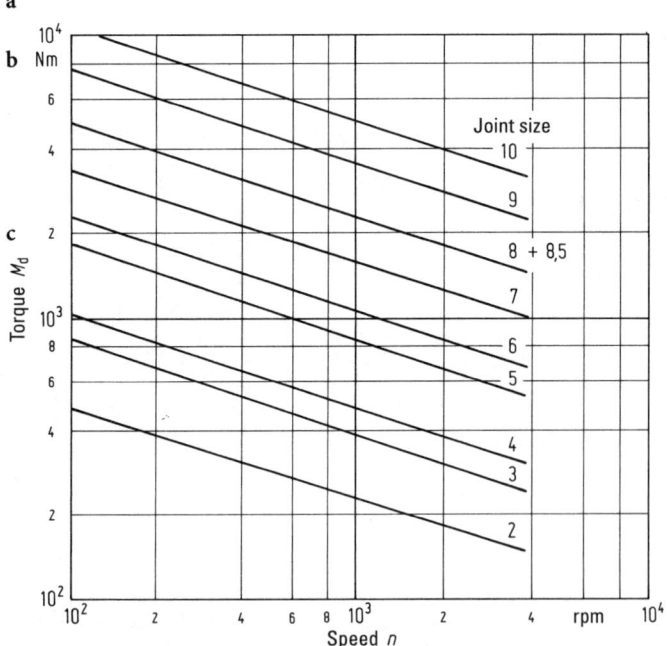

Fig. 5.29a–c. Hooke's jointed driveshaft, Mechanics design. **a** Principal dimensions in mm and torque capacity in Nm. M_n without plastic deformation; M_d values for $L_h = 3000$ hrs, $n = 100$ rpm, and $\beta = 3°$; **b** nomograph for Nm at rpm speed $n = 100$ to 4000 rpm, articulation angle $\beta = 3°$ and $L_h = 3000$ hrs; **c** for torque M_d

the 'twenties, the French firm Glaenzer-Spicer SA, of Poissy, and the English firm Hardy-Spicer & Co. Ltd., of Birmingham, began to produce them under licence. Between 1928 and 1930, Spicer introduced needle bearings for the trunnions thereby increasing the efficiency of the driveshafts to 98%. Their life was similar to that of the whole vehicle.

Figures 5.29 show the Mechanics driveshaft[1]. A typical feature of this is the hub attachment with wing bearings, which are externally centred. This driveshaft cannot be balanced very exactly because the axial play between the joint cross and the wing bush (arising from the build-up of tolerances) means that the external centring of the complete joint cross cannot be kept sufficiently tight. Thus the driveshaft only achieves a balance level of $Q \approx 40$, which is however found to be adequate in the USA. With the separate cross and centre part many drivetrain designs, including those with support bearings, can be readily manufactured. The bolted connections of the individual assemblies make servicing easier. A particular advantage is the compact nature of these Hooke's joints through which the connection bearing forces are reduced because they are directly proportional to the overhang (Figs. 4.24 and 4.25, Eqs. 4.48–4.52).

The Mechanics driveshaft is installed in forklift trucks, wheel loaders and dumper trucks.

5.2.3 Driveshafts for Industrial use

The industry wishes Hooke's driveshafts for several applications from 2,4 to 15 000 Nm.

Mechanical Engineering (Figs. 5.31; 5.33; 5.34; 5.37)	*Mechanical Drive*
Rolling mills	Roll pillars
Broad bond cases	Calendering machines
Tin-snips	Ships
Paper mills	Crane plants

Their principal differences are in the form of the bearing lug:

- closed for general machinery (GWB series 687),
- split for heavy and very heavy driving mechanisms, with flange connections through Hirth- and Klingelnberg serrations (GWB series 390/92 and 498).

These are custom shafts where the user must specify some design details such as the hole patterns – 6, 8, 10, 16, 20 or 24 – for the flanges. There are short enclosed shafts, shafts with double flange joints, and intermediate shafts with or without plunging elements.

---

[1] Rockford Powertrain Inc., Rockford, Illinois, USA.

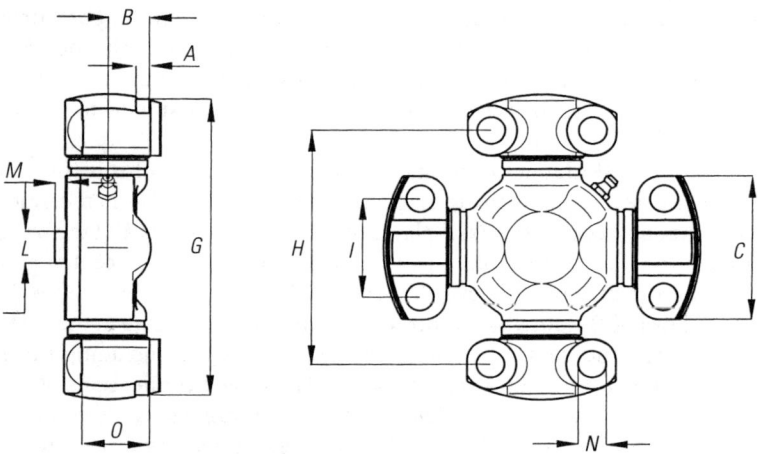

Specifications in mm

	2C	4C	5C	6C	7C	8.5C	8C	9C	10C	12.5C
G	79.35	107.92	115.06	140.46	148.38	165.10	206.32	209.52	212.70	280
A	4.50	5	5.50	5.50	7	7	7	8	9.50	10.50
B	13.10	15.50	17.50	17.50	20.60	25.40	20.60	25.40	32.50	38
C	48	51	62	62	72	96	72	96	120	128
H	59.53	87.32	88.90	114.30	117.48	123.83	174.60	168.25	165.10	227
I	33.32	36.52	42.90	42.90	49.20	71.44	49.20	71.44	92.10	92
L	9.50	9.50	14.26	14.26	15.85	15.85	15.85	15.85	25.35	35
M	3.10	3.10	4.1	4.1	5.10	5.20	5.10	5.20	8.50	8
N	8.40	8.75	10.50	10.50	13.50	13.50	13.50	13.50	17	19
O	8	25	29	29	33	40	33	40	49	62

Fig. 5.30. Driving dog and connection kit of a wing bearing driveshaft. Design from Mechanics/Borg Warner/Italcardano

5.2 The Cardan Compact 2000 series of 1989

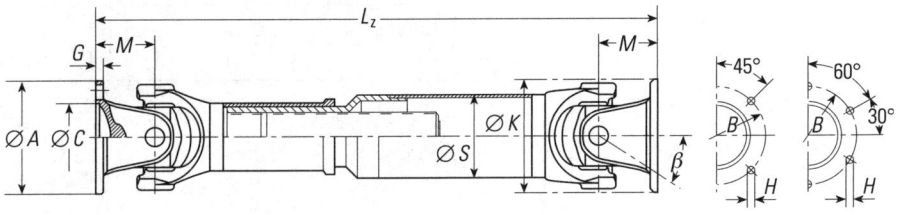

Joint size			687.15	687.20	687.30	687.45	687.65
M_N		Nm	1800	2700	5000	13000	27000
Flange-∅	A	mm	100	120	150	180	225
Specifications	B	mm	84	101.5	130	155.5	196
	C^a	mm	57	75	90	110	140
	G	mm	7	8	10	12	15
	H^c	mm	8.25	10.25	12.25	14.1	16.1
	i^b	–	6	8	8	8	8
	K	mm	90	98	127	172	204
	M	mm	48	54	70	95	110
	S	mm	63.5 × 2.4	76.2 × 2.4	90 × 3	120 × 4	142 × 6
	β	∢	25°	25°	25°	25°	25°
	L_z	mm	346	379	504	595	110
	L_a	mm	60	70	110	110	64.6
Physical data	G_W	kg	5.7	8.4	14.2	28	47.3
	G_R	kg	3.62	4.37	6.44	11.44	20.12
	Jm	kgm²	0.0043	0.0089	0.0245	0.1002	0.2614
	Jm_R	kgm²	0.0034	0.0059	0.0122	0.0385	0.0932
	$c_R \cdot 10^5$	Nm/rad.	0.34	0.6	1.25	3.93	9.5
Cross trunnion data	D	mm	27	30.2	34.9	52	65
	d	mm	2.6	2.5	3	5	6
	l	mm	9.8	11.8	14	17.8	24
	z	–	25	25	28	23	23
	R	mm	26.5	28.1	30.3	53.55	58.25
	$2CR$	Nm	540	928	1282	4832	8136

[a] Centering fit H 6, [b] number of flange holes, [c] fit C 12.

Fig. 5.31. Specifications, physical data and dimensions of the cross trunnion of a Hooke's jointed driveshaft for stationary (industrial) use. GWB design. c_R = torsional stiffness per 1000 mm tube; $2\,CR$ = dynamic transmission parameter; d = needle/roller-∅; D = trunnion-∅; G_W = weight of the driveshaft; G_R = weight per 1000 mm tube; i = number of rolling rows; Jm = mass moment of inertia for driveshaft without tube; Jm_R = moment of inertia per 1000 mm tube; l = length of needle/roller; L_z = min. collapsed length; L_a = max. plunge; $L_z + L_a$ = max. working length; R = effective joint radius; z = number of rollers in one row

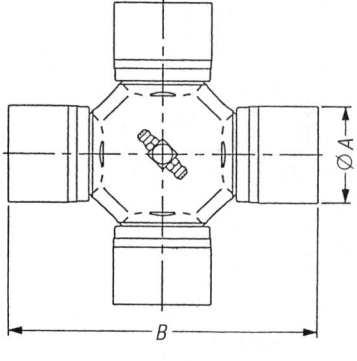

size of joint	⌀ A [mm]	B [mm]
687.15	27.0	74.5
687.20	30.2	81.8
687.25	34.9	92.0
687.30	34.9	106.4
687.35	42.0	119.4
687.40	47.6	135.17
687.45	52.0	147.2
687.55	57.0	152.0
687.65	65.0	172.0

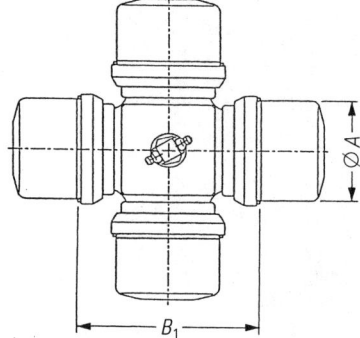

size of joint	⌀ A [mm]	B_1 [mm]
390.60	83	129
390.65	95	139
390.70	110	160
390.75	120	176
390.80	130	196
392.50	74	129
392.55	83	139
392.60	95	160
392.65	110	176
392.70	120	196
392.75	130	216
392.80	154	250
392.85	170	276
392.90	195	315

Fig. 5.32. Universal joints crosses for medium and heavy industrial shafts. GWB design

5.2 The Cardan Compact 2000 series of 1989

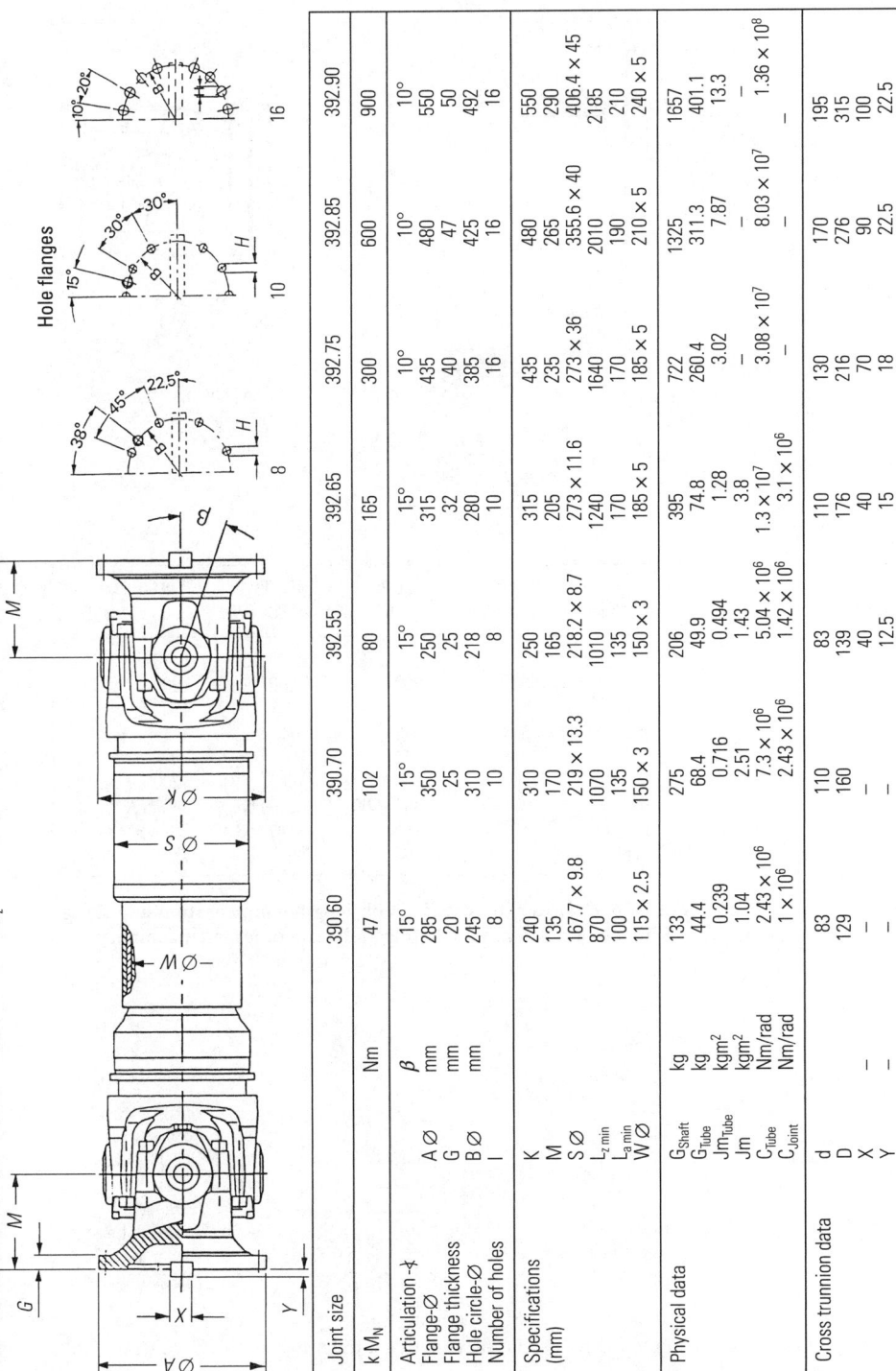

Joint size			390.60	390.70	392.55	392.65	392.75	392.85	392.90
k M_N		Nm	47	102	80	165	300	600	900
Articulation -$\sphericalangle$		β	15°	15°	15°	15°	10°	10°	10°
Flange-Ø	A Ø	mm	285	350	250	315	435	480	550
Flange thickness	G	mm	20	25	25	32	40	47	50
Hole circle-Ø	B Ø	mm	245	310	218	280	385	425	492
Number of holes	I		8	10	8	10	16	16	16
Specifications (mm)	K		240	310	250	315	435	480	550
	M		135	170	165	205	235	265	290
	S Ø		167.7 × 9.8	219 × 13.3	218.2 × 8.7	273 × 11.6	273 × 36	355.6 × 40	406.4 × 45
	$L_{z\,min}$		870	1070	1010	1240	1640	2010	2185
	$L_{a\,min}$		100	135	135	170	170	190	210
	W Ø		115 × 2.5	150 × 3	150 × 3	185 × 5	185 × 5	210 × 5	240 × 5
Physical data	G_{Shaft}	kg	133	275	206	395	722	1325	1657
	G_{Tube}	kg	44.4	68.4	49.9	74.8	260.4	311.3	401.1
	Jm_{Tube}	kgm²	0.239	0.716	0.494	1.28	3.02	7.87	13.3
	Jm	kgm²	1.04	2.51	1.43	3.8	–	–	–
	C_{Tube}	Nm/rad	2.43 × 10⁶	7.3 × 10⁶	5.04 × 10⁶	1.3 × 10⁷	3.08 × 10⁷	8.03 × 10⁷	1.36 × 10⁸
	C_{Joint}	Nm/rad	1 × 10⁶	2.43 × 10⁶	1.42 × 10⁶	3.1 × 10⁶	–	–	–
Cross trunnion data	d		83	110	83	110	130	170	195
	D		129	160	139	176	216	276	315
	X		–	–	40	40	70	90	100
	Y		–	–	12.5	15	18	22.5	22.5

Fig. 5.33. Specification, physical data and dimensions of the cross trunnion of Hooke's jointed driveshafts for industrial use, heavy type, with length compensation and dismantable bearing cocer. $L_z + L_a$ = max. working length (mm)

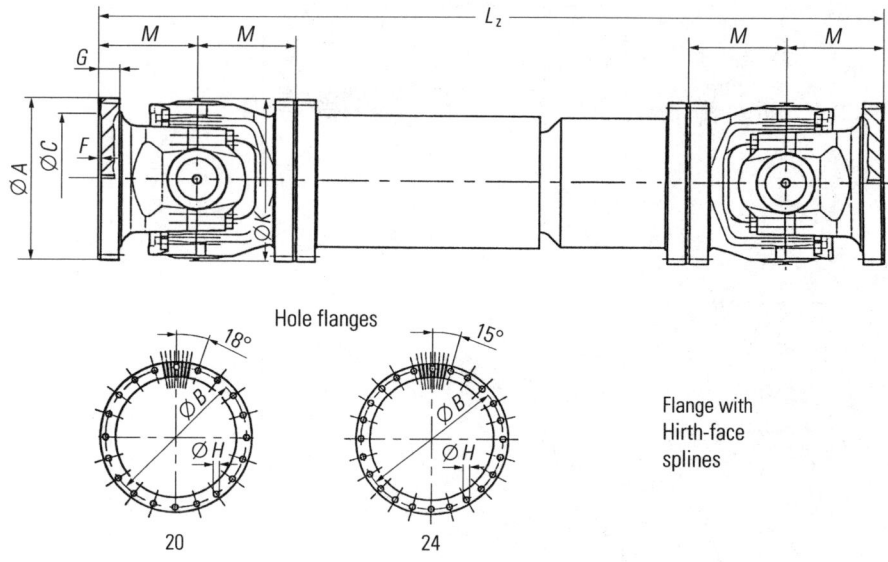

Joint size Nominal torque	M_0	k·Nm	498.00 1100–1450	498.20 2600–3400	498.40 5000–6700	498.60 8600–11 500
Articulation	β	n°	5, 10, 15°	5, 10, 15°	5, 10, 15°	5, 10, 15°
Flange-$\varnothing$	A$\varnothing$	mm	600	800	1000	1200
Flange-thickness	G	mm	75	100	125	150
Hole circle-$\varnothing$	B$\varnothing$	mm	555	745	925	1115
Number of holes	I		10	24	20	20
Rotation-$\varnothing$	K$\varnothing$	mm	1200	800	1000	1200
	M	mm	740–775	480–500	625–655	740–775
Coil-$\varnothing$	d	–	M 24	M 30	M 42 × 3	M 48 × 3
Screw-head	s	mm	36	46	65	75

Fig. 5.34. Specifications of Hooke's heavy driveshafts with length compensation and flange connections by Hirth- or Klingelnberg splines for rolling-mills and other big machinery

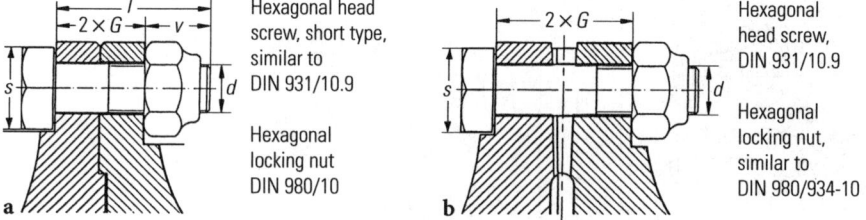

Fig. 5.35. Flange screw joints for industrial driveshafts. **a** joint size 392 with contrary bolt, **b** super heavy joint size 498

5.2 The Cardan Compact 2000 series of 1989

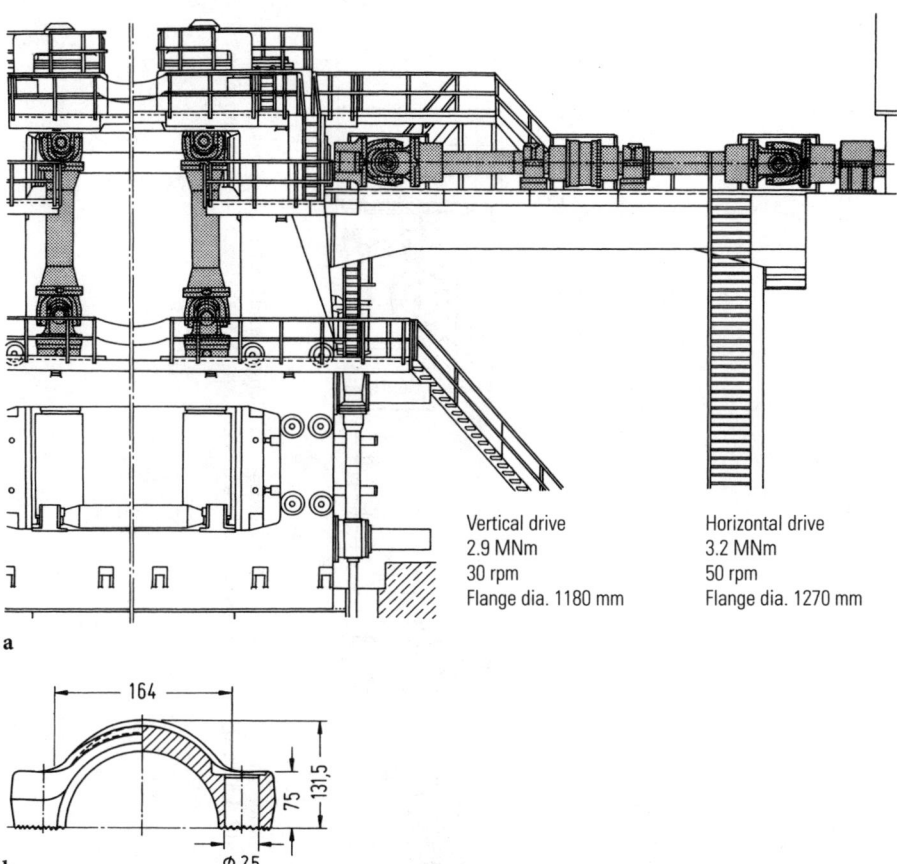

Vertical drive
2.9 MNm
30 rpm
Flange dia. 1180 mm

Horizontal drive
3.2 MNm
50 rpm
Flange dia. 1270 mm

Fig. 5.36a-c. Hooke's jointed driveshaft – heavy series. Made by GWB. **a** Universal slabbing mill train, articulation angle of the shafts $\beta = 15°$; **b** trapezoidal-serrated bearing cap, secured by four bolts (heaviest series)

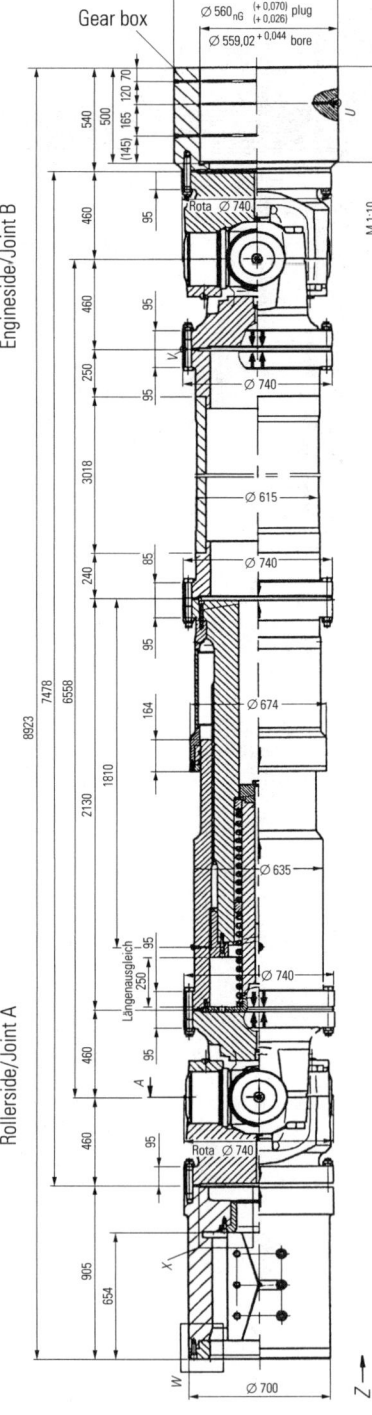

Fig. 5.37. Hooke's driveshaft for a Steckel-mill in reversal operation with tension spring for the plunge. All flanges with Hirth-splines (type 498). Flat tap on the roller side, compound oil pressure on the gearbox-side. $\beta = 7°$, $M_o = 2\,400\,000$ Nm, continuous torque = $1\,500\,000$ Nm, $L_z = 7228$ mm, $L_a = 250$ mm, $G = 14.1$ to. GWB design 1998

5.2.4 Automotive Steering Assemblies

In an accident, a rigid steering column will tend to act as a spear. Hence an articulated shaft gives safety to the driver. It also provides design flexibility. The shafts are simple being made up of two single joints (Fig. 5.38a), with interposed elastic parts to absorb vibration and noise (Fig. 5.38b). If Z- or W-configurations are not possible due to packaging requirements, centred double steering joints must be used to achieve uniform transmission of torque and rotation (Fig. 5.38c).

If a driveshaft which can plunge is fitted in a lorry between the steering wheel and the steering gear, the driver's cab can be tilted without disconnecting the driveshaft.

Steering joints must be light and cheap. The use of needle bearings in thin-walled bushes gives good freedom from play and high torque capacity. Universal joint bearing assemblies shown in Fig. 5.38 are used for DIN 808 shaft joints. The bearing assembly is held in the yoke by staking, which also preloads the needle bearings axially. The joint yokes are forged from aluminium alloys or pressed from extruded sections. The connection is via a splined shaft to DIN 5481 with pinch bolts. For passenger cars a 16 mm diameter stubshaft is used, with 25 mm for lorries. Steering joints are lubricated for life. For data on steering joints see Table 5.6.

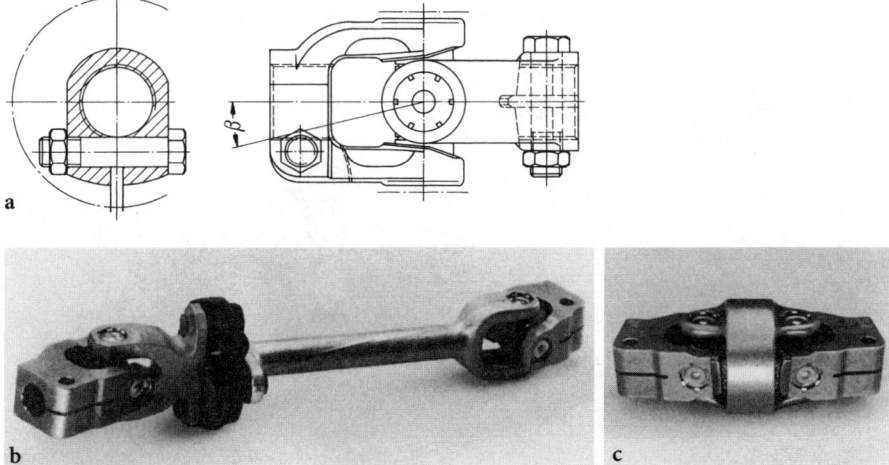

Fig. 5.38a–c. Steering joints. **a** Single steering joint, made by Willi Elbe. For lorries it is made in steel, for passenger cars in aluminium; **b** steering driveshaft for a car, forged in aluminium, with elastic coupling, made by Alusingen; **c** centred double steering joint for cars, in aluminium, made by Willi Elbe (German patent 2854232)

Table 5.6. Steering joint data

	Passenger cars		Lorries Single
	Single	Double	
	a	b and c	a
Material	Steel	Aluminium	Steel
Dynamic articulation angle β per joint	55°	42°	35°/55°
INA bush diameter D in mm (Fig. 5.31c)		15	19/22
Torque capacity M_0 in Nm		> 300	> 400
Yield point in Nm		> 250	> 300
Torque M_d in Nm for 100 000 cycles on rig tests		70	150
Attachment serrations		1 × 54 9/16″ 11/16″	1 × 79

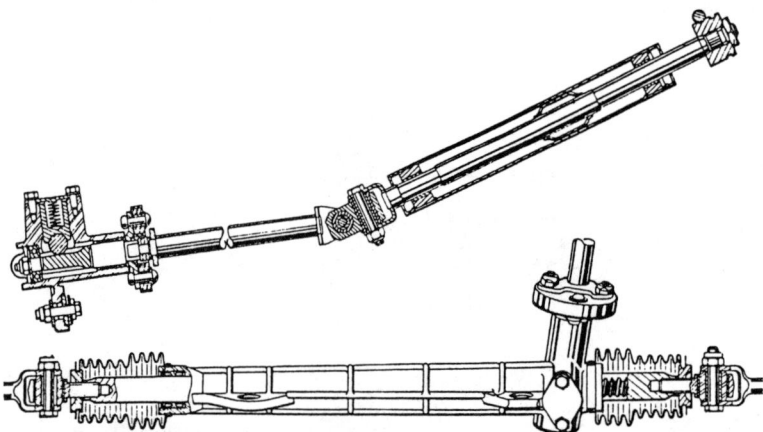

Fig. 5.39. Jointed steering column on a passenger car 1972 (Renault 5 L)

5.2 The Cardan Compact 2000 series of 1989

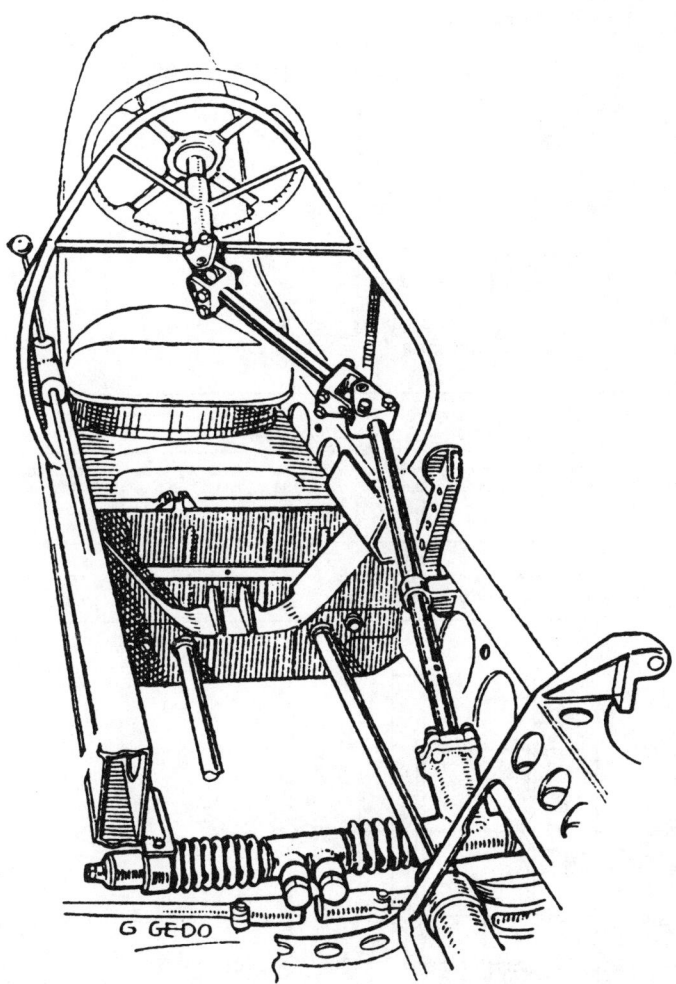

Fig. 5.40. Steering joint driveshaft with two single joints, on a race car 1948 (Deutsch-Bonnet)

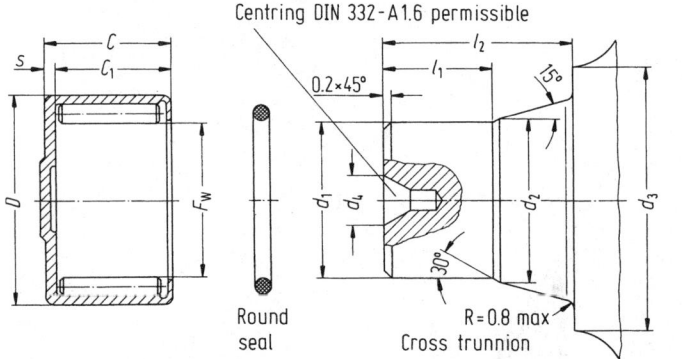

Logo-gram BU	Weight g	Dimensions in mm					Load carrying coeff.		Seal ring DIN 3770-NBR 70	Connecting dimensions at joint trunnion					
		F_W	D	C	C_1 min.	s ±0,05	dyn. C N	stat. C_o N		d_1 k6	l_1 min.	d_2 −0,1	l_2 ±0,2	d_3	d_4 max.
0407	0,8	4	7	4	3,4	0,63	1360	1310	4 × 1	4	3,4	4,3	5	5,7	2
0509	1,2	5	9	5	4,4	0,63	2440	2320	5 × 1,2	5	4,4	5,3	6,2	7,5	2
0711	1,8	7	11	6	5,4	0,63	3700	4200	7 × 1,5	7	5,4	7,3	7,3	9,5	2,5

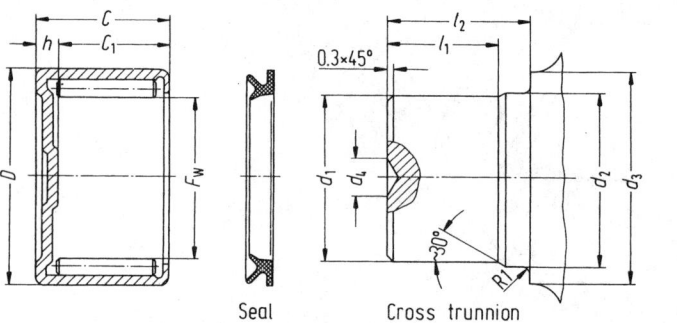

Logo-gram BU...A	Weight g	Dimensions in mm					Load carrying coeff.		Seal ring D	Connecting dimensions at joint trunnion					
		F_W	D	C	C_1 min.	h −0,15	dyn. C N	stat. C_o N		d_1 k6	l_1 min.	d_2 −0,1	l_2 ±0,2	d_3	d_4 max.
1015	6,8	10	15	9,35	7,60	1,75	6000	8800	1015	10	7,4	10,3	10,1	13,5	–
1319	12,5	13	19	11,85	9,60	2,25	10100	15000	1319	13	10	13,3	12,1	17,5	–
1622	16,5	16	22	12,85	10,60	2,25	12200	20700	1622	16	11	16,3	13,1	20,5	2,5
1824	19,3	18	24	13,50	11,25	2,25	13600	24500	1824	18	11,7	18,3	13,8	22,5	2,5
2127	25,0	21	27	15,50	13,25	2,25	17000	34500	2127	21	13,7	21,3	15,8	25,5	2,5

Fig. 5.41a–d. Full complement roller bearing, made by INA. Thin wall sheet metal bush. Misalignment of the needles is avoidable if tolerances are specified. **a** Small BU series, with inner diameter $F_w = 4$ mm and outer diameter $D = 7$ mm; **b** table of dimensions for BU series. Greased with roller bearing grease to DIN 51825 KP 2K for − 20° to + 120°C. Tolerance range of the bore in the steel housing F 7; **c** medium series BU ... A, with inner diameter $F_w = 10$ mm and outer diameter $D = 15$ mm; **d** table of dimensions for BU ... A series

5.2 The Cardan Compact 2000 series of 1989

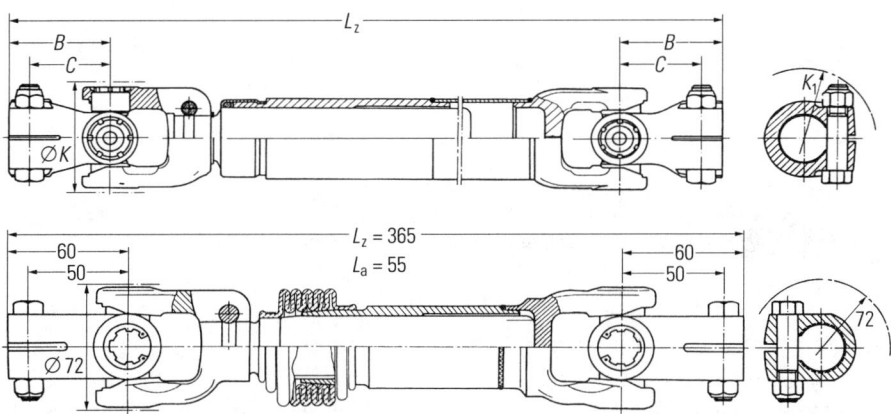

L_z = min. collapsed length
L_a = length compensation (plunge)

a

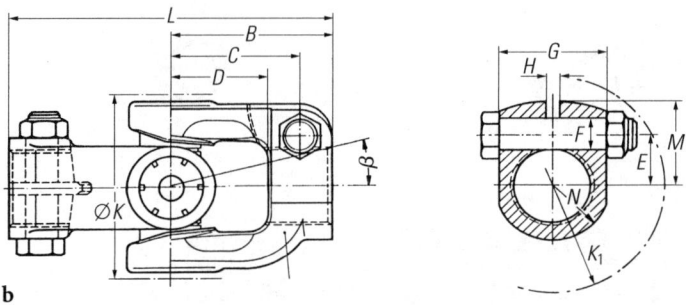

b

M_N Nm	β ·°	B mm	C mm	K mm	K_1 mm	E mm	F mm	G mm	H mm	D mm	M mm	N mm	Serration	Maintenance
300	90°	50	40	58	68	15.8	8.4	32	3	29	26	20	1 × 79	free
	90°	60	50	68	72	16.7	10.4	33	4	36.5	28	17	1 × 79	greasing
400	35	50	40	82	72	16.7	10.4	33	4	28.5	29	17.5	1 × 79	greasing

M_N = max. torque without deformation of joint parts
$L = 2 \times B$,
β = max. articulation angle in sag. Dynamic angle max. 55°

Fig. 5.42a, b. Steering joints with sliding serration ca. 1970 (DIN 5464). **a** Driveshaft with length compensation and vibration absorber, **b** clamp connection. Design ZF Schwäb. Gmünd

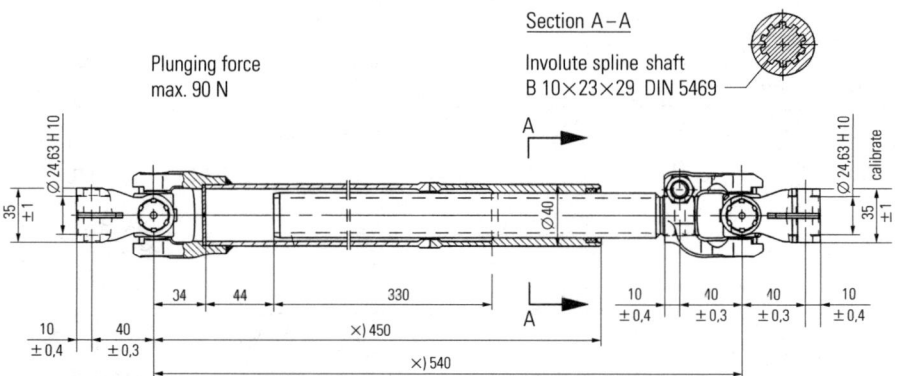

Fig. 5.43. Steering driveshaft with sliding serration for balls ca. 1980. Design ZF Schwäb. Gmünd

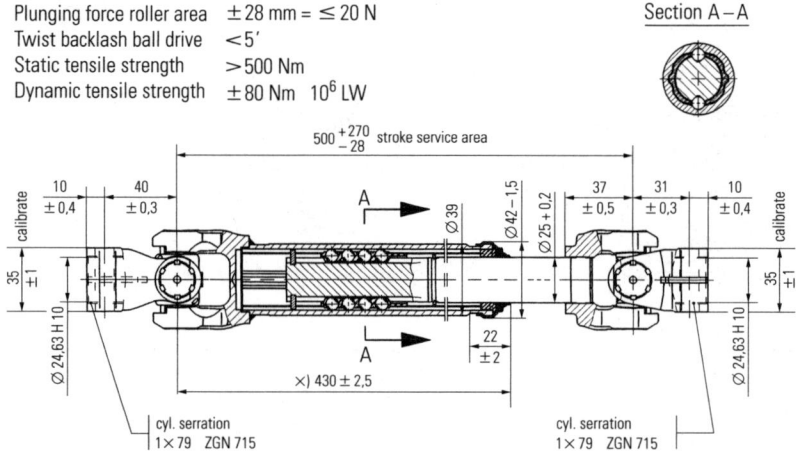

Fig. 5.44. Steering joints driveshaft with sliding serration through balls for movable driver's cab of commercial vehicles 1989. Design ZF Schwäb. Gmünd

5.2 The Cardan Compact 2000 series of 1989

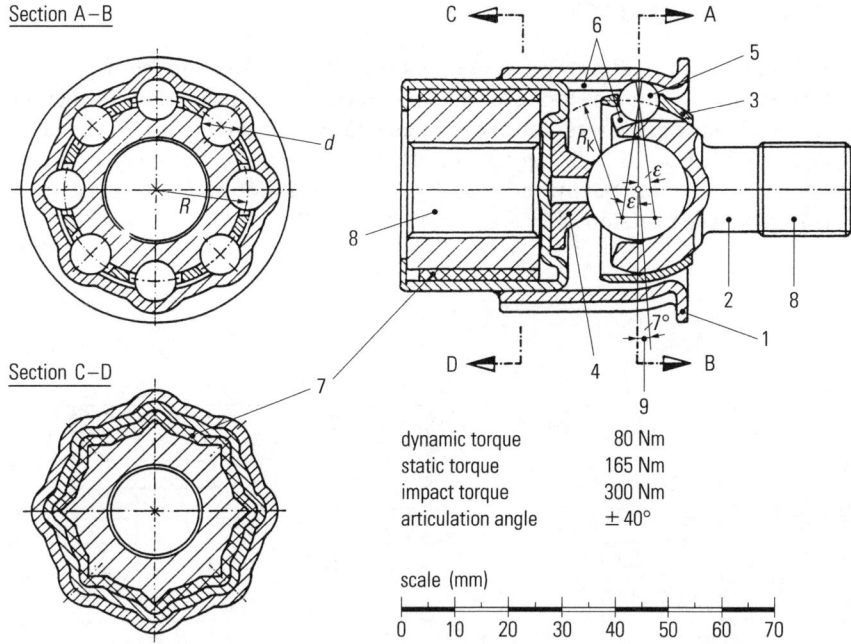

Fig. 5.45. Constant velocity steering mechanism joint, manufactured without machining, GKN Löbro design. Only the shaft connections (8) are formed by cutting operation. *1* outer part, *2* inner part, *3* axially supported cage, *4* radially adjustable ball centring, *5* balls with offset guidance from guided from undercut-free running tracks *6*, *7* vibration insulation, *8* shaft connection, *9* self-locking relief

Running ball dia. $d = 7.938$ mm Track curvature $R_K = 26$ mm
pitch circle radius $R = 17.3$ mm $R_K/d = 3.275$
offset angle $= 12°$

offset $\varrho/d = 2.17$

Opening angle	δ	β	
	10°	40°	Transition from curvature
	32°	+32°	R_K to a straight at $\beta = +32°$
	40°	+40°	

dynamic torque   80 Nm
static torque   165 Nm
impact torque   300 Nm
articulation angle ± 40°

5.2.5 Driveshafts to DIN 808

The best known joints still in general use are the single Hooke's joint shown in Fig. 5.46a, type E, and the double Hooke's joint derived from it, shown in Fig. 5.46b, type D [5.5]. Their basic design was standardised as DIN 808 as early as 1925. The standard was republished in 1957 and since then has been continuously updated. Although DIN 7551 was partly replaced by DIN 808, it continued to exist from 1943–1970. The latest version of DIN 808 dates from 1972.

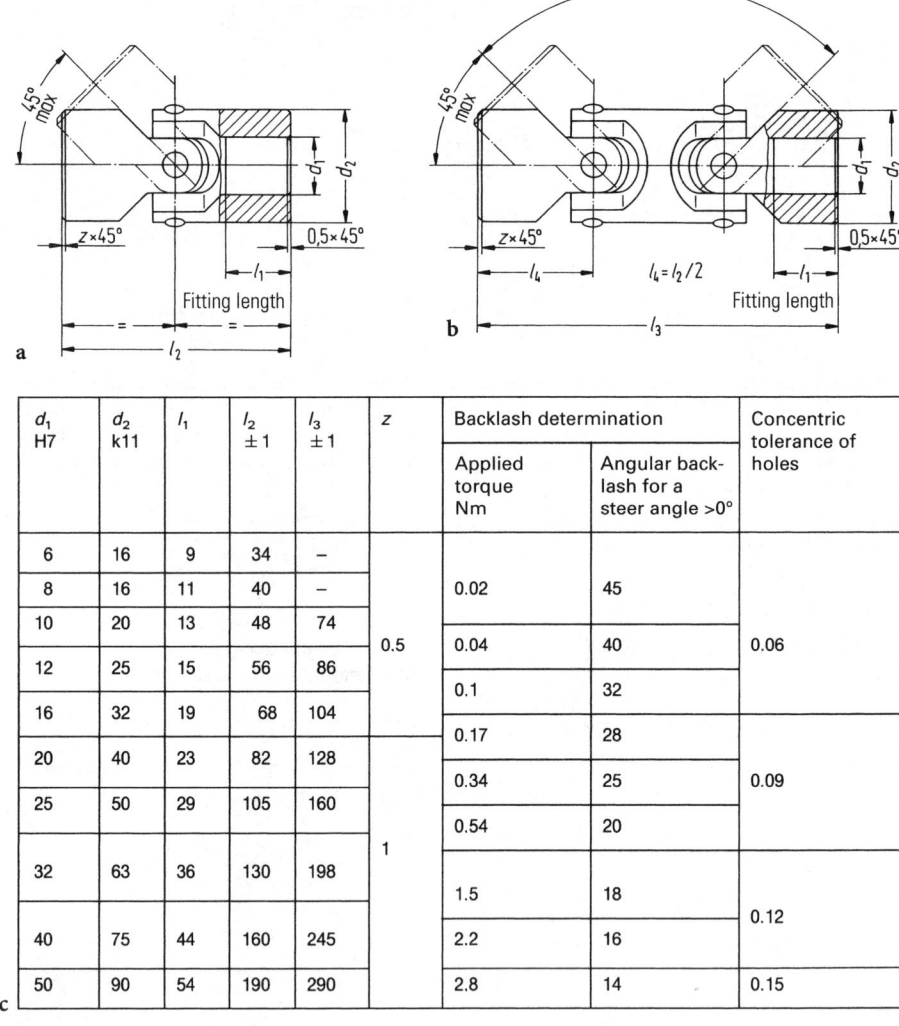

d_1 H7	d_2 k11	l_1	l_2 ±1	l_3 ±1	z	Backlash determination		Concentric tolerance of holes
						Applied torque Nm	Angular backlash for a steer angle >0°	
6	16	9	34	–	0.5	0.02	45	0.06
8	16	11	40	–				
10	20	13	48	74		0.04	40	
12	25	15	56	86		0.1	32	
16	32	19	68	104		0.17	28	0.09
20	40	23	82	128		0.34	25	
25	50	29	105	160		0.54	20	
32	63	36	130	198	1	1.5	18	0.12
40	75	44	160	245		2.2	16	
50	90	54	190	290		2.8	14	0.15

Fig. 5.46a–c. Single (E) and double (D) joint with plain (G) and needle (W) bearings according to DIN 808. **a** Single joint, $d_1 = 6$ to 50 mm; **b** double joint, $d_1 = 10$ to 50 mm; **c** principal dimensions in mm; for selection chart and correction values f_β for Hooke's with plain or needle bearings see DIN 808

5.2 The Cardan Compact 2000 series of 1989

Joint Sizes

DIN 808 specifies ten single joint sizes and eight double joint sizes for a maximum articulation angle of 45° per joint (Fig. 5.46c). Each joint consists of three parts:
- the cross which transmits the torque, and
- the two yoke shafts in which it is mounted.

The joint is designed with plain or needle bearings. The high precision bearings may have up to 45' torsional play for the smallest joint, and up to 14' for the largest joint. The eccentricity between the connecting bores is limited to 0.06 to 0.15 mm. Hardened and ground case-hardening steel, with a maximum strength of 600 N/mm² is used.

The joints are fixed on the mating shafts of 6 to 50 mm diameter by various means including tapered pins, dowels or notched pins, adjusting springs to DIN 6885, and square or hexagonal bores. The full complement needle roller bearing assemblies with thin-walled, drawn, cups give a particularly high load capacity. As an assembly with a seal, they allow a relatively large cross trunnion diameter: $d_1 = 4$ mm for a bush diameter of $D = 7$ mm, and $d_1 = 21$ mm for $D = 27$ mm.

Torque Capacity

Single joints with plain bearings. The maximum torque capacity is 2000 Nm; the speed should not exceed 1000 rpm. The steering torque for continuous operation, at an articulation angle of 10°, is a function of the speed in the selection chart in DIN 808. For larger articulation angles, the correction value f_β must be employed; the drive output is divided by the correction value f_β to give the steering torque N. For articulation angles less than 10°, the steering torque N from the DIN 808 chart may be increased. The increase is 25% between 0° and 5°, between 5° and 10° linear interpolation is required.

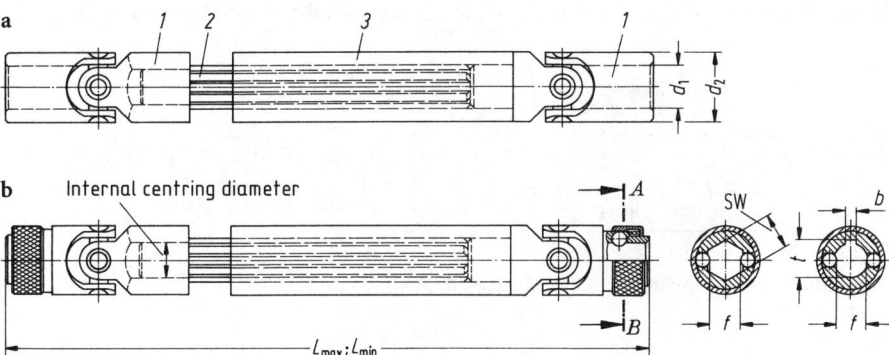

Fig. 5.47 a–c. Hooke's jointed driveshafts to DIN 808, made by Willi Elbe. **a** For general machine and machine tool construction (up to 4000 rpm). *1* single joints; *2* intermediate shaft; *3* muff with DIN 5463 splines. Maximum speed with needle bearings (W) – 4000 rpm, with plain bearings (G) – 1000 rpm; **b** with quick-change coupling for drill spindles (up to 6000 rpm). Plunge $L_{max} - L_{min} = 80$ to 200 mm for shaft diameters 10 to 50 mm; **c** section A–B through quick-change coupling **b**

Single joints with needle bearings. The maximum torque capacity and speed are 1000 Nm and 4000 rpm respectively. The relationship between the reduced torque M_{red} and the product of the life L and speed n, can be obtained from the chart in DIN 808.

Double joints. They may only be loaded up to 90% of the limits for single joints.

Fitting and Assembly

When two single joints are assembled to form a driveshaft, the yoke axes of the joints on the intermediate shaft must lie in one plane, as is the case for a double joint, otherwise the rotary movement would be non-uniform. A Z- or W-configuration should be employed. If the driving shaft is moved relative to the driven shaft, both shafts should only be displaced axially or parallel to their starting position or both. This is possible if the joints are fitted with a telescopic intermediate shaft and thus form an axially free driveshaft (Fig. 5.47a).

The drilling machine driveshaft with needle-bearings and a quick-change coupling (Fig. 5.47b) is a variation of the simple, telescopic driveshaft. The joint shank, fitted with a hexagon (Fig. 5.47c) or keyway to DIN 6885 is mounted on the drill spindle connection until the two locking balls engage in the retaining groove.

In order to achieve good durability of the joints, low operating temperatures are important; these should not exceed 60 °C. High temperatures for continuous operation, caused by internal friction of the joints, can be reduced by smaller articulation angles or lower shaft speeds.

5.2.6 Grooved Spherical Ball Jointed Driveshafts

The principal dimensions of grooved spherical ball joints (Figs. 5.48 and 5.49) are similar to those for DIN 808 joints. Instead of the cross, they work with cross-grooved balls which guide the shank yokes. The joint is particularly simple to take apart because it only needs to be articulated through 90°. It can then to be split into

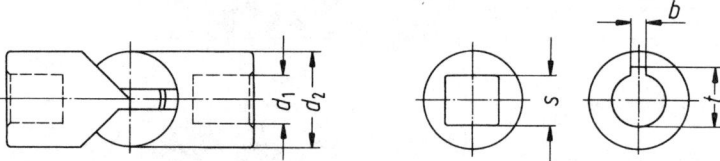

Fig. 5.48. Grooved spherical ball joint, made by Alfred Heyd

Fig. 5.49. Slow speed driveshaft (up to 1000 rpm for articulation between 3° and 5°) with two single grooved spherical ball joints

5.2 The Cardan Compact 2000 series of 1989

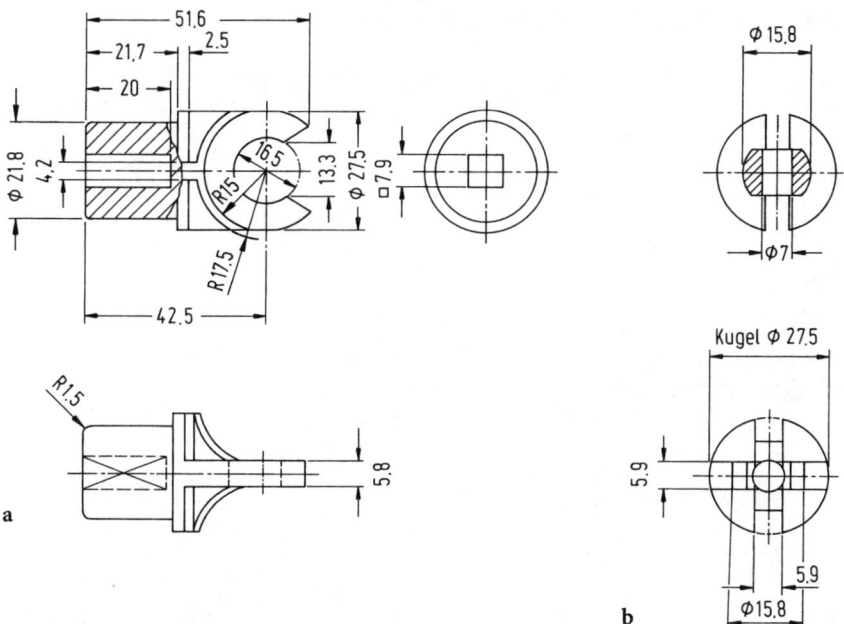

Fig. 5.50a, b. Individual parts of a plastic grooved spherical ball joint. **a** Joint shank, **b** ball

its three individual parts by means of the flats in the base of the ball groove. In addition to the single or double joint design, there are also driveshafts with plunging elements made out of stainless steel (St 1.4306) and a ball made of gunmetal for use in damp conditions.

In continuous operation grooved spherical ball joints transmit about the same torques up to an articulation angle of 35° as DIN 808 joints with plain bearings (Fig. 5.47); on the other hand they can only rotate at half the speed because of their greater joint friction. For an almost straight joint, they should only be used up to 1000 rpm; for articulation angles around 15°, only low speeds are acceptable. 35° articulation is only permissible for turning over by hand. The torque capacity is given by the connection bores for shaft diameters of 4 to 90 mm. Grooves spherical ball joints can withstand much lower dynamic loads than Hooke's joints.

The plastic joint is a further development of the grooved spherical ball joint. It is also made up of two identical joint shanks (Fig. 5.50a) and the cross-grooved centring ball (Fig. 5.50b). The parts are made simply by injection moulding and are mounted by snapping the joint yokes together. The spring-back of the yokes ensures that the joints hold together and also preloads the joint. Because of the precision forming the joint is cheap, light, maintenance-free and corrosion resistant. Hence it can be used in the control drives of window blinds, awnings, ventilation grilles, in chemical apparatus and in the open air.

5.3 Driveshafts for Agricultural Machinery

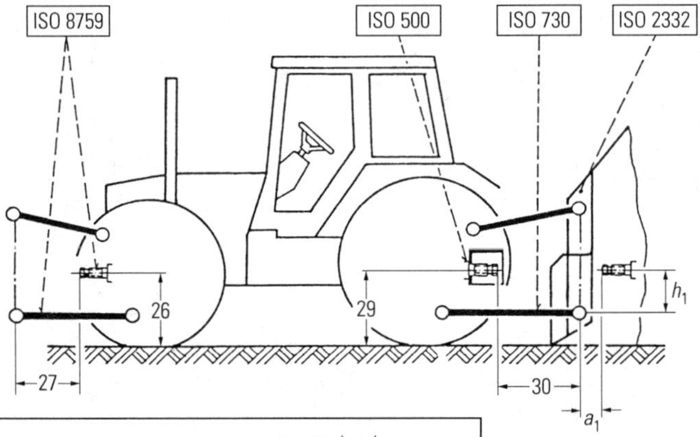

Fig. 5.51. Connection measurements for three tractor sizes from ISO standards

output kW size			tractor (mm)		
		up to	48	82	80–185
			1	2	3
three-point front	ISO 8759	T 27 26	550–675 550–850	550–675 550–850	
three-point rear	ISO 730 ISO 500	T 30 h 29	500–575 450–675	550–625 550–775	575–675 650–875
hitch coupling	ISO 6489/2	h 33	750–900	820–1100	820–1000
tow hook	ISO 6489/1	37	50–110	50–110	50–110
adjustable drawbar	ISO 6489/3	c 29 35 short 34 standard l long	min. 200 250 400 500	min. 220 250 400 550	min. 250 350 500 650

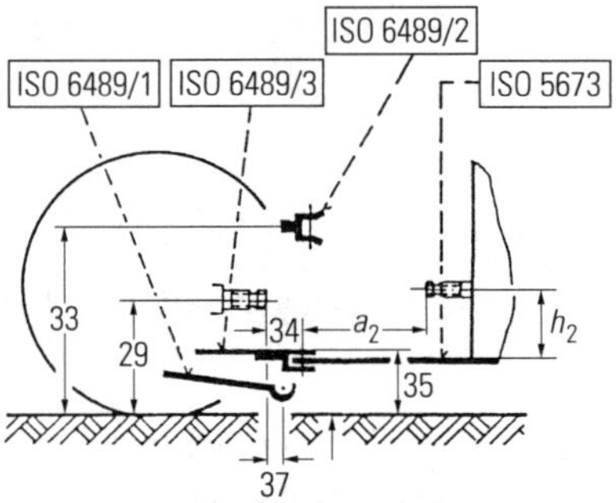

machinery (mm) size			1	2	3
three-point		a_1 h_1	150–180 100	150–180 100	300–330 100
hitch	ISO 5673	a_2 Standard long h_2	400 500 250	400 550 250	500 650 250

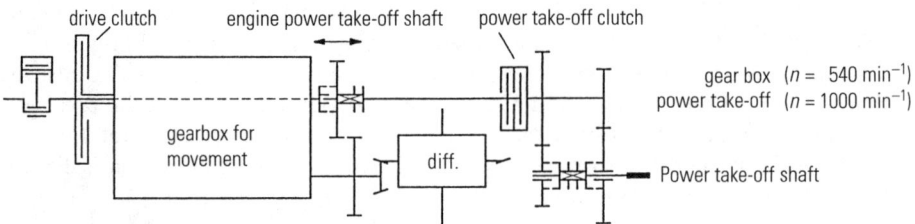

Fig. 5.52. Tractor drivetrain connection shaft. For loaders, farmyard manure spreaders and balers only one joint is articulated

5.3 Driveshafts for Agricultural Machinery

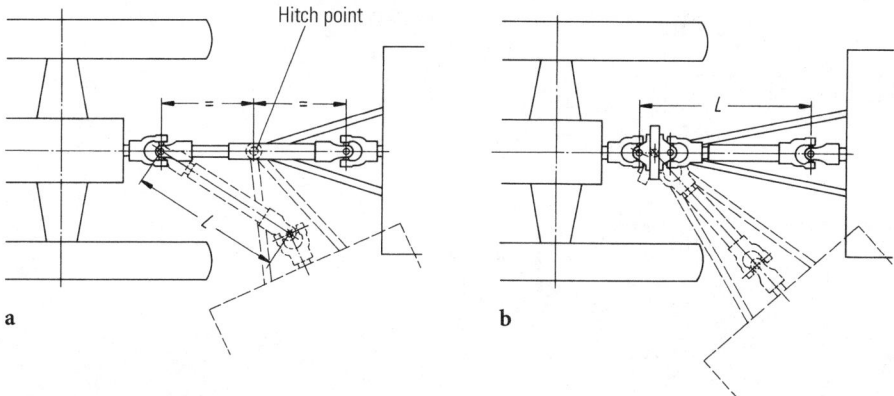

Fig. 5.53a, b. Attachment and PTO drive for implements. **a** Adjustable drawbar attachment with standard hitch. If the hitch point is half-way along the PTO shaft, the articulation angles are equal (W-configuration); **b** short hitch. If the joint centre and the hitch point coincide, only slight changes in length occur in the driveshaft even when going round bends. The non-uniformity can only be compensated by using wide-angle constant velocity joints. Articulation angles at the implement end are determined by the position of the attachment shaft

After the First World War, agriculture derived great economical benefits from the inniternal combustion engine-powered farm tractor. The introduction in 1918, by International Harvester Co., of mass produced power take-off shafts with their standardised attachment dimensions, opened up possibilities for driving additional implements. This made the farm tractor a source of mobile power in the field (Fig. 5.52).

Joints used in agriculture take up a special position because of their particular use. They vary a great deal therefore from machine and vehicle construction. In addition to rough conditions of use and little care and attention, they also stand idle for long periods in the course of a year. A characteristic of driveshafts used in the agricultural industry is the need for a large plunge and articulation angles of 45° to 75°.

When the joint articulates, the driving Hooke's joint steers the sliding centring disc into an eccentric position, forcing the driven joint to the same angle of articulation. Since it is open on both ends, because of the sliding centring disc, it must be greased and serviced very regularly and carefully.

5.3.1 Types of Driveshaft Design

Since the introduction of the power take-off drive, each of the three most common sizes of tractor shown in Fig. 5.52 has its own power take-off shaft (Fig. 5.54) and a specified operating area for the driveshaft (Figs. 5.57a, b). The driveshaft for the implement is coupled to the tractor by a spring loaded quick disconnect coupling (Fig. 5.57c). It snaps into the annular groove on the power take-off shaft.

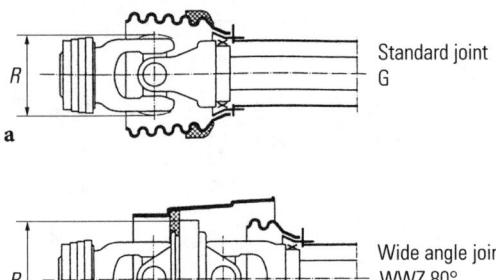

Standard joint G

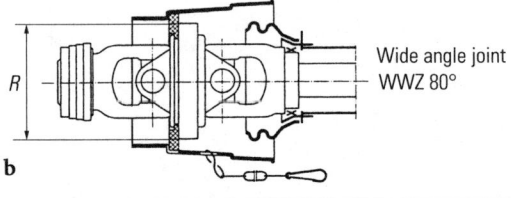

Wide angle joint WWZ 80°

Joint size	Rotary ⌀ R mm	Limits		Dynamic capacity K 540 min⁻¹	
		Ms [Nm]	Mp [Nm]	1000 min⁻¹	
				P [kW] (PS)	Mn [Nm]
G 2000	59	170 (1505)	900 (7965)	6 (8)	110 (970)
				9 (12)	90 (795)
G 2100	69	335 (2965)	1100 (9735)	12 (16)	210 (1860)
				18 (24)	175 (1550)
G 2200	76	565 (5000)	1750 (15 490)	20 (27)	355 (3140)
WWG 2280	142			31 (42)	295 (2600)
G 2300	92	840 (7435)	2350 (20 800)	28 (38)	500 (4420)
WWG 2380	155			44 (60)	415 (3675)
G 2400	95	1240 (10 975)	3800 (33 630)	39 (53)	695 (6150)
WWZ 2450	125	1165	2895	120	
WWG 2480				61 (83)	580 (5130)
WWZ 2480	145	1730 (10 600)	4500	178	
G 2500	108	2150 (19 030)	6000 (53 100)	66 (90)	1175 (10 400)
WWG 2580	180			102 (139)	975 (8630)
G 2600	125	2895 (25 620)	7800 (69 030)	79 (107)	1400 (12 390)
				122 (166)	1165 (10 310)
G 2700	145	4530 (40 090)	10 600 (93 810)	119 (162)	2095 (18 540)
				182 (248)	1740 (15 400)

Fig. 5.54a–d. Joint sizes for agricultural applications, made by J. Walterscheid GmbH (GK N Automotive AG). **a** single joint (*G*); **c** wide angle constant velocity joint (WWG); **b** principal dimensions in mm and torque capacity in Nm

5.3 Driveshafts for Agricultural Machinery

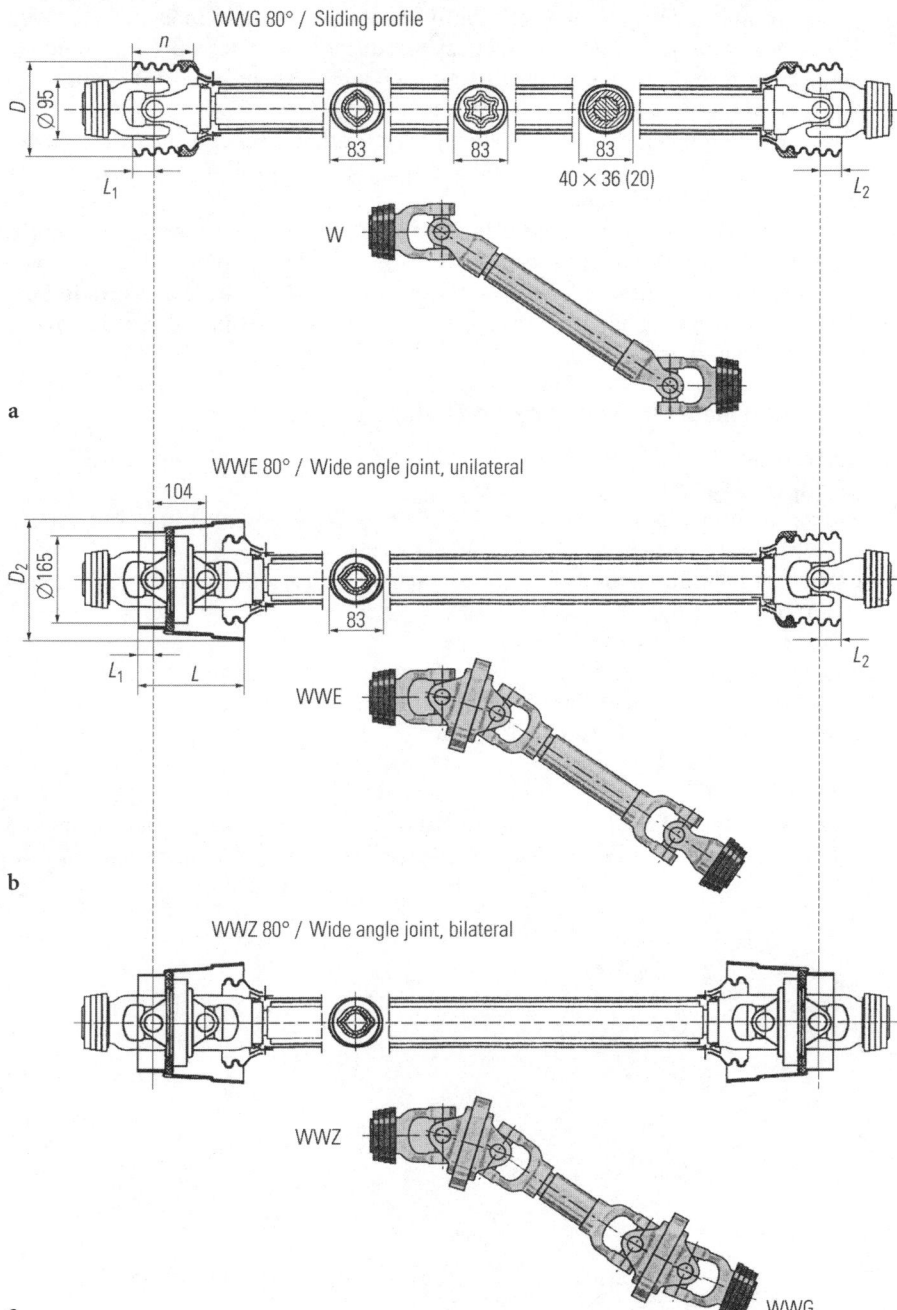

Fig. 5.55a–e. Designs of driveshafts for agricultural uses, made by Walterscheid. **a** WWG shaft for standard Z- and W-configuration; **b** WWE shaft for adjustable drawbar attachment; **c** WWZ shaft for articulation angles which cannot be compensated (1 WWZ = 2 WWG)

- Type G. Simplest design. Two single joints are connected by an intermediate shaft. Substantial changes in length are possible due to the telescopic intermediate shaft. The driveshaft itself transmits the rotary movement uniformly with compensated angles in Z- and W-articulation.
- WWE type. Wide angle constant velocity driveshaft. It transmits the rotary movement through a wide angle joint and a single joint in a uniform manner, as long as the single joint is straight.
- WWZ type. Wide angle constant velocity driveshaft with two wide angle joints. It transmits the rotary movement uniformly for articulation angles up to 36°, even with different and out-of-phase articulation angles. When the shaft is stationary, articulation angles of 80° are possible, e.g. for sharp turns; this reduces damage to the implement drive.

5.3.2 Requirements to be met by Power Take Off Shafts

- full transmission of power from the take-off shaft to the implement for the standard speeds 540 and 1000 rpm,
- plunge between the tractor and implement for all types of trailer and for the three-point attachment on the tractor (Fig. 5.56),
- a large articulation angle to allow the tractor to turn relative to its implements and to negotiate tight bends (Fig. 5.53),
- robust design with overload protection,
- protection against accidents by means of protective cones and cover tube (Figs. 5.57a, b) such as those of Kurt Schroeter (German pat. 948568/1953),
- simple handling, maintenance and repair.

Length Compensation

The sliding profiles of the intermediate shafts (Figs. 5.55a–d and 5.27) are designed for large changes in length and the lowest possible sliding resistance. Special tube

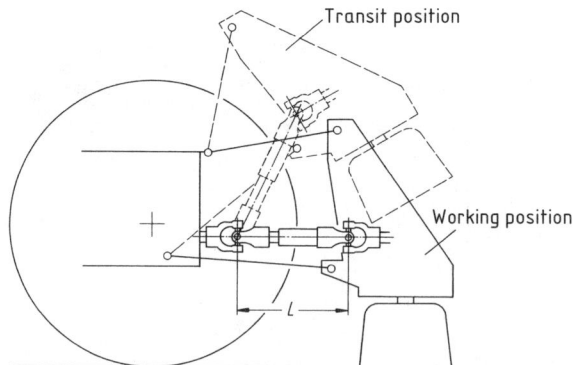

Fig. 5.56. Three-point attachment of implement piggy-backed onto the power take-off shaft. The driveshaft length is determined by the distance L for a horizontal arrangement (maximum overlapping of the sliding elements). Because of the three-point kinematics (no parallelogram effect), the articulation angles are only compensated by the Z-configuration in one position of the implement

5.3 Driveshafts for Agricultural Machinery

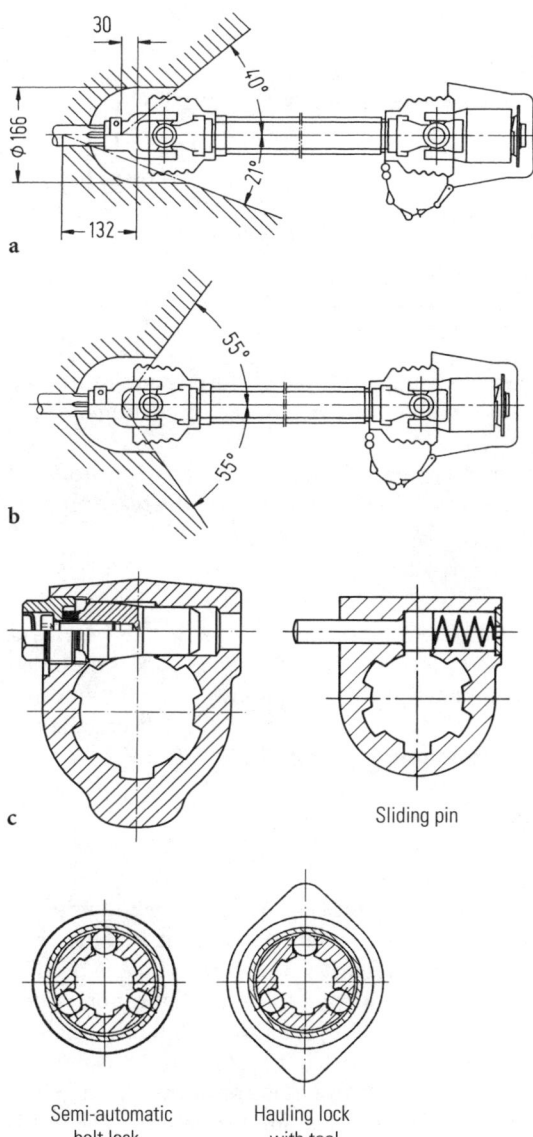

Fig. 5.57a–c. Operating area for power take-off shaft on tractor with safety shield on the implement. **a** Side view, **b** plan view, **c** driveshaft quick disconnect couplings. The clutches protect thw whole machinery, because the output of the tractor is higher. Design J. Walterscheid GmbH

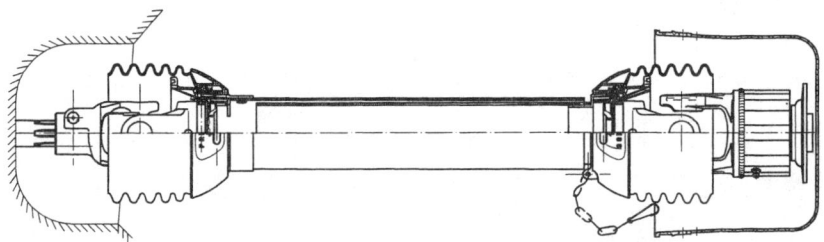

Fig. 5.58. Hooke's joints in agricultural use with protection covers. Design J. Walterscheid

shapes have been chosen for this; these distribute the load uniformly and centre the shaft so that it runs smoothly. The sliding forces can be reduced by coating the surfaces (e.g. with Rilsan).

For extreme changes in length there must be sufficient tube to allow the profile to support with its full length. All slip yokes are hardened and fitted with grease nipples.

Driveshaft Protection

The design of the driveshaft must include protection against contact and accident (Figs. 5.54 a–c). A telescopic plastic tube with elastic, funnel-shaped ends is mounted on the driveshaft. The conical ends completely enclose the joints, without hindering their movement, even in articulation, or their connection to the power take-off. A chain prevents the driveshaft guard from rotating (Figs. 5.57a, b, 5.58).

5.3.3 Application of the Driveshafts

For three-point attachment of the driven implement (Fig. 5.56) a true Z-configuration with compensation of the articulation angles is only possible for a single position of the implement. The W type of driveshaft with two single joints is used. If there are changes in stroke during operating however, the Z-configuration is disrupted. The non-uniform rotation of the shaft arising from the differential angle of the imperfect Z-configuration should not be too great; in general, +/– 7% is permissible if the vibrations are to remain acceptable. If the shaft is not working, the full articulation angle of the joint can be exploited. For drawn implements, like the drawbar attachment with a short hitch (Fig. 5.53a), the connecting shafts are positioned such that the hitch point is located midway between the two single joints of the driveshaft. An exact W-configuration thus occurs.

For attachments with a long hitch (Fig. 5.53b), a WWE driveshaft, comprising a wide angle joint which take up the whole articulation, and a single joint are required. They must be arranged such that the tractor hitch is near the wide angle joint centre. The single joint which is almost straight then only has to accommodate the differential angle resulting from the displacement of the hitch point.

If spatial displacements occur between the connecting shafts, that is, undefined arrangements with varying articulation angles of both joints, or if the articulation

5.3 Driveshafts for Agricultural Machinery

Tube profiles

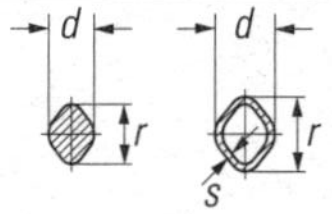

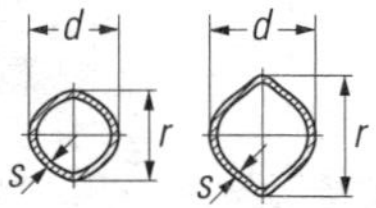

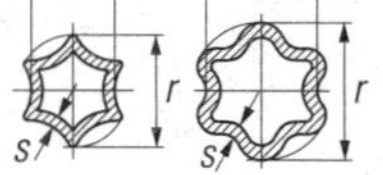

Profiles	00a/0a	00aGA*/0aG*	0v/1	0vGA*/1G*	1b/2a	1bGA*/2aG*	S4LH/35
r [mm]	31	31 39	40	40 48	49 57.5	49 57.5	51 61
d [mm]	23.5	23.5 30	34.5	34.5 41	39.5 48	39.5 48	37 47
s [mm]	2.8	2.6	4.0	4.0 2.7	4.5 4.0	4.5 3.5	6.0 4.5
Protection	SD05	SD05	SD15	SD15	SD25	SD25	SD25

Tube profiles

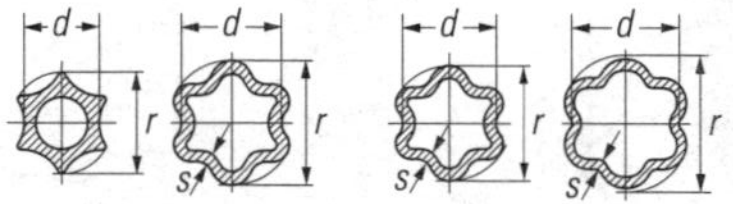

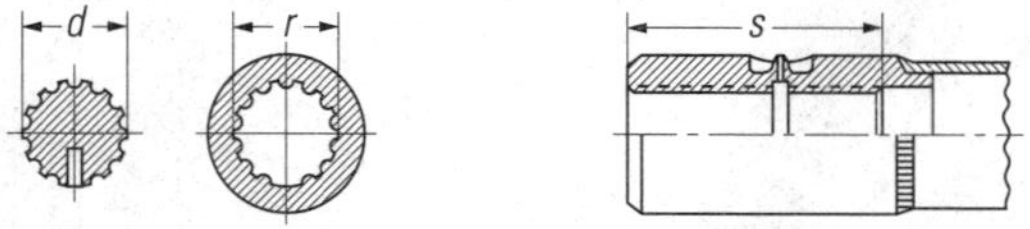

Profiles	S4/S5 S4GA*/S5	S5H/S6	25 × 22 (14)	30 × 27 (16)	35 × 31 (18)	40 × 36 (20)	52 × 47 (25)
r [mm]	51	61 71.5	40	51	51	52	75
d [mm]	37	57 57.5	25	30	35	40	52
s [mm]	4.5	4.5 5.0	50	80	90	100	120
Protection	SD25	SD35	SD05Z	SD10Z	SD15Z	SD25Z	SD35Z

GA Surface coated; G special outside profile for coated inside profile; H hardened

Fig. 5.59. Sizes for sliding profiles of driveshafts on agricultural machinery

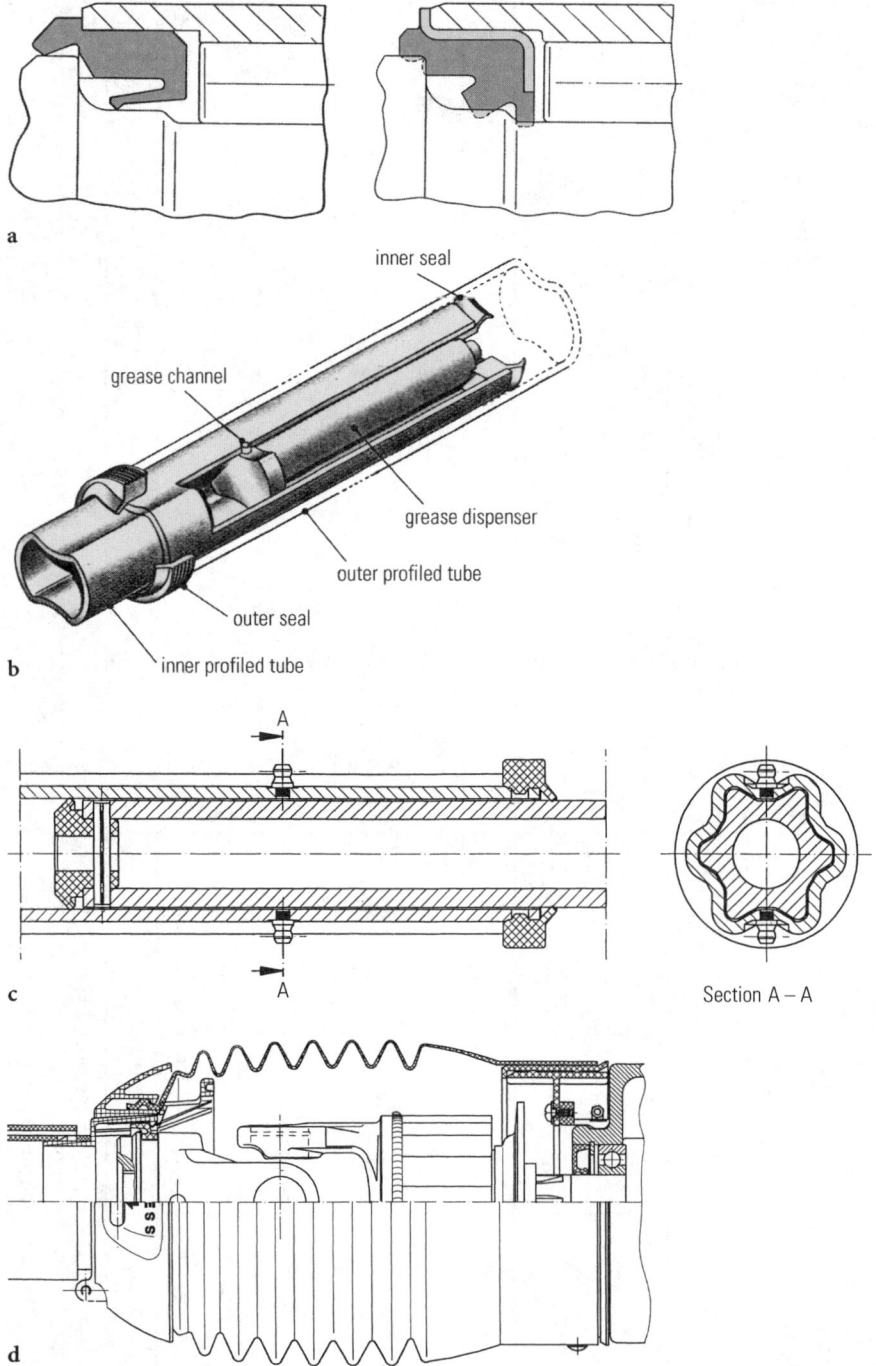

Fig. 5.60 a–d. a Plastic double lip seal on the Hooke's joint between tractor and machinery; **b** profiled tube system, **c** profiled tube in section together with scraper, **d** full protection of the Hooke's joint. GKN Walterscheid design

5.3 Driveshafts for Agricultural Machinery

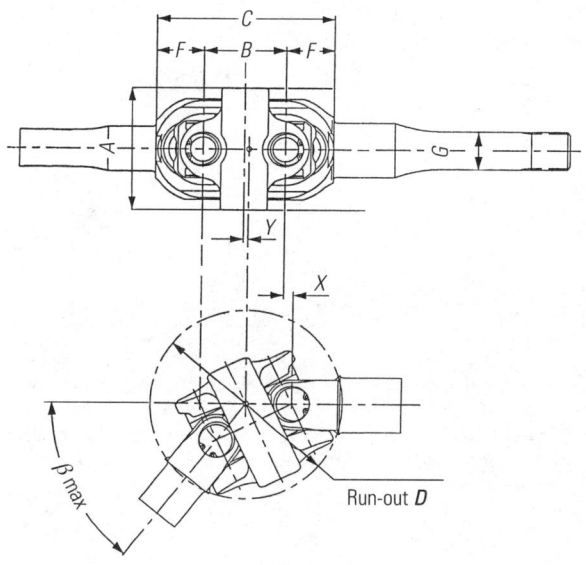

	Joint sizes							
M_{max}	1200	1650	2300	3000	4000		5000	
							42°	50°
$\varnothing$ A	85	94	106	114	122		142	
B	53	59	64	71	76		69	75
C_{min}	123	133	138	159	186		159	165
F_{min}	32	32	32	38	48		36	
$\varnothing$ G_{min}	26	28	30	34	38		38	
$\varnothing$ D	132	152	164	182	200		198	204
Y		3.2/3.6	3.5/4.1				5.2	6
X		6.3/7.3	7/8.2				10.4	12

Fig. 5.61. Double joints with misaligned trunnions. For fwd tractor with steering angle greater 50°. Articulation angle $\beta_{max} = 60°$. Shaft $\varnothing$ G from heat treatable steel $\geq$ 930 N/mm². $\varnothing$ D necessary install space at β_{max}. Y joint pledge for β_{max} (mm), X plunge. ITALCARDANO design

angles of the single joint are too big, then a WWZ wide angle driveshaft with two identical wide angle joints must be fitted (Fig. 5.54c). This is expensive but runs the most smoothly.

If the driveshaft is to work reliably in rough agricultural conditions, all lubricating points must be greased before starting work and every eight hours.

The joint crosses are supported by needle or roller bearings. They are forged, precision turned and case hardened. Double lip seals are fitted to the joint cross to protect against dirt and moisture, and grease dispensers are mounted in the inner profile for continuous lubrication (Fig. 5.60).

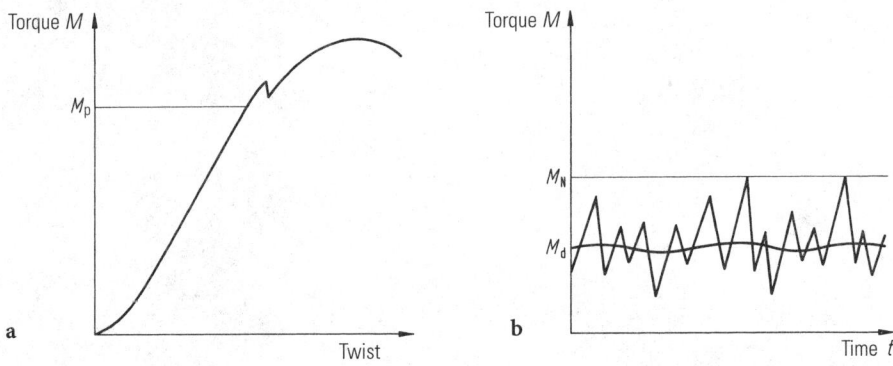

Fig. 5.62 a, b. Loading characteristics for a driven implement. **a** Static load, M_0 limit of static load carrying capacity; **b** dynamic load, M_N is the pulsating torque and gives the highest recurring torque (the dynamic limiting load for the needle bearings); M_d is the nominal or mean torque and defines the durability and permissible articulation angles

Joint Selection

The torque of the work implement is made up chiefly of the uniform or intermittent starting torque, the consistent pulsating or alternating fatigue loading torque, the acceleration torque and the overload or peak torque for a jammed implement. To define the torque variations for the operation of the driven device, the torque has to be recorded (Fig. 5.62). The rated torque is given by the maximum operating torque and the size of the joints is determined by the large articulation angle. If the peak torques are above the permissible, pulsating torque value of the joint size selected, overload protection is needed in the power take-off shaft.

Parts of the machinery that are swivelled away or folded over can articulate the joints through more than 90°. In these cases double joints on short driveshafts are required.

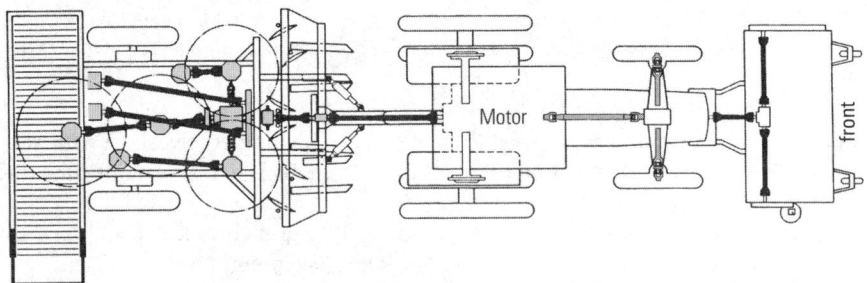

Fig. 5.63. Fully automatic sugar beet harvesting machine, diagram of drive mechanism. GKN Walterscheid design

5.4 Calculation Example for an Agricultural Driveshaft

A driveshaft is sought for a rotary mower attached using a three-point hitch. In order to protect the whole machine drive system from impact and jamming, an overload clutch should be used. The operating data are:

- output to be transmitted   P = 25 kW
- drive speed   n = 540 rpm
- operating articulation angle $\beta = 18°$
- connection profiles
  tractor side   1 3/8 6-part
  machine side   1 3/8 6-part
- required service life   1000 h

the following dates are to be determined:

a) the size of driveshaft
b) its calculated service life
c) the response torque of the overload coupling

From the specified output 25 kW and the speed 540 rpm, the torque can be calculated from Mn_x = 9550 P/n = 9550 × 25/540 = 442 Nm.

From experience, the load factor f_{ST} = 1.3 to 1.6 is used for the dynamic load in Table 4.2.

For a load factor f_{ST} = 1.6, a pulsating torque of Ms_x = 1.6 × Mn = 1.6 × 442 Nm = 707 Nm results.

From Fig. 5.54 you can select

Joint size G 2300
Mn 500 Nm > Mn_x 442 Nm
Ms 840 Nm > Mn_x 707 Nm

Safety factor from the static torque Mp:

Mp/Ms_x = 2350 Nm/707 Nm = 3.32.

There is then an adequate safety factor against collision impact.

The three-point attachment produces a Z-configuration (see Figure 5.56) and so a W-shape driveshaft is required.

Solution b)

The life L_h of the G 2300 joint is calculated for continuous operating with Mn_x 442 Nm and a speed of n = 540 rpm and an articulation angle of 18°.

The torques Mn given in Fig. 5.54c relate to speeds of 540 rpm and 1000 rpm, an articulation angle of 5° and a bearing life of 1000 hrs.

The conversion to the operating conditions is calculated from the following formulae:

$L_h = L'_h \cdot n_x/K \cdot \beta_x (Mn \cdot \cos \beta_x/Mn_x)^{10/3}$
$K = n/\beta \cdot (\cos \beta)^{10/3}$

$K_{540} = 540/5 \cdot (\cos 5°)^{10/3} = 106 \text{ l/min°}$
$K_{1000} = 1000/5 \cdot (\cos 5°)^{10/3} = 197 \text{ l/min°}$
$L_h = 1000 \cdot 540/106 \cdot 18 (500 \cdot \cos 18°/442)^{10/3} = 361 \text{ h}$

The life L_h is clearly the required life of 1000 hrs and so the next biggest size must be chosen.

Checking size 2400

$Mn = 695 \text{ Nm}$
$L_h - 1000 \cdot 540/106 \cdot 18 (695 \cdot \cos 18°/442)^{10/3} = 1082 \text{ h}$
Size 2400 meets the requirements.

Safety factors:

$Mn/Mn_x = 695 \text{ Nm}/442 \text{ Nm} = 1.57$
$Ms/Ms_x = 1240 \text{ Nm}/707 \text{ Nm} = 1.75$
$Mp/Ms_x = 3800 \text{ Nm}/707 \text{ Nm} = 5.37$

The driveshaft is therefore adequately protected against impact and jamming.

In order now to protect the machinery against these effects, the response torque of the overload clutch must be determined. The clutch factor K_k is in the range 1.3 to 2.5, depending on the type of clutch.

Friction clutches are used for mowing machines with large inertia. For these the factor $f_{ST} = 1.3$.

$M_{clutch} = Ms_x \cdot 1.3 = 707 \cdot 1.3 = 920 \text{ Nm}$

For the chosen driveshaft:

W 2400 with 2-rib tubular profile and a friction clutch 920 Nm.

5.5 Ball Jointed Driveshafts

Ball jointed driveshafts possess constant velocity characteristics. They run quietly and free the designer from restrictive Z- or W-configurations. The only requirements are to keep the articulation angle β and the product $n\beta$ as small as possible.

Table 5.7 compares the main features of Hooke's jointed and ball jointed driveshafts. The "Rzeppa" joint is used as a constant velocity joint especially in cars with front wheel drive. The technology of this joint seemed to have been brought to maturity by A. Rzeppa and B. Stuber (Figures 1.23, 1.26, 1.28), but the automotive industry demanded *a longer service life, a greater steering lock angle of the front wheels, and a better vibration and noise isolation of the engine from the vehicle*. The development needed for this was made possible by new non-machining methods: the more precise manufacturing and better surface quality have led to both a much longer service life and quieter running.

5.5 Ball Jointed Driveshafts

Table 5.7. Comparison of driveshafts

	Hooke's jointed driveshaft (fixed joint only)	Constant velocity driveshaft (fixed and plunging joint)	
Rotary movement	Not uniform	Uniform	
Constant velocity condition	Inflexible W- or Z-configuration; not constant velocity in the centre part of the shaft	Constant velocity is obtained in any configuration, in and out of plane, in all parts	
Articulation angle $\beta°$	< 43°	Fixed joint	< 50°
		Plunging joint	< 30°
Plunge	By means of shatt or shank splines	Via plunging joint	
Durability	High	Lower	
Recommend limit $n\beta$	25 000	Fixed joint	10 000
		Plunging joint	30 000
Speed limit n (rpm)	4000	Fixed joint	3 000
		Plunging joint	10 000
Balancing	Necessary	Not required	
Multi-piece, out-of-plane drive line	Not easy	Always possible	

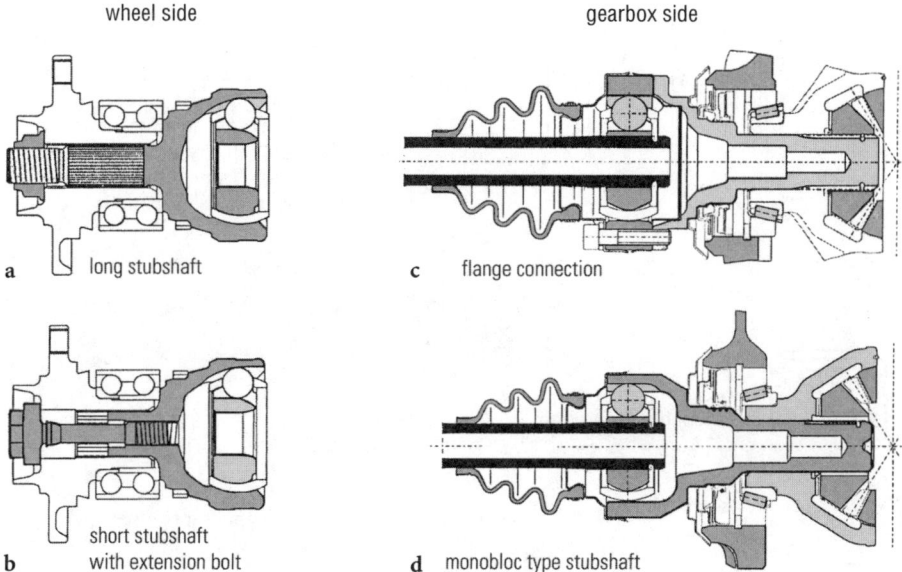

Fig. 5.63. Connecting the constant velocity driveshaft in the drive line. **a** long stubshaft, **b** short stubshaft with extension bolt, **c** flange connection, **d** monobloc type stubshaft [5.17]

5.5.1 Boots for joint protection

The service life of an exposed fixed or plunging joint depends on the boot. Firstly, the designer must ascertain the contrasting stresses in the stretched and compressed states: a stretched boot which is stable in shape has many folds and thick walls, a compressed one has fewer folds and thin walls.

In the sealing area, one or several sealing lines must be provided by lips or surface on the joint and shaft (Fig. 5.65).

The most important dimensions for the boot are (Fig. 5.64):

- outside diameter D of the joint housing
- diameter d of the seat on the shaft
- length L of the boot

The most favourable articulation behaviour of the boot is when $L/D = 1$ to 1.5. The speed is assumed to be 2500 rpm. Temperatures between $-40°$ and $+120°$ must be reckoned with. Boots made out of thermoplastic elastomers (TPE) are less sensitive and therefore preferred. Joints on the gearbox side, with their temperatures of up to 120°, need boots made from ether elastomers, which are, however, stiffer because of

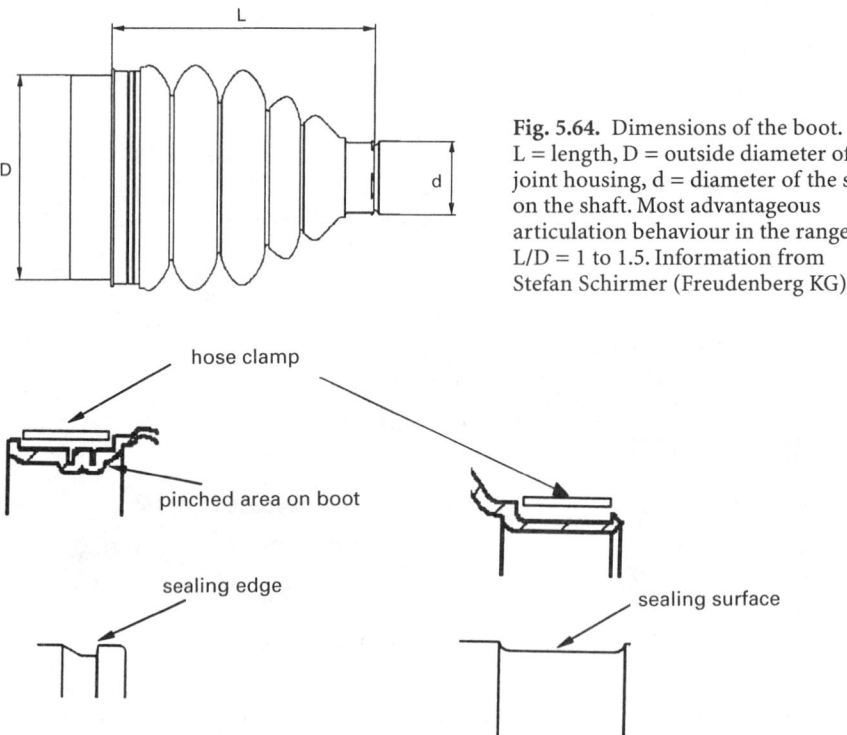

Fig. 5.64. Dimensions of the boot. L = length, D = outside diameter of the joint housing, d = diameter of the seat on the shaft. Most advantageous articulation behaviour in the range $L/D = 1$ to 1.5. Information from Stefan Schirmer (Freudenberg KG)

Fig. 5.65. Sealing areas of the boot. Several sealing lines through **a** lips, **b** sealing surface. Information from Stefan Schirmer (Freudenberg KG)

5.5 Ball Jointed Driveshafts

their greater hardness. The roller boots with radial support on propshafts are also dimensionally stable.

5.5.2 Ways of connecting constant velocity joints

Both ends of the driveshaft must be easy to connect. There are two conventional approaches for both the wheel-side and gearbox-side connections (Figures 5.63a–d). For connection on the wheel side the ball bearing on the wheel is often

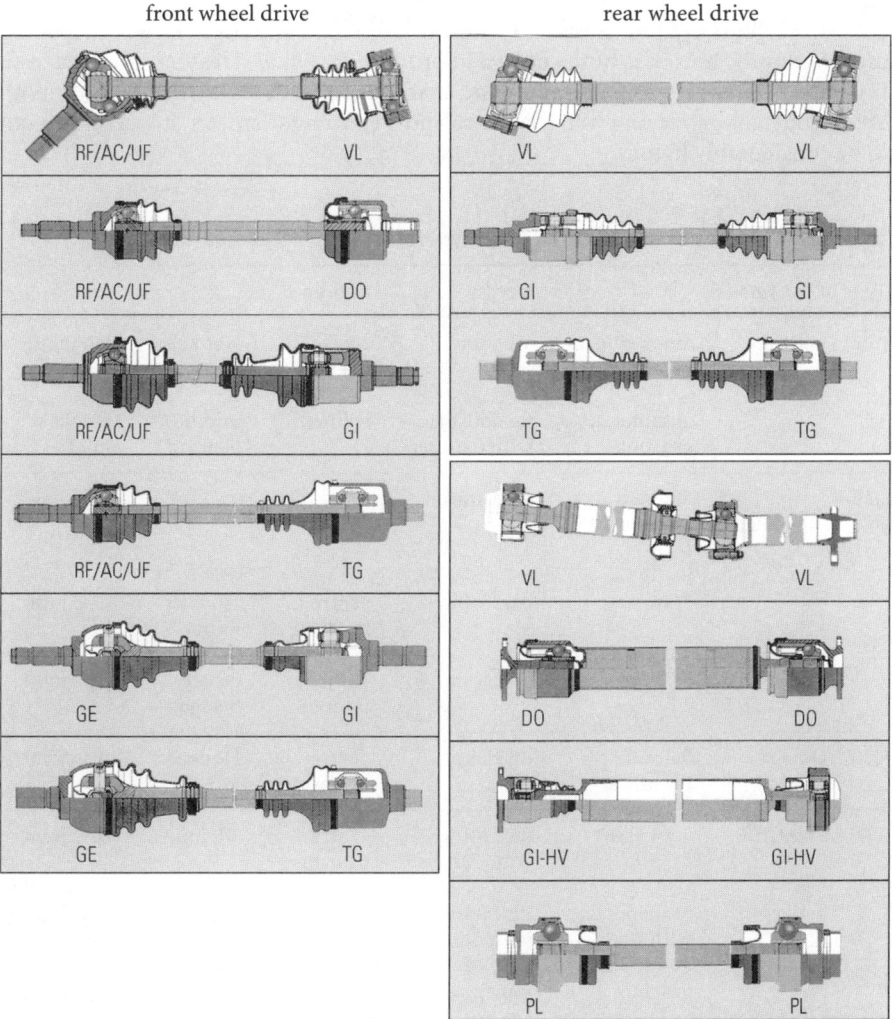

Fig. 5.66. Differing applications and designs of connections for front wheel and rear wheel ball-jointed driveshafts, for vehicles with independent wheel suspension and high speed propshafts, GKN Löbro design

axially preloaded. The use of a short stubshaft means that mounting without dismantling the wheel suspension is simple. Connecting at the gearbox-side by means of a bolted flange connection is simple; in production, it can be supplied with the attached flange directly onto the conveyor belt. The stubshaft connection is lighter and also simple to fit; it has no eccentric interfaces and therefore does not excite vibrations.

5.5.3 Constant velocity drive shafts in front and rear wheel drive passenger cars

Solid steel shafts are too heavy and transmit noise and vibration. However, making them hollow by boring is both costly and produces residual stresses in the material. Instead of the cold reshaped monobloc shaft has been developed (Fig. 5.68). With either constant or varying wall thickness and a constant diameter, it has turned out to be considerably lighter.

Table 5.8. Examples of standard long shaft systems for passenger cars around 1998

Type of car (*model*)		front	middle		rear
Audi Quattro	automatic gearbox	VL joint	centre bearing	Hooke's joint	VL joint
	automatic gearbox	flexible disc with damper	splined	middle bearing	Hooke's joint
BMW	automatic gearbox E 36/7	disc joint			VL joint
	3/94	disc joint	centre bearing	Hooke's joint	disc joint
	5-Series	disc joint	Hooke's joint	centre bearing	disc joint
OPEL Omega	Saloon Estate car	disc joint	centre bearing	Hooke's joint	disc joint
VOLKSWAGEN	4 ×4 short	disc joint with damper	centre bearing	VL joint	disc joint
	Transporter with automatic gearbox	disc joint with damper and VL joint	centre bearing	centre bearing and VL joint	disc joint

5.5 Ball Jointed Driveshafts

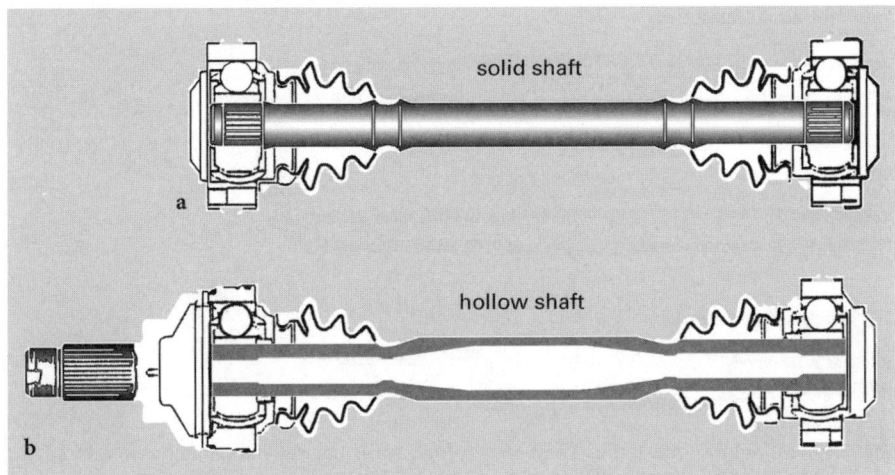

Fig. 5.67. Comparison of passenger car halfshafts. **a** Solid shaft with wide joints, bolted at both ends, **b** Hollow shaft with narrow joints, inserted at wheel end. GKN Löbro design/BMW [5.17]

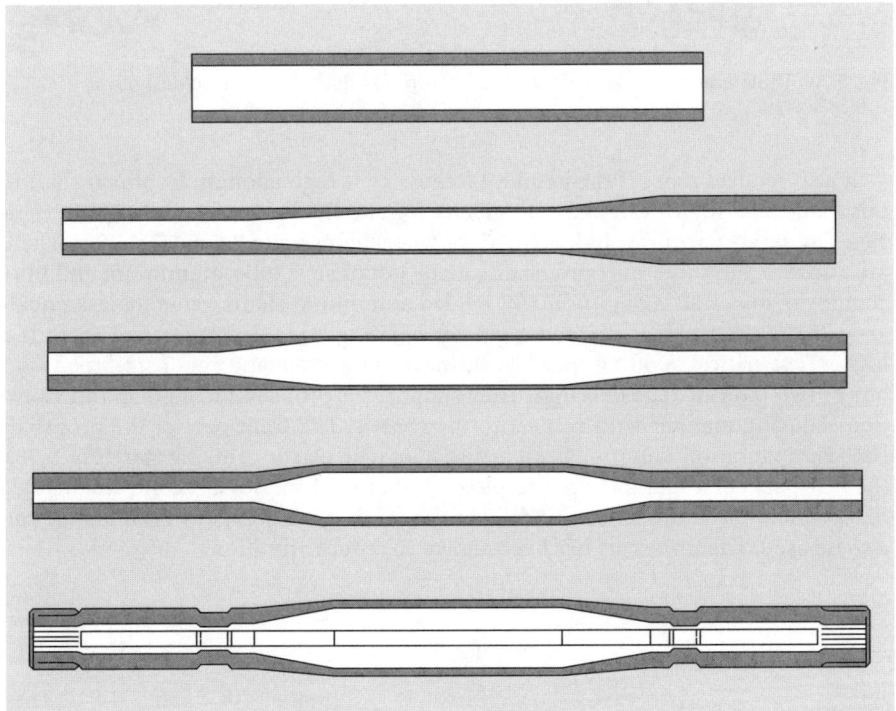

Fig. 5.68. Monobloc sideshaft reshaped of steel

160% torsional resistance

40% weight saving

160% bending resistance

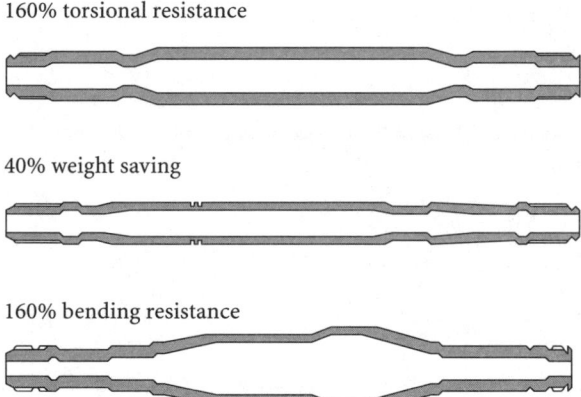

Fig. 5.69. Monobloc halfshafts. They have a high torsional resistance and a weight saving of around 40%. Their bending frequency is over 200 Hertz

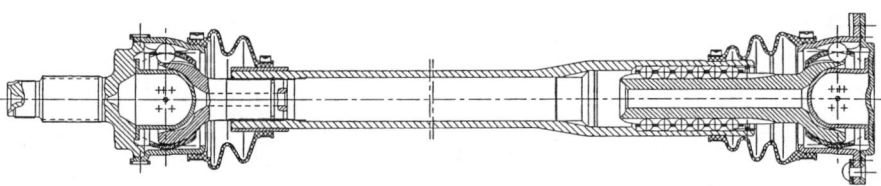

Fig. 5.70. Unilateral, noise- and vibrationless propeller shaft, GKN Löbro design

It also reduces noise if the bending frequency is high enough. A splined shaft of this kind can also be designed with greater longitudinal plunge for all-terrain vehicles. The single part propshaft is fitted for wheel bases up to 2.6 m (Fig. 5.70). There are also two and three-part propshafts made out of steel tube, aluminium and fibre composite materials. Longitudinally welded aluminium shafts generate less imbalance. If the propshaft is divided a greater bending stiffness results and there is a higher first critical bending speed. Multi-part shafts are made out of steel tube with one or two intermediate bearings. They support the propshaft in a noise and vibration-reducing manner with respect to the chassis. The front part of the propshaft runs via a vibration-damping flexible disc joint with elastic centring at 0° (Fig. 5.70); the rear part corresponds to the single-part shaft and carries a second small angle fixed joint at the end. Narrow, modified fixed joints or VL joints without plunge can also be used. The front part has been shown to reduce vibration.

5.5 Ball Jointed Driveshafts

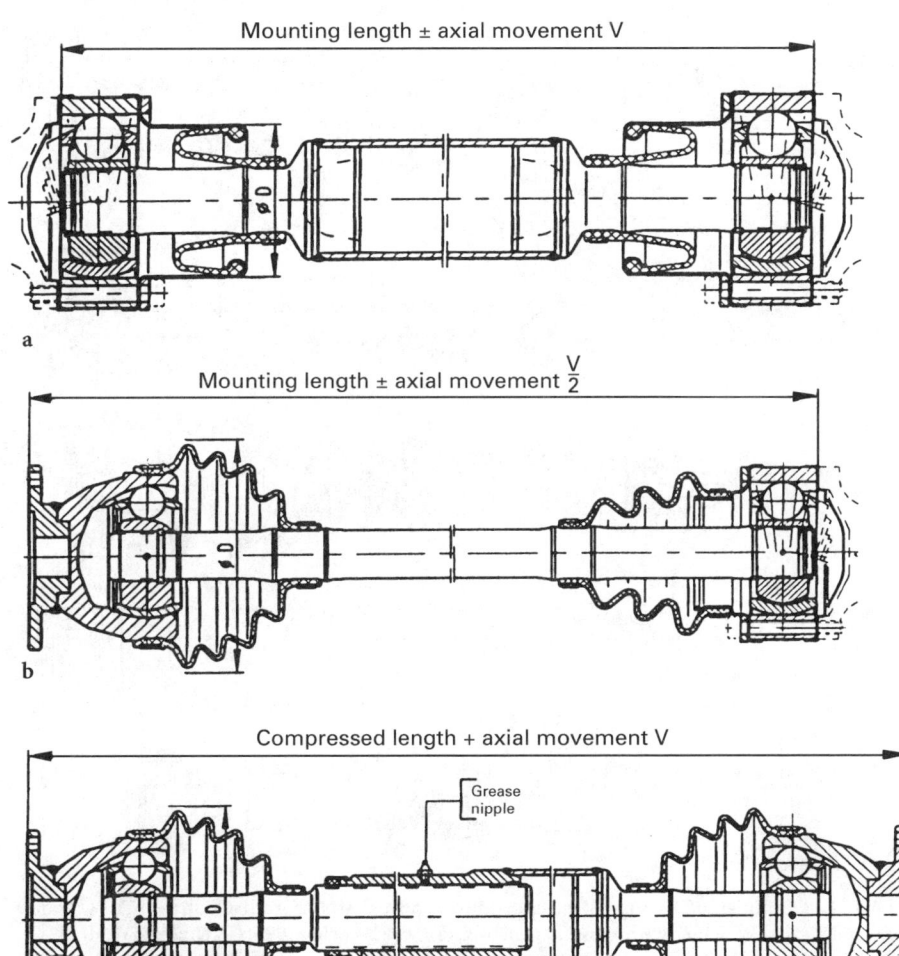

Fig. 5.70a–c. Drive shafts with CV joints for Commercial and Special vehicles, Industry, made by GKN Driveline. **a** 2 VL, high speed design, integrated length compensation; **b** RF + VL, wheel and differential side; **c** 2 RF with slip spline connection. Dates (mm):

	2 VL	RF + VL	2 RF
Mounting length	130–200	250–420	–
Compressed length	–	–	423–750
Plunge V	16–25	12–25	70–105
⌀ Tube/Shaft K	50 × 3	24–45	40 × 2
			70 × 4
⌀ D	90–188	88–160	94–165
No. of holes	–	–	4–8

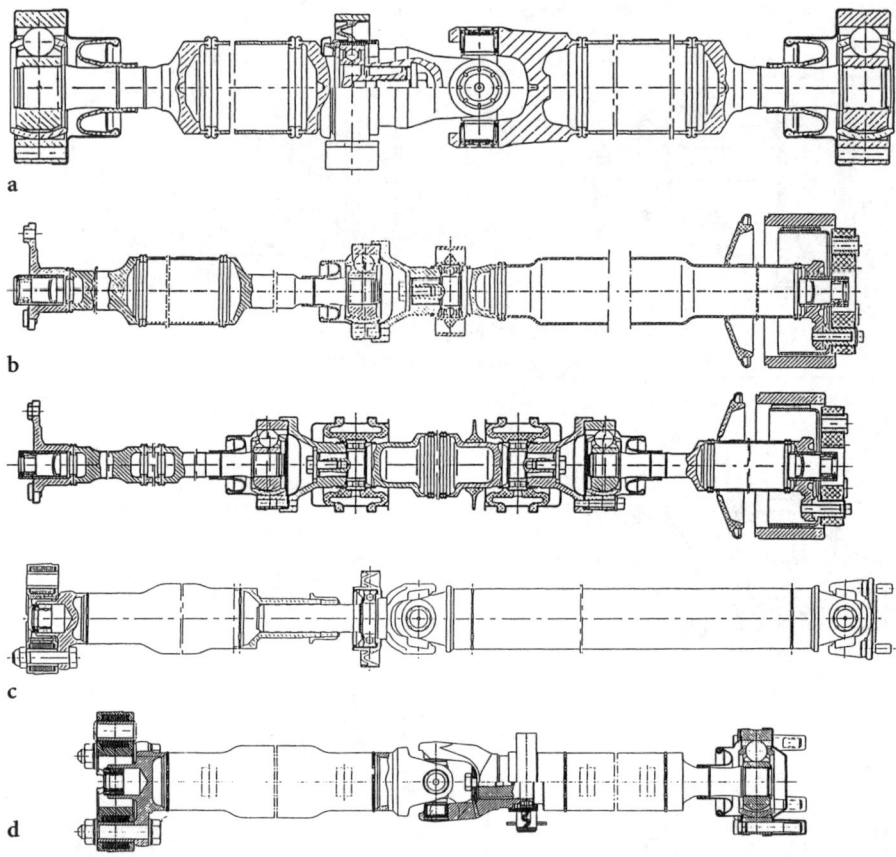

Fig. 5.71. Examples of two-part, mass produced propshafts. **a** Audi-Quattro, **b** VW Transporter T4, **c** BMW-E36/7 with manual gearbox, **d** BMW 5 series 1995. GKN Löbro design

5.5.4 Calculation Example of a Driveshaft with Ball Joints

A ball jointed driveshaft as shown in Fig. 5.72 is loaded by an electric motor with an output torque $M_x = 1000$ Nm ($f_{ST} = 1.0$ from Table 4.2). It comprises a fixed and a plunging joint with the following dimensions:

	UF 95 fixed joint	VL 107 plunging joint
Ball diameter d in mm	17.462	22.225
Effective radius R in mm	30.25	31.95
Track radius r_L in mm	$R/\cos 9°$	∞
Pressure angle α	45°	40°
Skew angle γ	0°	16°
Reciprocal conformity value ψ	1,04	1,04
Number of balls z	6	6

5.5 Ball Jointed Driveshafts

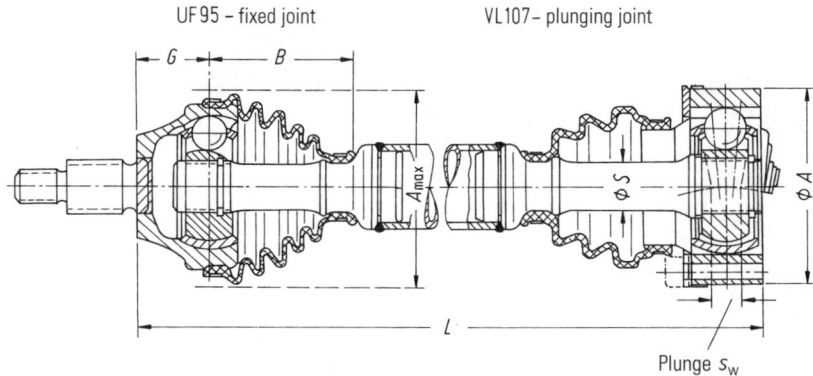

Fig. 5.72. Rzeppa ball jointed driveshaft for the calculation example 5.5.4. Löbro design (GKN Automotive AG)

The compressive force P on the individual balls is:

for the RF 95 joint, from (4.16a)

$$P = \frac{1000}{6 \cdot 0.03025 \cdot 0.7071} = 7792 \text{ N},$$

for the VL 107 joint, from (4.16b)

$$P = \frac{1000}{6 \cdot 0.03195 \cdot 0.6428 \cdot 0.9613} = 8442 \text{ N}.$$

The following have to be calculated:
a) the conformity $\varkappa_L$ along the track (Fig. 5.72)
b) the Hertzian coefficient cos τ,
c) the sum of the curvatures $\Sigma \varrho$
d) the elliptical coefficients μ, ν, $\mu\nu$ and $2K/\pi\mu$,
e) the axes of the contact ellipse $2a$ and $2b$,
f) the total deformation δ_0 at the inner and outer contact points,
g) the Hertzian stress p_0

1. Solution for UF 95 Fixed Joint (Fig. 5.73a)
a) the conformity $\varkappa_L$ along the track from (4.54),
inner:

$$\varkappa_L = \frac{17.462}{\dfrac{2\,(30.25 + 0)}{0.7071 \cdot 0.9877} - 17.462} = 0.2525$$

outer (Fig. 5.73b): [2]

$$\varkappa'_L = \frac{17.462}{\dfrac{-2\,(30.25 + 0)}{0.7071 \cdot 0.9877} + 17.462} = -0.1678$$

---

[2] The values for the outer contact points have a prime suffix, e.g. $\varkappa'_L$.

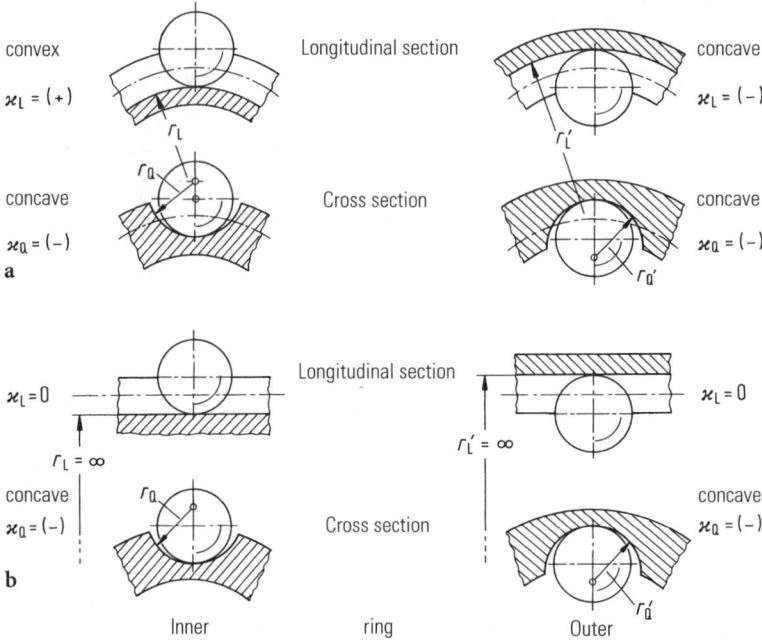

Fig. 5.73 a, b. Shapes of tracks in the ball joint. **a** Fixed joint, **b** plunging joint

b) The Hertzian coefficient $\cos \tau$ from (4.57),

inner:
$$\cos \tau = \frac{0.2525 + 0.9615}{2 + 0.2525 - 0.9615} = \frac{1.2135}{1.2905} = 0.9403,$$

outer:
$$\cos \tau' = \frac{-0.1678 - (-0.9615)}{2 - 0.1678 - 0.9615} = \frac{0.7934}{0.8707} = 0.9112,$$

c) The sum of the curvatures $\Sigma \varrho$ from (4.56),

inner:
$$\Sigma \varrho = \frac{2}{17.462}(2 + 0.2525 - 0.9615) = \frac{2}{17.462} 1.291 = 0.1479,$$

$$d\Sigma \varrho = 2.5820$$

outer:
$$\Sigma \varrho' = \frac{2}{17.462}(2 - 0.1678 - 0.9615) = \frac{2}{17.462} 0.8707 = 0.0997$$

$$d\Sigma \varrho' = 1.7410 \text{ durch Interpolieren oder aus [3.11],}$$

5.5 Ball Jointed Driveshafts

d) The elliptical coefficients μ, ν, $\mu\nu$ and $2K/\pi\mu$ in Table 3.2 follow through linear interpolation

inner:

$\cos\tau$	μ	ν	$\mu\nu$	$2K/\pi\mu$
0.9403	3.8334	0.4112	1.5762	0.6022

outer:

$\cos\tau$	μ	ν	$\mu\nu$	$2K/\pi\mu$
0.9112	3.2519	0.4485	1.4584	0.6615

e) The axes of the contact ellipse $2a$ and $2b$ from (3.45 and 3.46)

inner:
$$2a = \frac{4.72}{10^2} \, 3.8334 \underbrace{\sqrt[3]{\frac{7792}{0.1479}}}_{37.4881} = 6.7830 \text{ mm}$$

$$2b = \frac{4.72}{10^2} \, 0.4485 \cdot 37.4881 = 0.7936 \text{ mm},$$

outer:
$$2a = \frac{4.72}{10^2} \, 3.2519 \underbrace{\sqrt[3]{\frac{7792}{0.0997}}}_{42.7548} = 6.5624 \text{ mm}$$

$$2b = \frac{4.72}{10^2} \, 0.4485 \cdot 42.7548 = 0.9051 \text{ mm},$$

f) The total deformation δ_0 at the contact point, inner and outer, from (3.53) and (d) above,

inner:
$$\delta_0 = \frac{2.78}{10^4} \, 0.6022 \, \sqrt[3]{7792^2 \cdot 0.1479} = 0.0348 \text{ mm} \approx 35 \text{ μm}$$

outer:
$$\delta_0 = \frac{2.78}{10^4} \, 0.6615 \, \sqrt[3]{7792^2 \cdot 0.0997} = 0.0335 \text{ mm} \approx 34 \text{ μm},$$

g) The Hertzian stress p_0 from (3.54).

inner:
$$p_0 = \frac{858}{1.5762} \, \sqrt[3]{7792 \cdot 0.1479^2} = 3018 \text{ N/mm}^2$$

outer:
$$p_0 = \frac{858}{1.4584} \, \sqrt[3]{7792 \cdot 0.0997^2} = 2508 \text{ N/mm}^2.$$

2. Solution for VL 107 Plunging Joint (Fig. 5.73b)

a) $\varkappa_L = \dfrac{22.225}{\dfrac{2(31.95+\infty)}{0.7071 \cdot 1} - 22.225} = 0$

inner: $\varkappa_L = 0\ (\psi_L = \infty)$,
outer: $\varkappa_L = 0\ (\psi_L = \infty)$.

Hence the curvatures are the same inside and outside as are the contact ellipse, the deformation and the Hertzian stress.

b) $\cos \tau \dfrac{0 - 0.9615}{2 + 0 - 0.9615} = \dfrac{-0.9615}{1.0385} = 0.9259$

c) $\Sigma\varrho = \dfrac{2}{22.225}(2 + 0 - 0.9615) = \dfrac{2}{22.225} \cdot 1.0385 = 0.0935;\quad d\Sigma\varrho = 2.0770$

d) The elliptical coefficients in Table 3.3 can be read without interpolation

cos τ	μ	ν	μν	2K/πμ
0.9259	3.5065	0.4309	1.5109	0.6340

e) The axes of the contact ellipses in (3.45) and (3.46) are

$2a = \dfrac{4.72}{10^2} 3.5065 \sqrt[3]{\dfrac{\overbrace{8442}^{44.8619}}{0.0935}} = 7.4250\ \text{mm}$

$2b = \dfrac{4.72}{10^2} 0.4309 \cdot 44.8619 = 0.9124\ \text{mm}$,

f) $\delta_0 = \dfrac{2.78}{10^4} 0.6340 \sqrt[3]{8442^2 \cdot 0.0935} = 0.0332 \approx 33\ \mu\text{m}$,

g) $p_0 = \dfrac{858}{1.5109} \sqrt[3]{8442 \cdot 0.0935^2} = 2382\ \text{N/mm}^2$.

Discussion of the results: The very high conformity of the balls $\varkappa_Q = -0.9615$ ($\psi = 1.04$) in the cross-section of the track results in a long, narrow and curved contact ellipse (Fig. 4.32). The second condition of the Hertzian theory (Chap. 3) requires that the surfaces of the bodies touch only over very small areas. In 1928 Helmut Stellrecht [4.26, p. 8] extended the Hertzian limit derived for two touching balls [1.21, p. 167] $a \leqq 0.1r$ to cover the case of contact between a ball and a track,

$$\dfrac{\text{Contact area}}{\text{Surface area of ball}} = \dfrac{\pi a^2}{4\pi r^2} \leqq \dfrac{1}{4}\left(\dfrac{1}{10}\right)^2 \leqq \dfrac{1}{400}.$$

For the VL 107 plunging joint

$$\dfrac{\text{Contact area}}{\text{Surface area of ball}} = \dfrac{3.7 \cdot 0.45}{4\pi \cdot 22.225^2} = 0.00084 \leqq \dfrac{1}{1190}$$

Hence the new condition is satisfied.

5.5 Ball Jointed Driveshafts

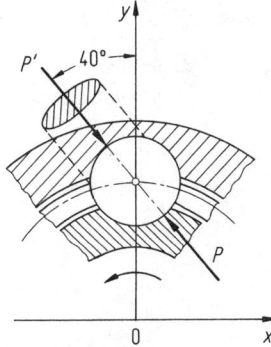

Fig. 5.74. Contact ellipse in the ball joint with very close conformity

But if the limit relates to the length of the smi-axis a, this is more than four times bigger than $0.1r$. Moreover, in Heinrich Hertz's 1882 analysis [1.21, p. 158] only small curvatures of the contact area are allowed. Since high conformities have proved very successful in practice, it would be a good idea to clarify this matter and to widen the theory to cover these conditions. Figure 5.74 also shows that the contact ellipse goes over the edges of the track. This should always be avoided, even in the case of radiussed edges. The contact becomes shorter and wider if a conformity which is not too close is chosen. The Hertzian stress can be increased for shorter duration loading up to 3200 N/mm².

5.5.5 Tripode Jointed Driveshafts Designs

The tripode jointed driveshaft system consists of the end connections, the joints and the intermediate shaft [5.1, 5.2]. For pode jointed driveshafts there is no separate plunging element because this function is performed by the plunging joint. When the tripode driveshaft is fitted in a front wheel drive passenger car, a fixed joint GE with a large articulation angle is needed outboard (Fig. 5.75a). For rear wheel drive the inboard and outboard conditions are similar, so both sides can be fitted with plunging joints GI.

The tripode plunging joint is characterised in the straight position by a particularly small plunging force because the compressive load is transmitted predominantly by rolling friction. Ball joints on the other hand need greater plunging forces for small articulation angles. The forces only become relatively greater than those for the tripode joint as the angle increases (Fig. 5.75). Hence the straight GI joint is particularly suitable for transmitting large torques with small plunging loads, e.g. for rail vehicles.

In the French high speed train (TGV), the electric motors are rigidly mounted in the bogie. The torque of 3000 Nm is transmitted to the sprung drive axles via a driveshaft with two Hooke's joints and an intermediate GI joint for length compensation (Fig. 5.77).

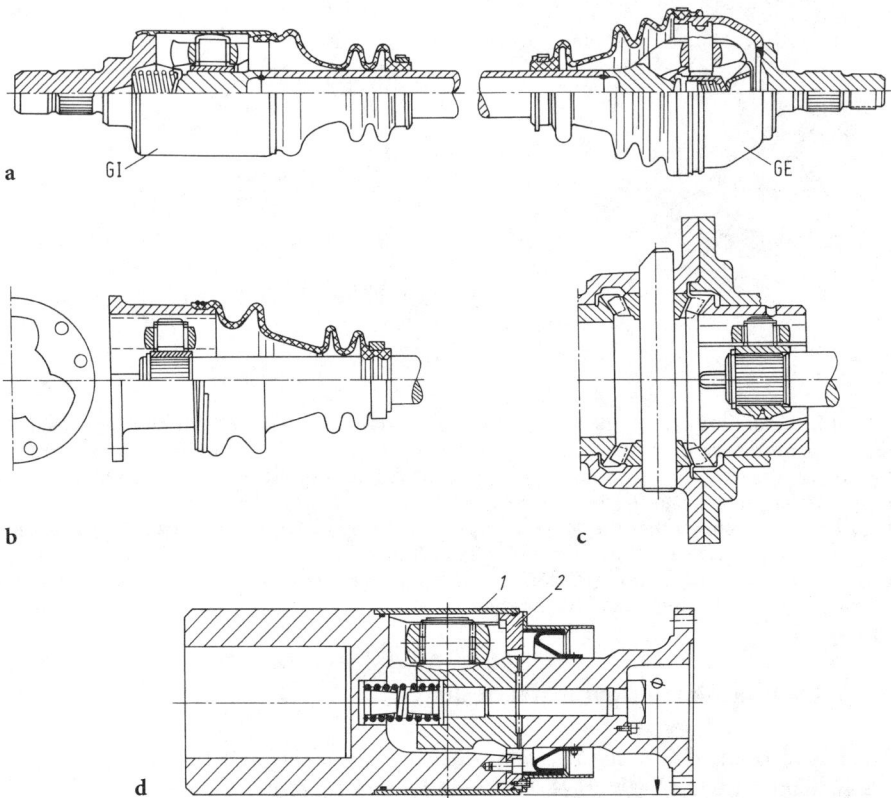

Fig. 5.75a–d. Applications of the tripode joint. **a** Halfshaft with GI and GE tripode joints for a front wheel drive passenger car. Made by Glaenzer-Spicer; **b** tripode plunging joint. Made by Citroën. The outer joint body is tulip with a closed a flange. Fixed by bolts on the flange; **c** tripode fitting. The side gear of the differential acts as the joint tulip. FIAT 127; **d** heavy duty tripode GI plunging joint for driving rolling mill stands, GWB design. The joint and flange diameters are the same. The middle part is floating and spring loaded. *1* Protection tube; *2* support ring

5.5.6 Calculation for the Tripode Jointed Driveshaft of a Passenger Car

The example is based on a passenger car with front engine, five speed gearbox, rear wheel drive, using the following operating data and assumptions:

- maximum engine output P_{eff} = 106 kW at n_M = 5500 rpm,
- maximum torque M_M = 217 Nm at n_M = 3000 rpm,
- GVW G = 19 228 N,
- permissible rear axle load G_R = 10 448 N
- height of centre of gravity, laden h = 0.55 m,
- wheelbase l = 2.80 m,
- static rolling radius R_{stat} = 0.288 m,
- dynamic rolling radius R_{dyn} = 0.310 m,

5.5 Ball Jointed Driveshafts

- gearbox ratios i_a

1st	2nd	3rd	4th	5th
3.862	2.183	1.445	1.000	0.844

  axle ratio $i_A = 3.70$,
- mean articulation angle $\beta = 7°$.

Starting takes place at maximum torque with the coefficient of friction $\mu = 1$ and the shock factor $f_{ST} = 1.2$. The vehicle is driven at 2/3 of the maximum engine torque and constant speed in the individual gears. Gears are used, 1, 5, 27, 40 and 27% of the time respectively (Fig. 5.76).

The joints should achieve a life of at least 100 000 km. The size of tripode joint required is obtained from the stresses arising from the starting torque M_A and the adhesion torque M_H. The durability of the selected joint is then checked.

Solution

The starting torque from (4.30)

$$M_A = 1.2 \cdot 0.5 \cdot 217 \cdot 3.70 = 1860 \text{ Nm/shaft}.$$

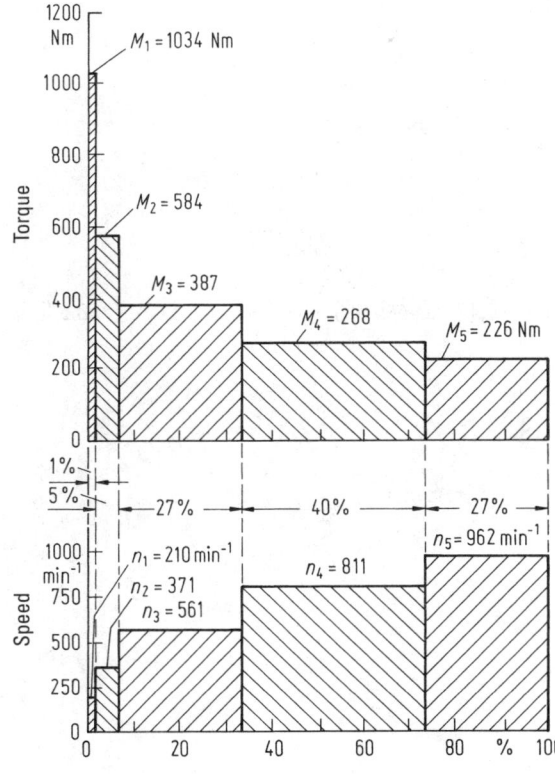

Fig. 5.76. Block programme for calculating the life of a driveshaft for a passenger car with five speed gearbox and rear wheel drive. Such drive programmes show the varying, irregular loading of a vehicle

Table 5.9. Durability values for the selected tripode joint on rear drive

Line	Eq. No		Gear				
			1st	2nd	3rd	4th	5th
1		a_x	1%	5%	27%	40%	27%
2		i_x	14.29	8.08	5.35	3.70	3.12
3	(4.70)	n_x	210	371	561	811	962
4	(4.71)	V_x	24.5	43.4	65.6	94.8	112.4
5	(4.72)	M_x	1034	584	387	268	226
6	(4.73)	L_h	143.2	450.1	1022.9	2130.6	2995.2

The adhesion torque (without differential lock) from (4.33a)

$$M_H = 1.2 \, \frac{1065 \cdot 9.81 \cdot 2.80}{2 \, (2.80 - 1 \cdot 0.55)} \, 0.288 = 2247 \text{ Nm/shaft}.$$

The AAR joint (M_N = 2600 Nm, M_d = 450 Nm) is selected from Fig. 4.89. In order to check this joint for durability, the values for the individual gears are listed in Table 5.9.

The life in hours from (1.20) is obtained

$$\frac{1}{L_h} = \frac{0.01}{143.3} + \frac{0.05}{450.1} + \frac{0.27}{1022.9} + \frac{0.40}{2130.6} + \frac{0.27}{2995.2}$$
$$= 10^{-5} \cdot (6.9783 + 11.108 + 26.395 + 18.774 + 9.0144) = 10^{-5} \cdot 72.270$$

L_h = 1383.7 hrs.

The average driving speed from (4.71)

$$V_S = 0.01 \cdot 24.5 + 0.05 \cdot 43.4 + 0.27 \cdot 65.6 + 0.40 \cdot 94.8 + 0.27 \cdot 112.4 = 88.3 \text{ kph}$$

The calculated life in kilometres from (4.73)

$$L_S = 1383.7 \cdot 88.3 = 122\,318 \text{ km}.$$

The AAR joint meets the durability requirements.

5.6 Driveshafts in railway carriages

5.6.1 Constant velocity joints

The tripode plunging joint is characterised in the straight position by a particularly small plunging force because the compressive load is transmitted predominantly by rolling friction. Ball joints on the other hand need greater plunging forces for small articulation angles. The forces only become relatively greater than those for the tripode joint as the angle increases (Fig. 5.78). Hence the straight GI joint is particularly suitable for transmitting large torques with small plunging loads, e.g. for rail vehicles.

In the French high speed train (TGV), the electric motors are rigidly mounted in the bogie. The torque of 3000 Nm is transmitted to the sprung drive axles via a driveshaft with two Hooke's joints and an intermediate GI joint for length compensation (Fig. 5.77).

In 1998 GKN Löbro GmbH started to supplying UF/VL double joints in sizes 32/32, 160/160 or 16000/13200, for the individual drive units of the wheels of a driverless underground railway; +/– 7° articulation angles were required.

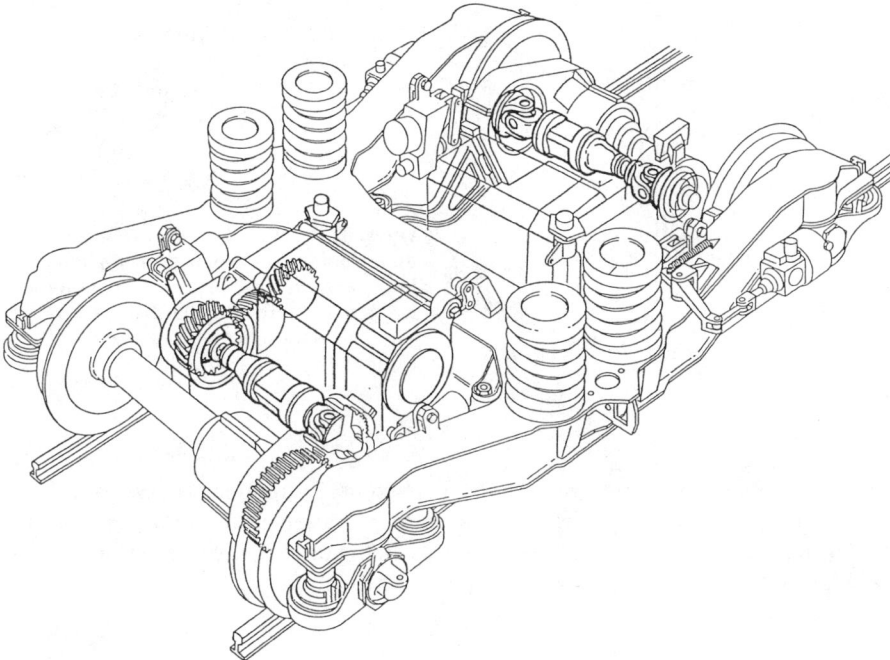

Fig. 5.77. Driveshafts in the bogie of the high speed train (TGV), built by Alstom, Hooke's jointed drive-shaft with a GI joint for plunge, built by GWB/Glaenzer-Spicer (GKN Automotive AG)

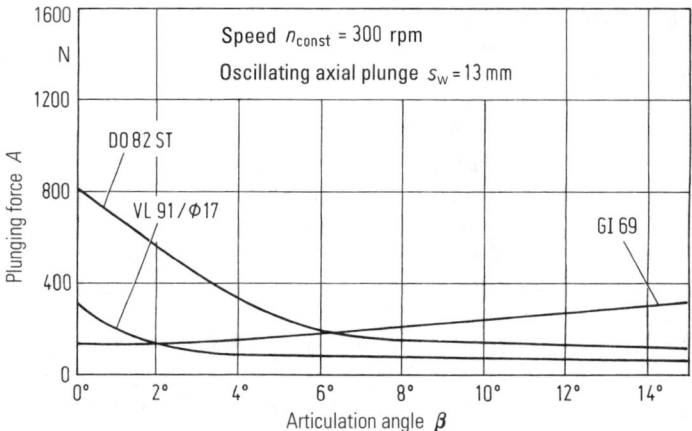

Fig. 5.78. Comparison of the axial plunging forces for CV joints at 100 Nm derived by Hardy-Spicer (GKN Automotive). For $\beta = 6°$ the plunging forces of DO and GI joints are similar

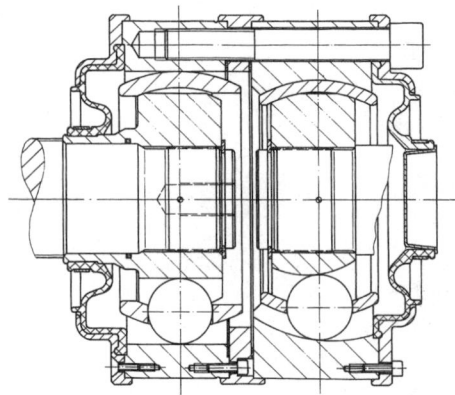

Fig. 5.79. UF/VL 32 constant velocity double joint with articulation angle $\beta \pm 7°$, in the driveline between electric motor and wheel drive of the Bombardier/Vevey VAL 208 driverless metro rail vehicles (Lille 1998). GKN Löbro design

Large constant velocity joints up to 23,000 Nm (fixed joints) and 67,500 Nm (plunging joints) have partly replaced Hooke's joints in stationary heavy machinery and in commercial and special vehicles where vibration-free running is required. For example constant velocity driveshafts are being used in the propulsion systems of ships.

5.7 Ball jointed driveshafts in industrial use and special vehicles

Constant velocity driveshafts are ideally suited as well in the industrial applications where unequal joint angles are encountered and low vibration generation is needed. Examples of applications are:

Filling Machines	–	Printing Machines
Soil Compactors	–	Marine Propulsion
Dynamometers	–	Mill Drives
Pump Drives	–	Machine Tools
Glass Manufacturing	–	Textile Machines
Commercial Vehicles	–	Railroad Equipment
Wind Turbines	–	Military Vehicles

Based on the physics of a cardan joint, cardanshafts have the inherent disadvantage of generating vibrations. Even with an aligned cardan shaft due to the non-uniform motion of the center section a variety of unwanted forces and vibrations is generated. Furthermore the sinusoidal radial forces generated by the cardan joint have a negative influence on the bearings.

GKN constant velocity driveshafts solve the vibration problems generated by cardanshafts and significantly reduce the radial and axial forces to the connecting bearings. They have no torsional or inertial excitations which are inherent in cardanshafts. The smooth torque transmitted from a GKN constant velocity driveshaft occurs even when the operating angles are unequal.

Main advantages of GKN constant velocity driveshafts in brief

- no oscillations and vibrations (low noise)
- easy installation, no alignment necessary for fit-in
- high angle capability
- maintenance-free
- no additional slip device necessary in the intermediate shaft, plunge achieved within the joint
- soft running plunge by rolling friction of the balls, even when torque is applied
- axial vibration can be easily absorbed as compared to slide spline arrangement
- reduced loads on bearings and on the supporting structure
- high speed applications
- compact design, small rotation diameter, short length
- proven in thousands of applications around the world.

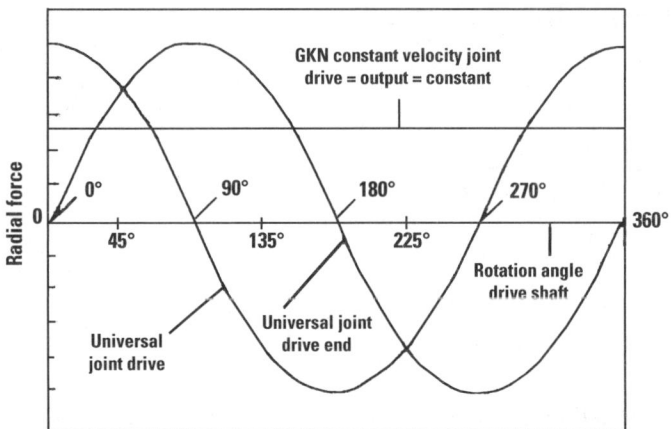

Fig. 5.80. Generated radial force at the universal joint and the constant velocity joint

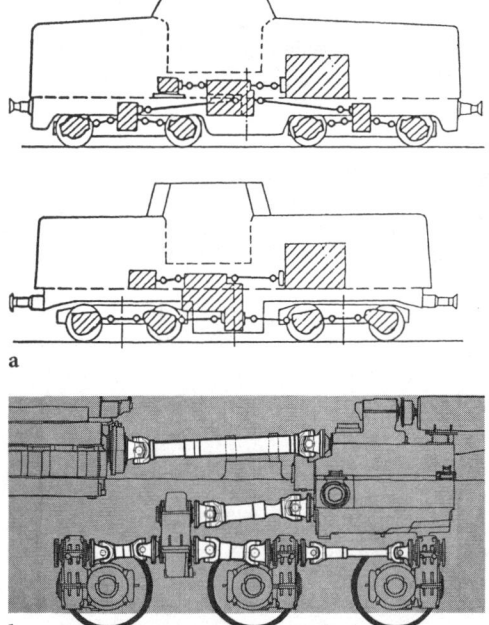

Fig. 5.81 a, b. Driveshafts in locomotive train engine. **a** 4-axled, **b** 6-axled. GWB design

5.8 Hooke's jointed high speed driveshafts

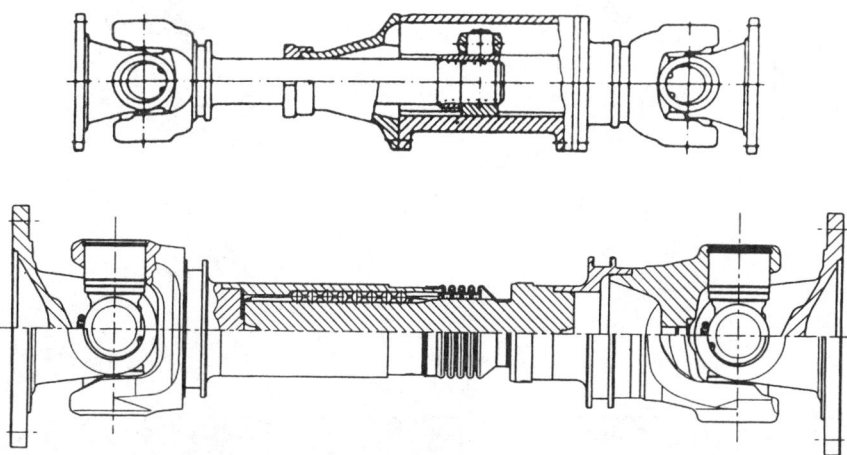

Fig. 5.82. Driveshafts for rail-cars. **a** Hooke's jointed driveshaft with GI joint for plunge, built by Glaenzer Spicer/GKN Driveline; **b** Hooke's jointed drive-shaft with ball spline, built by Spicer GWB Essen (DE-PS 19952245 A1).

5.8 Hooke's jointed high speed driveshafts

The Hooke's jointed driveshaft is a drive element that is well suited for transmitting high torques with high radial and axial displacements. In the case of heavy duty drives, such as in rail vehicles, cold rolling mill stands or test rigs, they must be carefully designed. Angle compensation between the input and output and adequate plunge must to be provided.

The dynamic and the alternating loads are critical for Hooke's jointed drive shafts, between the gearbox and axle or between the wheelsets of rail vehicles. $s_0 >$ = 1.5 must be assumed as the safety factor for a service life of at least 50,000 operating hours. The 1st and 2nd order critical bending speeds must be determined, and the radial and axial loads of the connection bearings known. Driveshafts in electrically driven power cars reach over 4,000 rpm.

High centrifugal forces arise at high speeds of the Hooke's jointed driveshaft. They expand the joint yoke which increases the bearing play on the trunnion cross (Fig. 5.83a). This deformation impairs the balancing quality G of the driveshaft because the connected middle part runs eccentrically due to axial play in the trunnion cross bearing. The yoke expansion is a function of the rotational diameter of the joint (Fig. 5.83b). The same happens if the centring via the front splines no longer works or the profile "kinks" under low torque. Measures to combat expansion of the yoke are a yoke bridge (Fig. 5.84a, b) or a bearing cap that goes right through, if there is a split bearing lug. A self-centring trunnion cross is also possible.

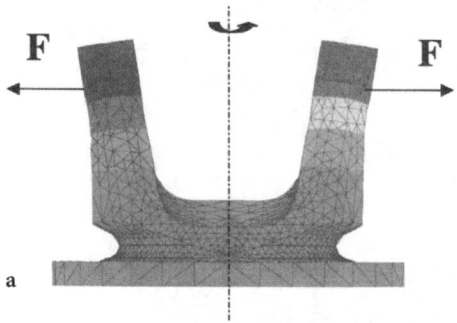
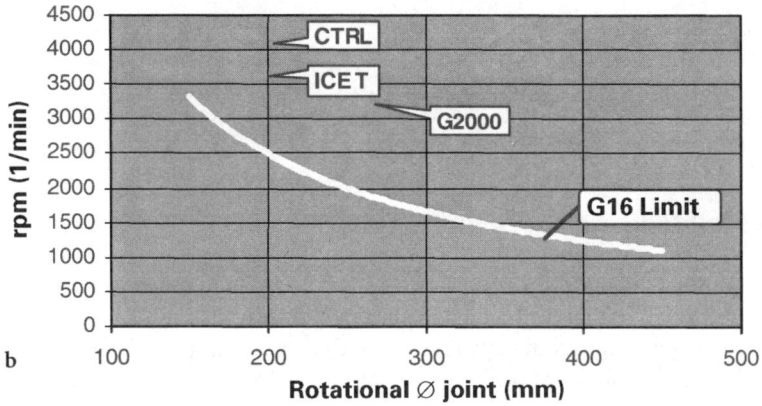

Fig. 5.83 a, b. Reaction of the centrifugal force on the joint. **a** Extension of the driver fork at 4000 rpm results in 0,11 mm per side of the GWB-type 390.65; **b** Cinetic energy values G dependent on rpm and joint's rotation diameter. Investigated by SPICER GWB Essen

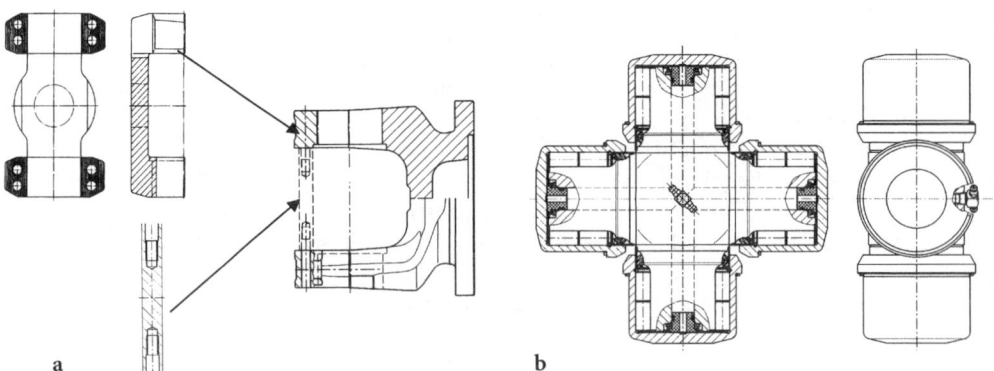

Fig. 5.84a,b. Measures against the extension of the joint fork under centrifugal force, patented for SPICER GWB Essen. **a** fork bridge or transmitted bearing cover (DE-PS 3617459.9); **b** closed self centered cross pin (DE-PS 10036203.6)

5.8 Hooke's jointed high speed driveshafts

A standard driveshaft at 3200 rpm achieves an unbalance G 24, a high speed shaft on the other hand achieves G 12 (Fig. 5.85b).

The design of a high speed shaft requires:

- rigid yokes to combat expansion
- high profile overlap against "kinking"
- bulk lubrication and a double lip seal
- use of a hollow shaft for length compensation. Voith Transmit GmbH supplies a driveshaft with the particularly large length compensation of 530 mm for offset rolling mills.
- use of balancing rings

Table 5.10. Comparison of a standard and a high speed drive-shaft (Fig. 5.82, 5.85), built by SPICER GWB Essen

GWB shaft	Lok G 2000 390.65	Rail-car ICE-T 688.65
Joint's Rotation-∅ (mm)	265	204
Flange-∅ (mm)	315	225
Mounting length (mm)	2270	1870
Length compensation (mm)	130	160
Weight (kg)	314	95
Speed$_{max}$ (rpm)	3150	3600
Torque dynamic	36000	11000
(Nm) static	90000	35000

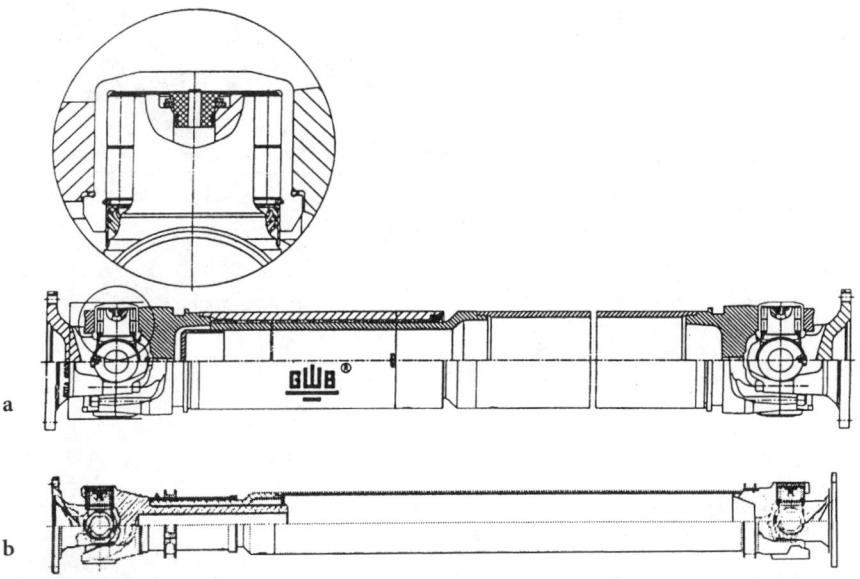

Fig. 5.85. High-speed drive shafts, built by SPICER GWB Essen. **a** with cross pin and grease chamber; **b** for the German bend over rail-car ICE-T (Alstom)

5.9 Design and Configuration Guidelines to Optimise the Drivetrain

Requirements			Recommendations			Illustrated in Fig.
Static load carrying capacity	< 1500 Nm		Hooke's and CV joints			4.10, 5.8, 5.9
	> 15000 Nm		Hooke's joints with Rilsan coated plunging elements			
Lubricated for life			Hooke's and CV joints			5.5, 5.11
High durability (low static load)			Hooke's joints with roller bearings			4.24–4.28
High static load (lower durability)			CV joints			
Fixed layout	W- or Z-configuration		Well aligned Hooke's joints			4.24–4.25
	Unequal articulation angles		CV joints without alignment			Tab. 4.5
Flexible layout			CV joints			4.24–4.28
Oscillating and vibrating longitudinal movement under torque			Articulation angle			
			β = 0 to 4°	GI joints		4.87–5.76
			β = 2° to 20°	VL joints		4.69–4.71
Small axial plunge	straight		Bouchard-Enke joint with needle bearings			1.25, 1.36
	angled		β = 0 to 4°	GI joints		5.75
			β = 2° to 20°	VL joints		4.69–4.71
			β = 10° to 25°	DO joints		4.66, 4.67
High speed (5000 to 12000 rpm)			VL and DO joints			4.66–4.71
High refinement	β = 0 to 4°		GI	CV joints:		4.85
	b = 2 to 15°		VL	zero clearance and		4.69–4.71
	b = 0 to 25°		VL 4 + 4	with slight preload		4.69
				Hooke's joints: equal flange, swing and shaft tube diameter		5.2–5.4
Boot seals			Speed n_{max} rpm	Articulation angle β_{max}	Type of boot	
Halfshaft	Car	FWD (outboard)	50	50°	Unsupported	5.63, 5.66
		RWD and FWD (inboard)	2500	20°		
Propshaft	Car		6000	10°		5.67–5.71
	Truck		2500	25°	Supported	5.25, 5.26
	Machinery		12000	5°		5.25–5.27
	Articulated vehicle		100 to 5000	50° to 5°	Stationary	5.23, 5.27
Operation in water			CV joints, with oil-tight boot seal			5.64, 5.65

5.9.1 Example of a Calculation for the Driveshafts of a Four Wheel Drive Passenger Car

Figure 5.86 shows the passenger car, with front engine and five speed gearbox, and gives the working data used in the example. The ball and Hooke's jointed driveshafts needed for front wheel, rear wheel and four wheel drive variants of the car are specified and their service lives are calculated. The joints are based on Löbro and GWB designs.

Working Data:
maximum engine power	P_{eff}	100 kW at 5900 rpm
maximum torque	M_M	176 Nm at 4500 rpm
GVW	G	16187 N
permissible front axle load	G_F	7279 N
permissible backaxle load	G_R	8908 N
axle ratio	i_A	4.111
height of centre of gravity, laden	h	0.5 m
static rolling radius	R_{stat}	0.296 m
dynamic rolling radius	R_{dyn}	0.301 m
torque distribution		
front : rear		36 : 64
Value of the differential lock		
front		100%
rear		100%

Gear Ratios	1	2	3	4	5
Gearbox i_s	3.600	2.125	1.458	1.071	0.829

Mean articulation angle β: outboard   7°, $A_x = 0.865$
 inboard   4°, $A_x = 0.926$
 propshaft 4°

Assumptions: coefficient of friction $\mu = 1$; shock factor $f_{ST} = 1.2$; load carrying factor $k_t = 1.33$.

Assumptions:
Starting occurs at maximum torque with the coefficient of friction $\mu = 1$ and the shock factor $f_{ST} = 1.2$. The car is driven at 2/3 of the maximum engine torque and constant speed in the individual gears. As shown in Table 4.14/15 the utilisation of gears 1 to 5 are: 1, 6, 18, 30 and 45%. The car should have a life of at least 100 000 km.

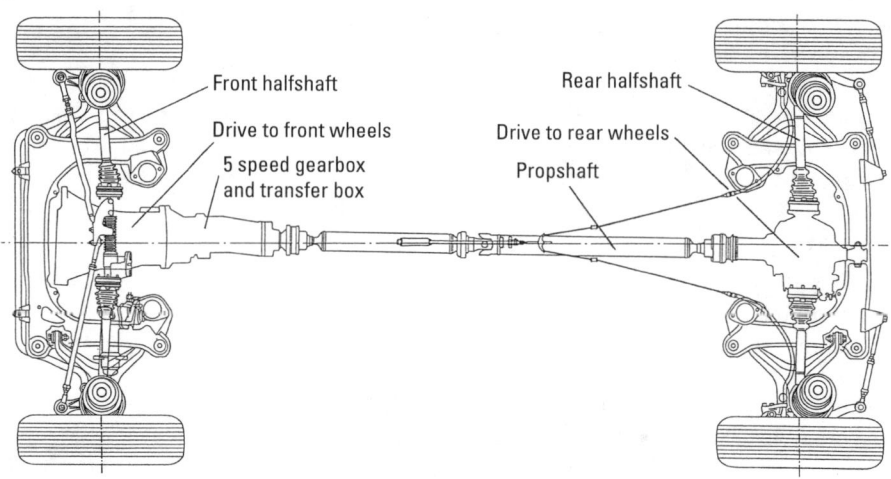

Fig. 5.86. Passenger car with four wheel drive and 3 differentials. 11 joints are fitted altogether: fixed joints, 6 VL plunging joints and 1 Hooke's joint. Audi design

Available CVJ Sizes

Front axle
Fixed joints on the wheel side M_0 (Nm) M_d (Nm)

UF 1300	1300	200
UF 2600	2600	360
UF 3300	3300	460

Plunging joints on the differential side

VL 1500	1500	200
VL 2600	2600	300
VL 3700	3700	522

Rear axle

VL 2300	2300	267
VL 2600	2600	302
VL 3700	3700	522

Longitudinal shaft

    VL-joints counted up above

5.9 Design and Configuration Guidelines to Optimise the Drivetrain

Solution:
- Calculate the stresses from the starting torque M_A and the adhesion torque M_H and select joint sizes from the smaller of the two torques, see Section 4.2.6 (Equs. 4.30 to 4.33) and Table 5.26
- Check the service life

1. *Front wheel drive*

With the equations from Sect. 4.1.3, the life L_{hx} is calculated for the individual gears in Table 5.9, 5.11.

Calculation of L_{hx} for line 6

$$L_{h1} = \frac{25\,339}{304^{0.577}} \left(\frac{0.865 \cdot 200}{868} \right)^3 = 7.4 \text{ hrs}$$

$$L_{h2} = \frac{25\,339}{515^{0.577}} \left(\frac{0.865 \cdot 200}{512.7} \right)^3 = 26.5 \text{ hrs}$$

$$L_{h3} = \frac{25\,339}{751^{0.577}} \left(\frac{0.865 \cdot 200}{351.4} \right)^3 = 66.3 \text{ hrs}$$

$$L_{h4} = \frac{470\,756}{1023} \left(\frac{0.865 \cdot 200}{258.1} \right)^3 = 138.6 \text{ hrs}$$

$$L_{h5} = \frac{470\,756}{1320} \left(\frac{0.865 \cdot 200}{200.1} \right)^3 = 230.5 \text{ hrs}.$$

Service life in hours as in (1.20)

$$\frac{1}{L_h} = \frac{0.01}{7.4} + \frac{0.06}{26.5} + \frac{0.18}{66.5} + \frac{0.30}{138.6} + \frac{0.45}{230.5}$$

$$= 10^{-3} (1.351 + 2.264 + 2.707 + 2.165 + 1.952) = 10^{-3} \cdot 10.44.$$

Hence

$$L_h = \frac{10^3}{10.44} = 95.79 \text{ h}.$$

Table 5.11. Durability values for front wheel drive with UF 1300

Line	Equ. No	Gear					
			1st	2nd	3rd	4th	5th
1		a_x	1%	6%	18%	30%	45%
2		i_x	14.80	8.74	5.99	4.40	3.41
3	(4.70)	n_x	304	515	751	1023	1320
4	(4.71)	V_x	34.4	58.2	84.9	115.7	149.3
5	(4.72)	M_x	868	512.7	351.4	258.1	200.1
6	(4.73)	L_{hx}	7.4	26.5	66.3	138.6	230.5

Table 5.12. Calculation of the starting and adhesion torques in the calculation example (Section 5.5.4)

Starting torque M_A from (4.30)	Adhesion torque M_H from (4.33)	Joint size required	
		Starting	Durability
1. Front wheel drive $M_A = 1/2 \cdot 176 \cdot 3.6 \cdot 4.11 = 1302$ Nm[a] 2604 Nm[b]	$M_{HV} = 1.2 \dfrac{742.5 \cdot 9.81 \cdot 2.525}{2(2.525 + 1 \cdot 0.5)} \cdot 0.3 = 1094$ Nm/shaft	UF 1100i outboard VL 1500 inboard	UF 3300 VL 3700
2. Rear wheel drive $M_A = 1302$ Nm/shaft[a] 2604 Nm/shaft[b]	$M_{HH} = 1.2 \dfrac{907.5 \cdot 9.81 \cdot 2.525}{2(2.525 - 1 \cdot 0.5)} \cdot 0.3 = 1998$ Nm/shaft	VL 2300 outboard and inboard	VL 3700
Longitudinal shaft $M_A = 1/1 \cdot 176 \cdot 3.6 = 633.6$ Nm	$M_H = 1.2 \dfrac{907.5 \cdot 9.81}{4.11} \cdot 0.3 = 780$ Nm/shaft	VL 2300 GWB C2015	VL 2600 GWB C2015
3. Four wheel drive Front axle $M_A = 0.36/2 \cdot 176 \cdot 3.6 \cdot 4.11$ $= 0.36 \cdot 1302 = 468.7$ Nm[a] 937.5 Nm[b]	$M_{HV} = 1.2 \dfrac{742.5 \cdot 9.81 \cdot 2.525}{2(2.525 + 1 \cdot 0.5)} \cdot 0.3 = 1094$ Nm/shaft	UF 1300i outboard VL 1500 inboard	UF 1300 VL 1300
Rear axle $M_A = 0.64/2 \cdot 176 \cdot 3.6 \cdot 4.11$ $M_A = 0.64 \cdot 1302 = 833$ Nm/shaft[a] 1666 Nm/shaft[b]	$M_{HH} = 1.2 \dfrac{907.5 \cdot 9.81 \cdot 2.525}{2(2.525 - 1 \cdot 0.5)} \cdot 0.3 = 1998$ Nm/shaft	VL 2300 outboard and inboard	VL 2300
Longitudinal shaft $M_A = 0.64/1 \cdot 176 \cdot 3.6 = 405.5$ Nm	$M_H = 1.2 \dfrac{907.5 \cdot 9.81}{4.11} \cdot 0.3 = 780$ Nm	VL 1500 GWB C2015	VL 1500 GWB C2015

[a] Without differential lock.
[b] With 100% differential lock.

5.9 Design and Configuration Guidelines to Optimise the Drivetrain

The mean driving speed V_m from (4.63)

$$V_m = 0.01 \cdot 34.4 + 0.06 \cdot 58.2 + 0.18 \cdot 84.9 + 0.30 \cdot 115.7 + 0.45 \cdot 149.3$$
$$= 121.0 \text{ km/h}$$

The calculated life in kilometres from (4.73a)

$$L_{s1} = 95.79 \cdot 121 = 11591 \text{ km.}$$

The UF 1300 joint, chossen from the starting torque requirements, does not achieve the required durability. The next largest joints must therefore be investigated.

The durability L_{s1} can be calculated for the same speeds and articulation angles, using the 3rd power of the torque ratio (4.73, a, b)

$$L_{s2} = L_{s1} \left(\frac{M_2}{M_1}\right)^3.$$

After this the distance durability for the joints is

$$\text{UF 2600} \quad L_{s2} = 11591 \left(\frac{360}{200}\right)^3 = 67599 \text{ km,}$$

$$\text{UF 3300} \quad L_{s2} = 11591 \left(\frac{460}{200}\right)^3 = 141028 \text{ km.}$$

Only the joint UF 3300 ($M_0 = 3300$ Nm, $M_d = 460$ Nm) fulfills the required life durability.

The VL 3700 joint ($M_0 = 3700$ Nm, $M_d = 3700$ Nm) is the selected plunging joint; it can only run at an average articulation angle $\beta_m = 4°$ but has a high M_d. Thus, for the VL 3700 joint the durability from (4.73) is

$$L_{s2} = 11591 \left(\frac{0.926 \cdot 522}{0.865 \cdot 200}\right)^3 = 252729 \text{ km.}$$

2. *Rear wheel drive*

Ball joints for the halfshafts.

The VL 2300 joint is chosen as shown in Table 5.10. The only differences compared with Table 5.9 are line 6 and the value $A_x = 0.926$ from the function of the articulation angle:

Line	Equ. No		Gear				
			1st	2nd	3rd	4th	5th
6	4.73a, b	L_{hx}	21.7	77.4	193.4	404.6	672.7

The life in hours from (1.20)

$$\frac{1}{L_h} = \frac{0.01}{21.7} + \frac{0.06}{77.4} + \frac{0.18}{193.4} + \frac{0.30}{404.6} + \frac{0.45}{672.7}$$
$$= 10^{-4}(4.608 + 7.752 + 9.307 + 7.415 + 6.689) = 10^{-4} \cdot 35.77.$$

$$L_h = \frac{10^4}{35.77} = 279.56 \text{ hrs}.$$

The average driving speed $V_m = 121$ kph remains unchanged.
The life in kilometers calculated from (4.73a, b)

$$L_{s1} = 279.56 \cdot 121 = 33\,827 \text{ km}.$$

The VL 2300 joint is insufficiently durable. The next bigger joints are therefore considered:

VL 2600

$$L_{s2} = 33\,827 \left(\frac{302}{267}\right)^3 = 48\,950 \text{ km},$$

VL 3700

$$L_{s2} = 33\,827 \left(\frac{522}{267}\right)^3 = 252\,779 \text{ km}.$$

Only a driveshaft with two VL 3700 plunging joints has the required durability.

Ball joints in the longitudinal shaft to the rear axle.

The whole engine torque, multiplied by the gear ratio, is transmitted by the propshaft. In addition its speed is 411% higher than that of the halfshafts. The VL 2300 joint ($M_0 = 2300$ Nm, $M_d = 267$ Nm) is selected on the basis of the starting torque (Table 5.13 and Fig. 4.69)

$$\frac{1}{L_h} = \frac{0.01}{75.5} + \frac{0.06}{215.8} + \frac{0.18}{459.3} + \frac{0.30}{855.8} + \frac{0.45}{1420.2}$$
$$= 10^{-4}(1.325 + 2.780 + 3.919 + 3.169 + 2.817) = 10^{-4} \cdot 14.698,$$

$$L_h = \frac{10^4}{14.698} = 680 \text{ hrs}.$$

The life in kilometres from (4.73)

$$L_s = 680 \cdot 121 = 82\,280 \text{ km}.$$

The VL 2300 joint does not have the required durability (4.60) shows that the next size, VL 2600 ($M_0 = 2600$ Nm, $M_d = 302$ Nm) is satisfactory

$$L_{s2} = 82\,280 \left(\frac{302}{267}\right)^3 = 119\,064 \text{ km}.$$

5.9 Design and Configuration Guidelines to Optimise the Drivetrain

Hooke's joint in the propshaft to the rear axle.

Joint size GWB C 2015 from Fig. 5.26 was checked for durability. From (4.28) one obtains line 6 of Table 5.9

Line	Eq. No	Gear					
			1st	2nd	3rd	4th	5th
6	(4.28)	L_{hx}	327.2	1113.8	2688.6	5557.4	10035.4

The life in hours from (1.20)

$$\frac{1}{L_h} = \frac{0.01}{327.2} + \frac{0.06}{1113.8} + \frac{0.18}{2688.6} + \frac{0.30}{5557.4} + \frac{0.45}{10\,035.4}$$

$$= 10^{-5}\,(3.056 + 5.387 + 6.695 + 5.398 + 4.484) = 10^{-5} \cdot 25.020$$

$$L_h = 3996.8 \text{ h}$$

The life in kilometres from (4.70)

$$L_{s1} = 3996.8 \cdot 121 = 483613 \text{ km}$$

The durability requirement is fulfilled by the GWB C 2015 joint.

Table 5.13. Values for the life of the propshaft for rear wheel drive

Line	Equ. No		Gear				
			1st	2nd	3rd	4th	5th
1		a_x	1%	6%	18%	30%	45%
2		i_x	3.6	2.13	1.46	1.07	0.83
3	4.70	n_x	1250	2113	3082	4206	5422
4	4.71	V_x	34.4	58.2	84.9	115.7	149.3
5	4.72	M_x	422.4	249.9	171.9	125.5	97.4
6	4.73a	L_{hx}	75.5	215.8	459.3	855.8	1420.2

3. *Four wheel drive*

Ball joints for the front halfshafts.

Lines 5 and 6 alter in Table 5.14. The input torque at the front axle is only 36% of the previous values of line 5.

Line	Equ. No		Gear				
			1st	2nd	3rd	4th	5th
5	(4.72)	M_x	312.6	1.84.6	126.5	92.9	72.0
6	(4.73)	L_{hx}	158.6	568.2	1420.4	2971.7	4947.6

The life in hours from (1.24)

$$\frac{1}{L_h} = \frac{0.01}{158.6} + \frac{0.06}{568.2} + \frac{0.18}{1461.2} + \frac{0.30}{2971.7} + \frac{0.45}{4947.2}$$

$$= 10^{-4}(0.631 + 1.056 + 1.267 + 1.232 + 0.909) = 10^{-4} \cdot 5.095$$

$$L_h = \frac{10^4}{5.095} = 1963 \text{ hrs}.$$

The life in kilometres from (4.74) and the same $V_m = 121$ kph

$$L_s = 1963 \cdot 121 = 237\,523 \text{ km}.$$

The UF 1300 joint ($M_0 = 1300$ Nm, $M_d = 200$ Nm) from Table 4.12 has the required durability. If the drive to the rear wheels could be disconnected, the resulting higher loading of the front drivetrain would have to be considered.

The life in kilometres of the inboard VL 2600 joint is greater in the ratio $\left(\frac{A_{4°}}{A_{7°}}\right)^3$, i.e.

$$L_{\text{inboard}} = L_{\text{outboard}} \left(\frac{0.925}{0.865}\right)^3 = 237\,523 \cdot 1.223 = 290\,457 \text{ km}.$$

The torques here are only 64% of the values in line 5 in Table 5.13

Line	Equ. No		Gear				
			1st	2nd	3rd	4th	5th
5	(4.72)	M_x	555.7	328.2	224.9	165.2	128.0
6	(4.73)	L_{hx}	82.5	295.1	737.8	1542.6	2570.2

The life in hours from (1.20)

$$\frac{1}{L_h} = \frac{0.01}{82.2} + \frac{0.06}{294.2} + \frac{0.18}{735.4} + \frac{0.30}{1537.6} + \frac{0.45}{2561.8}$$

$$= 10^{-4}(1.217 + 2.039 + 2.448 + 1.951 + 1.757) = 10^{-4} \cdot 9.412,$$

$$L_h = \frac{10^4}{9.412} = 1071.9 \text{ hrs}.$$

The life in kilometres from (4.73/74)

$$L_s = 1071 \cdot 121 = 129\,700 \text{ km}.$$

The VL 2300 joint ($M_0 = 2300$ Nm. $M_d = 267$ Nm) from Fig. 4.74 sufficiently durable. For a part-time front wheel drive the greater loading on the rear drivetrain would have to be taken into account.

5.9 Design and Configuration Guidelines to Optimise the Drivetrain

For a part-time front or rear wheel drive, Helmut Stellrecht's findings of 1928 apply [4.26], namely that the durability changes with the third power of the torque. Hence in the case of a passenger car which is subsequently fitted with a more powerful engine, the durability of the joints often becomes inadequate.

Ball joints in the longitudinal shaft.

For four wheel drive, the propshaft to the rear axle is only subjected to 64% of the engine torque. For this reason the VL 1500 joint ($M_0 = 1500$ Nm, $M_d = 200$ Nm) was chosen in Fig. 4.66. Because of this lines 5 and 6 in Table 5.14 change:

Line	Equ. No	Gear					
			1st	2nd	3rd	4th	5th
5	(4.72)	M_x	270.3	159.9	109.6	80.3	62.3
6	(4.28)	L_{hx}	121.1	248.1	737.0	1373.1	2280.8

The life in hours from (1.20)

$$\frac{1}{L_h} = \frac{0.01}{120.7} + \frac{0.06}{345.0} + \frac{0.18}{734.6} + \frac{0.30}{1368.7} + \frac{0.45}{2273.5}$$
$$= 10^{-4}(0.828 + 1.739 + 2.450 + 2.192 + 1.979) = 10^{-4} \cdot 9.188,$$

$$L_h = \frac{10^4}{9.188} = 1088.4 \text{ h}.$$

The life in kilometres from (4.74)

$$L_s = 1088{,}4 \cdot 121 = 131\,696 \text{ km}$$

The durability of the VL 1500 joint is satisfactory.

Hooke's joint in the propshaft.

Lines 1 to 4 in Table 5.14 and line 5 apply for the Hooke's joint as previously

Line	Equ. No		Gear				
			1st	2nd	3rd	4th	5th
6	(4.62)/(4.63)	L_{hx}	1450	4934	11914	24621	44506

The life in hours from (1.20)

$$\frac{1}{L_h} = \frac{0.01}{1450} + \frac{0.06}{4934} + \frac{0.18}{11914} + \frac{0.30}{23621} + \frac{0.45}{44506}$$

$$= 10^{-5}(0.670 + 1.216 + 1.511 + 1.218 + 1.011) = 10^{-5} \cdot 5645$$

$$L_h = \frac{10^5}{5645} = 17715 \text{ h}.$$

The life in kilometres from (4.74)

$$L_S = 17715 \cdot 121 = 2143515 \text{ km}.$$

Conclusions:
- If Hooke's joints and ball joints are fitted in the same shaft, the weakness of the ball joint is shown clearly by the relative durabilities, in this case 1338495/131696 km.
- It is important to recognize that changing the maximum engine torque alters the life in kilometres by the 10/3 power:

$$\frac{L_{s2}}{L_{s1}} = \left(\frac{M_1}{M_2}\right)^{10/3} = \left(\frac{1}{0.64}\right)^{10/3} = 4.428$$

$$L_{s2} = 4.428 \cdot 483613 = 2141438 \text{ km}.$$

Since the life in kilometres is more than twenty times greater than that required, one can try going back to the next smallest joint, GWB 2015 from Fig. 4.10. Its life in kilometres is given by

$$\frac{L_{S2}}{L_{S1}} = \left(\frac{210}{289}\right)^{10/3} = 0.345 \Rightarrow L_{S2} = 0.345 \cdot 2141438 = 738796 \text{ km}.$$

The static torque capacity must be checked using (4.21)

$$M_0 = 2 \cdot 38 \cdot 1 \cdot 24 \cdot 2 \cdot 8.80 \cdot 0.022 = 706 \text{ Nm}$$

According to (4.30 and 4.33) this joint should be assessed using the smaller of the two torques M_A or M_H. From Table 5.13 it can be seen that $M_A = 405.5$ Nm. Although the rated torque $M_N = 400$ Nm (Fig. 4.10) is exceeded, there is a high safety margin of $M_0/M_A = 711/405.5 = 1.75$ and so we can choose the smaller GWB C 2015 joint.

5.10 Literature to Chapter 5

5.1 Pahl, G.; Beitz, W.; Wallace, K. (editor): Engineering Design. Berlin: Springer 1988, Chap. 6

5.2 Pahl, G.; Kuettner, K. H.: Fundamentals (of design), in: Dubbel, Handbook of Mechanical Engineering, vol. 1. Berlin: Springer 1990

5.3 Reinecke, W.: Konstruktions-Richtlinien für die Auslegung von Gelenkwellenantrieben (Design guidelines for the layout of drivetrains). Motortech. Z. 19 (1958) p. 349–352, 421–425

5.4 Reinecke, W.: Gelenkwellen (Driveshafts), in: Bussien, Automobiltechnisches Handbuch (Automotive Manual), 18th edition, vol. 2, p. 389–406. Berlin: Techn. Verlag Herbert Cram 1965

5.5 Reuthe, W.: Untersuchung von Kreuzgelenken auf ihre Bewegungsverhältnisse, Belastungsgrenzen und Reibungsverluste (Study of the internal movements, load limits and friction losses of Hooke's joints). Diss. TH Berlin 1944. Konstruktion 1 (1949) p. 206–211, 234–239; 2 (1950) p. 305–312

5.6 Spicer, C. W.: Action, Application and Construction of Universal Joints. Automotive Industries 19 (1926), No. 4, p. 625–634

5.7 Cheng, H.: A numerical solution to the elastohydrodynamic film thickness in an elliptical contact. J. ASME Lubric. Techn. (1970) p. 155–162

5.8 Mazziotti, P. J.: Dynamic characteristics of truck driveline systems. SAE-Paper SP-262 (1964)

5.9 ISO 500, 1st edition, 1979

5.10 Miller, F. F.: Constant velocity universal ball joints – their application in wheel drives. SAE-Paper 958 A (1965)

5.11 Enke, K.: Konstruktive Überlegungen zu einem neuen Gleichlaufschiebegelenk (Ideas for designing a new CV plunging joint). Automobil. Ind. 15 (1970) p. 33–38

5.12 Krude, W.: Theoretische Grundlagen der Kugel-Gleichlaufgelenke unter Beugung (Theoretical principles of articulated ball CV joints) IV. Eppan-Tagung GK N Automotive AG 1973 (MS)

5.13 Kutzbach, K.: Quer- u. winkelbewegliche Wellenkupplungen (Shaft couplings for angle and translation). Kraftfahrtech. Forschungsarb. 6. Berlin: VDI-Verlag 1937

5.14 Stall, E.; Bensinger, J.; van Dest, J.-Cl: Neue Gelenke zur Isolation von Motoranregungen beim Frontantrieb (New joints for isolating engine excitations with front wheel drive). Automobiltechn. Zs. 95 (1993), No. 4, p. 204–209

5.15 Sedlmeier, R.; Schlonski, A.; Hedborg, J.: Wartungsfreie Gelenkwellen für schwere Nutzfahrzeuge (Maintenance-free driveshafts for heavy commercial vehicles). Automobiltechn. Zs. 98 (1996), No. 9, p. 400–405

5.16 Nienhaus, Cl.; Wilks, E.: Antriebssysteme in der Landtechnik (Drive systems in agricultural engineering). Landsberg/Lech: Verlag Moderne Industrie, 1997, Die Bibliotechnik der Technik, Vol. 156

5.17 Pierburg, B.; Amborn, P.; Gleichlaufgelenkwellen für Personenkraftfahrzeuge (Constant velocity drive shafts for passenger cars). Landsberg/Lech: Verlag Moderne Industrie, 1998, Die Bibliothek der Technik, Vol. 170

5.18 Young, John R.: Antrifriction Ball or Roller Units offer many Material Advantages in obtaining Truck Driveline Slip. SAE-Journ. 1968, January, p. 68–70

5.19 Rouillot, M.: Transmissions en matériaux composites (Transmissions made of composite materials). SIA-Journ. 1987, April, p. 105–108

5.20 Guimbretière, P.: Matériaux composites et liaison sol (Composite materials and connections). SIA-Journ. 1987, nov./dec., p. 74–83 (1st part)

5.21 Papendorf, J.: Einsatz von Kreuzgelenkwellen in hochtourigen Antrieben (Use of Hooke's jointed drive shafts in high speed drive systems). Tagungsband Dresdner Maschinenelemente Colloquium 2003, Chapter 9, p. 693–709. Aachen: Verlag Mainz 2003

5.22 idem: Wuchtverfahren bei Hochgeschwindigkeits-Gelenkwellen (Balancing methods for high speed driveshafts). Essen: Spicer GWB 2000

5.23 Gold, P.W.; Schelenz, R.; Haas, R.; Untersuchungen zur Antriebsstrang-Dynamik der Lok G 2000 BB (Investigations into the drivetrain dynamics of the G 2000 BB locomotive). Aachen: RWTH 2000

5.24 Derda, Th.; Weidner, U.: Drehmomentmessungen im Antriebsstrang der dieselhydraulischen Lokomotive MaK G 2000 (Torque measurements in the drivetrain of the MaK G 2000 diesel-hydraulic locomotive). Hamburg: Germanischer Lloyd 2002

5.25 Ritter, N.; Papendorf, J.: Konstruktive Maßnahmen zur Erhöhung zulässiger Betriebsdrehzahlen von Kreuzgelenkwellen in Schienenfahrzeuge (Design measures to increase permissible operating speeds of Hooke's jointed drive shafts in rail vehicles). Tagungsband 5, Intl. Schienenfahrzeugtagung (Rail Vehicle Conference), Dresden 2002. Aachen: Verlag Mainz 2002

5.26 Ritter, N.; Bertels, Th.; Peters, R.; Millet, P.: Betriebsnahe Erprobung von Antriebskupplungen in Schienenfahrzeugen (Testing drive couplings in railway vehicles in simulated operating conditions). Glasers Ann. 124 (2000), No. 2/3

Name Index

Amicus, 16th century engineer 1, 3
Anderson, Edmund B. 35, 36, 213, 260, 277
Aucktor, Erich (1913–1998), Dipl. Ing., Designer at Löbro 23, 31, 161, 162, 165, 172, 175, 179, 195, 202, 203

Bach, Carl Julius von (1847–1931), German professor of mechanical engineering 43, 45, 212
Balken, Jochen, Dr.-Ing. research engineer 55
Barlow, Peter W. (1776–1862), English professor of mathematics and strength of materials 41
Bendix-Weiss 19, 23
Bengisu, Oezdemir, Dr.-Ing., Rumanian scientist 61
Bernoulli, Jakob (1654–1705), Swiss professor of mathematics 40
Birfield Transmissioni SpA, Brunico (founded 1963), part of GKN Automotive 18
Bochmann, Hellmuth, Dr.-Ing. 105
Bouchard, Robert, French Head of Design at Nadella 14, 16, 29, 30, 258, 332
Boussinesq, Joseph-Valentin (1842–1929), French professor of physics and mathematics 89
Bratt, Erland, Swedish research worker at SKF 111
Braune, G., Dipl.-Ing., research worker at FAG 105
BTB, *see* Birfield Trasmissioni
Bussien, Richard (1888–1976), German senior engineer and designer 29, 33

Cardano, Geronimo (1501–1576), Italian doctor and mathematician 1–3, 49
Charles, V. (1500–1558), German emperor 1
Chenard & Walcker SA Gennevilliers (1899–1939) 14
Cheng, H. S. 219
Clement, Paul Robert 36

Cole, S. J. 106, 107
Coulomb, Charles-Augustin de (1736–1806), French engineer officer and physicist 40
Cull, William (1901–1994), British engineer and research worker at Scott u. Birfield Transmissions Ltd (now part of GKN Automotive) 174, 187, 189, 260, 262

Daimler Motoren 29
Dana Corp., Toledo/CDN (founded 1946) 49, 202, 204
Danielson, Axel, Swedish research worker at SKF 46
Dest, Jean Claude van, designer at Loebro 228
Devos, Gaston, French inventor 21, 24, 27, 32, 177
Diaconescu, D., Rumanian engineer 78
Duditza, Florea, Dr.-Ing., Rumanian lecturer 10, 78
Dunkerley, S., English engineer 140

Elbe, Willi, Gelenkwellen KG Tamm (founded 1951) 285, 294
Emerson, Victor Lee 36, 38, 39, 260
Enke, Kurt (1927–1984), Dr.-Ing., designer at Daimler-Benz 19, 23, 35, 36, 260, 332
Eschmann, Paul, Dr.-Ing., Head of Design at FAG 111
Euler, Leonhard (1707–1783), Swiss mathematician, physicist and astronomer 40

Faure, Henri, French inventor 29–31
Fénaille, Pierre († 1967), French inventor 12, 13
Fischer, Wilhelm, Dipl.-Ing., mathematician and FAG research worker 109, 119
Flick, John B. 33
Foeppl, August (1854–1924), Dr. phil., German professor of mathematics and engineering science 231
Freudenberg, KG 310

Galilei, Galileo (1564–1642), Italian mathematician and physicist 40
Geisthoff, Hubert (1930–2002), German engineer at J. Walterscheid GmbH (part of GKN Automotive) 12, 239
Girguis, Sobhy Labib. Egyptian Dipl.-Ing., development engineer 201, 202
GKN Transmissions Ltd., Birmingham (founded 1972), now part of GKN Automotive 158, 190, 194, 195, 263, 266
Glaenzer-Spicer SA, Poissy (founded 1838/1930), part of GKN Automotive 36, 38, 66, 201, 202, 211, 214, 221, 263, 265, 322, 327, 329
Gough, H. J., English metallurgist 112
Grashof, Franz (1826–1893), German professor of strength of materials, hydraulics and engineering science 56
Gruebler, Martin (1851–1935), German professor of mechanics and kinematics 55
GWB: Gelenkwellenbau GmbH Essen (founded 1946), part of GKN Automotive 126, 128, 133, 138, 139, 255, 256, 261, 268, 269, 271, 274, 275, 279, 280, 284, 322, 325, 328–331, 333, 336, 339, 342

Hardy-Spicer (& Co) Ltd., Birmingham (founded 1925), now part of GKN Automotive et seq. 189, 273, 326
Hardt, Arthur, German inventor 9
Harris, T. A., 232, 233
Herchenbach, Paul, German engineer at J. Walterscheid GmbH (part of GKN Automotive) 11, 12
Hertz, Heinrich Rudolf (1857–1894), German professor of physics 46, 81–112, 128, 154–158, 185, 188–206, 201, 204, 206, 211, 212, 215, 217–219, 231, 233, 235, 244, 317, 320, 321
Hesse, Ludwig Otto (1811–1874), Phil. D. German professor of mathematics 91
Hevelius, Johannes (1611–1687), astronomer from Danzig 1, 3, 4
Heyd, Alfred, driveshaft manufacturer 294
Hirschfeld, Fritz, 33–35
Honda Motors 187
Honnecourt, Villard de, French church builder in the 13th century 1
Hooke, Robert (1635–1703), English physicist 1–19, 34–41, 53, 54, 70–73, 81, 109–121, 123–154, 207–214, 244, 277–284, 292, 293, 295, 297, 300, 304, 308, 309, 312, 321, 325, 326, 329–334, 339, 342

Horch, August (1868–1951), Dr.-Ing. E. h., German mfg. engin., automotive designer 27
Hueckel, Fritz (1884–1973), Austrian hat manufacturer 49

INA Wälzlager Schaeffler KG, Herzogenaurach 255, 286
International Harvester Co. 297
Iveco-Magirus BV/AG, Ulm 136

Jacob, Werner, techn. man. at INA+Loebro 160, 161, 163, 165, 168, 182, 186, 195, 228
Jonkhoff, Henri Wouter 36, 37, 39
Jung, Hanns, DKW designer 14, 15

Karas, Franz (1906–1947), Dr.-Ing., Ob.-Ing., German lecturer 110
Kirchner, W., FAG engineer 105
Kittredge, John W. 33, 35, 36, 213
Kleinschmidt, Hans-Joachim, Dr.-Ing., GWB development engineer 140
Kommers, J. B., US metallurgist 112
Korrenn, Heinrich (1924–1992), Dr.-Ing., Managing Director of FAG 105
Krude, Werner, Dipl.-Ing., GKN Automotive development engineer 39, 157, 165, 166, 184, 185
Kunert, Karl-Heinz, Dipl.-Ing. 105
Kutzbach, Karl (1875–1942), Professor of Mechanical Engineering and Theory of Kinematics, TH Dresden 33, 34

Laval, Carl Gustav Patrik de (1845–1913), PhD. 140
Legendre, Adrien Marie (1752–1833), French professor of mathematics 94, 96
Leonardo Da Vinci (1452–1519), Italian painter, designer and inventor 1
Lloyd, R. A., Phil. D., GKN 228, 229
Löbro: Löhr & Bromkamp GmbH, Offenbach a. M. (founded 1948) 31, 182, 203–205, 208, 216, 230, 242, 260, 263, 266, 291, 311, 313, 314, 316, 317, 326, 333
Lundberg Gustaf (1882–1961), Swedish professor and research worker at SKF 105, 111, 184

Macielinski, J.W. (1910–1972), research engineer at Birfield and GKN Tansmissions Ltd. (now part of GKN Automotive) 158, 164, 190–193, 201, 207, 208
Metzner, Eberhard (1932–2000) Dipl. Math. Audi Design Department 58, 59, 61, 63, 65, 70

Name Index

Miner, Milton, A., Douglas Aircraft Co. test engineer, see also Palmgren 49
Mohr, Christian Otto (1835-1918), prof. of mechanics + graphostatics 1887 110, 231, 236
Molly, Hans, Dipl. Ing., development engineer 59, 61
Monge, Gaspard (1746-1818), French professor at the Ecole Polytechnique 5
Moore, Herbert Fisher (1875-1960), US professor of metallurgy 112
Mundt, Robert (1901-1964), Dr.-Ing. professor, SKF Director 84, 112

Niemann, Gustav (1899-1982), Dr.-Ing., professor of Mechanical Engineering TH Brunswick and Munich 112
Nugué, Charles, French engineer and technical author 10

d'Ocagne, Maurice (1862-1938), French professor of mathematics and kinematics 10, 12, 55
Orain, Michel (1920-1998), Dr.-Ing., Head of Development at Glaenzer-Spicer SA 35, 36, 38, 39, 66, 67, 69, 71, 73-75, 77, 78, 191, 214, 221

Paland, Ernst Günther, Prof. TU Hannover 182, 184, 195, 196, 231, 234, 235
Palmgren, Nils Arvid (1890-1971), Dr.-Ing., SKF Director and development engineer 46, 47, 49, 109, 111-114, 184, 192, 207, 209
Philips, I. R., Australian professor of kinematics 155
Philon of Byzantium, engineer in about 230 BC 1
Planiol, André (1893-1955), French engineer and technical author 10
Poisson, Simeon Denis (1781-1840), professor of Analysis and Mechanics, Paris 100
Poncelet, Jean-Victor (1788-1867), French engineering officer, mathematician and hydraulics specialist, Professor of Applied Mechanics, Head of École Polytechnique, Paris 5, 6, 40, 214

Retel, Julie-Marie-René, French inventor 14
Reuleaux, Franz (1829-1905), Professor of Kinematics, Berlin 53, 56
Rheinmetall: Rheinische Metallwaren- u. Maschinenfabrik Soemmerda AG 42

Roessler, Heinrich, Dipl.-Ing., Chief Designer at UNIMOG 255
Rubin, Wolfgang, designer at Löbro 195
Rzeppa, Alfred Hans (1885-1965), US engineer and designer 19, 21-29, 49, 153, 154, 160, 162, 163, 166, 170, 176, 177, 179, 187, 189, 201, 203, 308, 317

Sager, Johann, German mathematician, doctor and philosopher of the 16th c. 2
Salzenberg, Wilhelm, lecturer, Allgem. Bauschule Berlin 27, 40, 41
Sarrus, Pierre-Frédéric (1798-1861), French professor of mathematics, Strasbourg 69
Schott, Caspar (1608-1666), Jesuit father 1, 3, 49
Schroeter, Kurt (1905-1973), Dipl.-Ing., engineer at J. Walterscheid GmbH (part of GKN Automotive) 298
Schwenke, Robert (1873-1944), Berlin designer and inventor 27, 28, 33, 213
SKF: Svenska Kullager-Fabriken AB, Gøteborg (founded 1907) 47, 113
Spicer, Clarence Wilfred (1875-1939), US inventor and entrepreneur 46, 273
Steeds, William, OBE BSc. (Eng.), lecturer at the Royal Military College of Science and Technology, author 55
Stellrecht, Helmut, Dipl.-Ing., Schweinfurt 320, 341
Straub, Theodor, PhD., Director of Studies, Ingolstadt 2
Stribeck, Richard (1861-1950), German professor of mech. engineering 45, 46, 117, 130, 210
Stuber, Bernard K., US engineer and inventor 23-25, 32, 160, 179, 184, 187, 308
Suczek, Robert, US inventor 29, 30
Sundberg, Knut, SKF research worker 46

Taniyama, K. 162
Tracta joint, see also Fénaille 10, 12, 13

Vanderbeek, Herbert Calvin, US engineer 27, 28
Villard, Marcel, French designer 14, 15

Wahlmark, Gunnar A. 35, 36
Walterscheid KG/GmbH, Jean, Lohmar (founded 1919), part of GKN Automotive 299, 303-306
Warner, Archibald A. 33

Weber, Constantin Heinrich (1885–1976), Dr.-Ing., designer, Studienrat, Masch.-Bauschule Dortmund, Professor of Strength of Materials, TH Dresden and Brunswick 46, 81, 90, 91, 95

Weibull, Waloddi (1887–1979), PhD., Swedish professor and SKF research engineer 46

Weiss, Carl William, US inventor and entrepreneur 19–22, 153, 154, 170, 176, 177, 187

Welschof, Heinrich, German engineer at J. Walterscheid GmbH and Löbro (parts of GKN Automotive AG) 161, 162, 195

Whitney, William A., US inventor 17–19, 21, 25, 260

Willimek, Walter (1908–1982), Dipl.-Ing. Technical Manager at Löbro 175

Willis, Robert (1800–1875), English. professor of applied mechanics and kinematics 10

Wingquist, Sven G. (*1876), founder of the SKF 1907 38, 39

Winter, Hans (1921–1999), Dr.-Ing., Professor of Mechanical Engineering at TH Munich 155

Woehler, August (1819–1914), German railway engineer 47, 112, 113

Wooler, John 23

Subject Index

Agricultural equipment 296
angle of tracks 62, 155
angular acceleration 8, 151
– difference 7–9
– relationship 5–8
– velocity 7 ff., 66–70
articulation angle 6, 7, 10, 15, 19, 21, 216, 221, 245, 250, 253, 260, 297, 298, 309
auxiliary steering 22–26, 177

Balancing quality 252, 257 ff.
ball centring 23–26
ball joint, grooved spherical 294, 295
ball jointed driveshaft 33–36, 309
ball steering 177
ball track 17, 18, 24, 153 ff., 318
ball track, circular 64–66
–, circular cross-section 174, 175, 189 ff.
–, elliptical cross-section 174, 175, 189 ff.
–, flank 174
–, helical 200 ff.
–, ogival (gothic arch) 174, 175
–, parallel axes (straight) 200 ff.
ball trunnion 45, 46
bar linkage 53–57
bearing, block 28–32
–, needles 12, 46, 121, 129, 209 ff., 220, 256, 273, 286, 294, 331
–, roller 116, 117, 121, 122, 129, 209, 286
bending moment 41–43
Bendix-Weiss joint, see plunging joint
bipode joint, see plunging joint
boots 310

Cage, clearance 181 ff.
–, control 177, 181, 186
cardan error 5–8
centre of ball tracks 22, 155, 187, 188
circlip 12
coefficient of conformity 19–21, 110, 158, 189–191, 209, 211
compressive force, cyclic 115, 124, 126
–, mean equivalent 119–125

configuration W+Z of joints 9
conformity, along track 21, 189, 210–212
–, across track 189, 201, 202
connecting shaft 250
connection, cross serration 251–253
–, DIN/SAE 252
–, flange 251–253
–, inner race 238, 243
–, Mechanics 273, 276, 278
–, wing bearing 251, 253
constant velocity, conditions for 8–12, 23, 54
–, demonstration 82 ff.
–, fixed joint 57–62, 187 ff.
–, plunging joints 17–21, 29–32, 200–206, 220, 221
contact, area 88
–, ellipse 90, 91
–, rectangular 103, 104
critical bending speed 140–144
cross bearing, dimensions 129
–, trunnion 40–42

Design guidelines for joints 244, 332
double Hooke's joint, see fixed joint driveshaft
–, agricultural machinery 206 ff.
–, ball jointed 308, 309
–, design 109 ff.
–, Hooke's jointed 116
–, tripode jointed 321, 322
Duditza's equation 78
durability calculation 323, 324

Family tree of joints 176
Fischer's equation 119
fixed joint
– Bouchard 14–16
– double Hooke's 8 ff., 144–147, 271, 285, 287
– Glaenzer Spicer 189, 201, 211, 217, 218, 220, 221
– Honda 187, 215
– Hooke 5, 116

fixed joint
- Loebro 188, 190, 202–205, 216, 230
- Rzeppa 21, 22, 26, 187–189
- Tracta 12–14
- tripode 214 ff.
- Weiss 19–22
four-wheel drive passenger car 333–342
- front-wheel drive 8, 14

Geometry coefficient 119, 120
- generating centre 23

Hardness curve 236, 237
- conditions 238
heat treatment 237
Hertzian coefficient 88
- theory 46, 81 ff., 106, 107
Hooke's joint, see fixed joint
- kinematic chain 10, 53 ff.

Interaction of forces 148 ff.
intermediate shaft 9, 54

Joint centre 12, 15, 42 ff., 57 ff.

Kinematic chain 53 ff.

Lemon play 174
length compensation 303
lip seal 129
load spectrum 208
loading of joints, dynamic 81, 106, 112
- static 111

Materials and manufacture 215 ff.
Miner's rule, see Palmgren-Miner rule
mirror image and symmetry 61
monopode joint 70
motor vehicle construction and design 12 ff., 130–133

Non-uniformity (of rotation or torque transmission) 5 ff.

Offset steering 19, 24, 177
opening angle 177
operating factor 125, 127

Pairs of rolling elements 17, 19
Palmgren-Miner rule 49, 207, 209
Palmgren's equation 47, 112, 113
pilot lever 22
plane of symmetry 23, 54, 55, 61
plunge resistance 258, 259
plunging elements 17, 18, 258

plunging joint
- Bendix-Weiss 19–22
- bipode 213
- Bouchard 29, 30
- Bussien 29
- Devos 37
- Enke (star) 19, 23
- Faure 30, 31
- Rzeppa (principle) 160
- Schwenke 27, 28
- Suczek 29, 30
  tripode 214 ff., 322, 324
- VL 202 ff., 205, 206
pode joint, see also monopode and polypode joint 209–213
-, bipode 213
polypode joint 71 ff.
Poncelet's equation 5–8, 122
power take-off shaft 298–300
pressure angle 200, 201
profile, lemon 259, 301
-, splined shaft 296
-, sliding 258, 259, 301
propeller shaft 134, 262, 269–273

Radius of curvature
- rolling surface 46, 82, 83
- track 17, 18, 24, 155 ff., 174–176, 318
Rilsan coating 261
roller, effective length 121
rolling contact fatigue strength 47, 232–234

Sealing 254
shock factor, see operating factor
skew angle 62–64, 115
slip yoke connection 251–253
star joint, see plunging joint
steer drive axle 268, 302
steering ball joint 23, 285–291
Stribeck's equation 117, 130
support bearing 270
- force 148 ff.
surface pressure 42–46
symmetry, see mirror image and symmetry

Tongue and groove (Tracta) joint, see plunging joint
torque capacity, dynamic 121 ff., 154 ff.
- static 116 ff., 153
track, see ball track
transmission elements 56, 57
tripode joint, see also fixed and plunging
- joints 73–78, 214 ff., 223–228

– plane 214ff.
– trunnion 209–212
track, *see* ball track
–, crossing angle 26, 57–59, 177
–, inclination angle 61, 64, 115
trunnion stress 40–42, 209–212

Unit pack 12, 255, 256

W-configuration 9, 150–152, 270
Weber's analogy 46, 81, 91
wide angle joint 11, 12, 298, 299, 303
Woehler's function 47, 112

Yoke forces 10, 19, 41, 42, 148ff.

Z-configuration 9, 10, 16, 17, 153, 249, 268, 270, 299, 301, 303, 332